Oak Origins

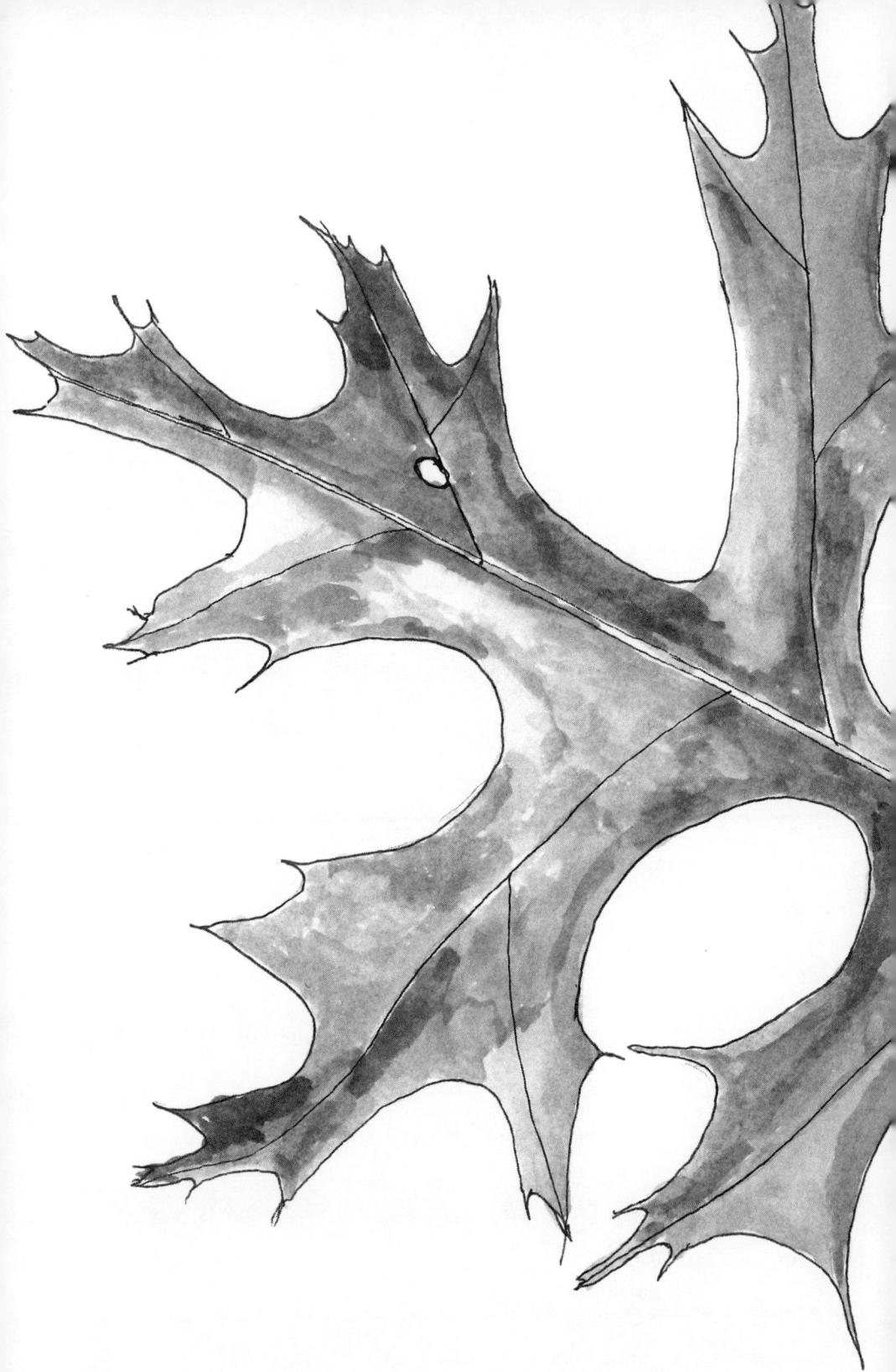

Oak Origins

From Acorns to Species
AND THE *Tree of Life*

ANDREW L. HIPP

Illustrations by Rachel D. Davis
Foreword by Béatrice Chassé

The University of Chicago Press   Chicago and London

The University of Chicago Press, Chicago 60637
The University of Chicago Press, Ltd., London
© 2024 by Andrew L. Hipp
Illustrations © 2024 by Rachel D. Davis
Foreword © 2024 by Béatrice Chassé
Published 2024
Printed in the United States of America

33 32 31 30 29 28 27 26 25   3 4 5

ISBN-13: 978-0-226-82357-7 (cloth)
ISBN-13: 978-0-226-82358-4 (e-book)
DOI: https://doi.org/10.7208/chicago/9780226823584.001.0001

Library of Congress Cataloging-in-Publication Data

Names: Hipp, Andrew, author. | Davis, Rachel D., illustrator. | Chassé,
    Béatrice, writer of foreword.
Title: Oak origins : from acorns to species and the tree of life / Andrew L.
    Hipp ; illustrations by Rachel D. Davis ; foreword by Béatrice Chassé.
Description: Chicago : The University of Chicago Press, 2024. | Includes
    bibliographical references and index.
Identifiers: LCCN 2024011381 | ISBN 9780226823577 (cloth) |
    ISBN 9780226823584 (ebook)
Subjects: LCSH: Oak—Origin. | Oak—Evolution. | Oak—Ecology.
Classification: LCC SD397.O12 H55 2024 | DDC 634.9/721—dc23/
    eng/20240423
LC record available at https://lccn.loc.gov/2024011381

♾ This paper meets the requirements of ANSI/NISO Z39.48-1992
(Permanence of Paper).

For Paul Manos and Alan Whittemore

Contents

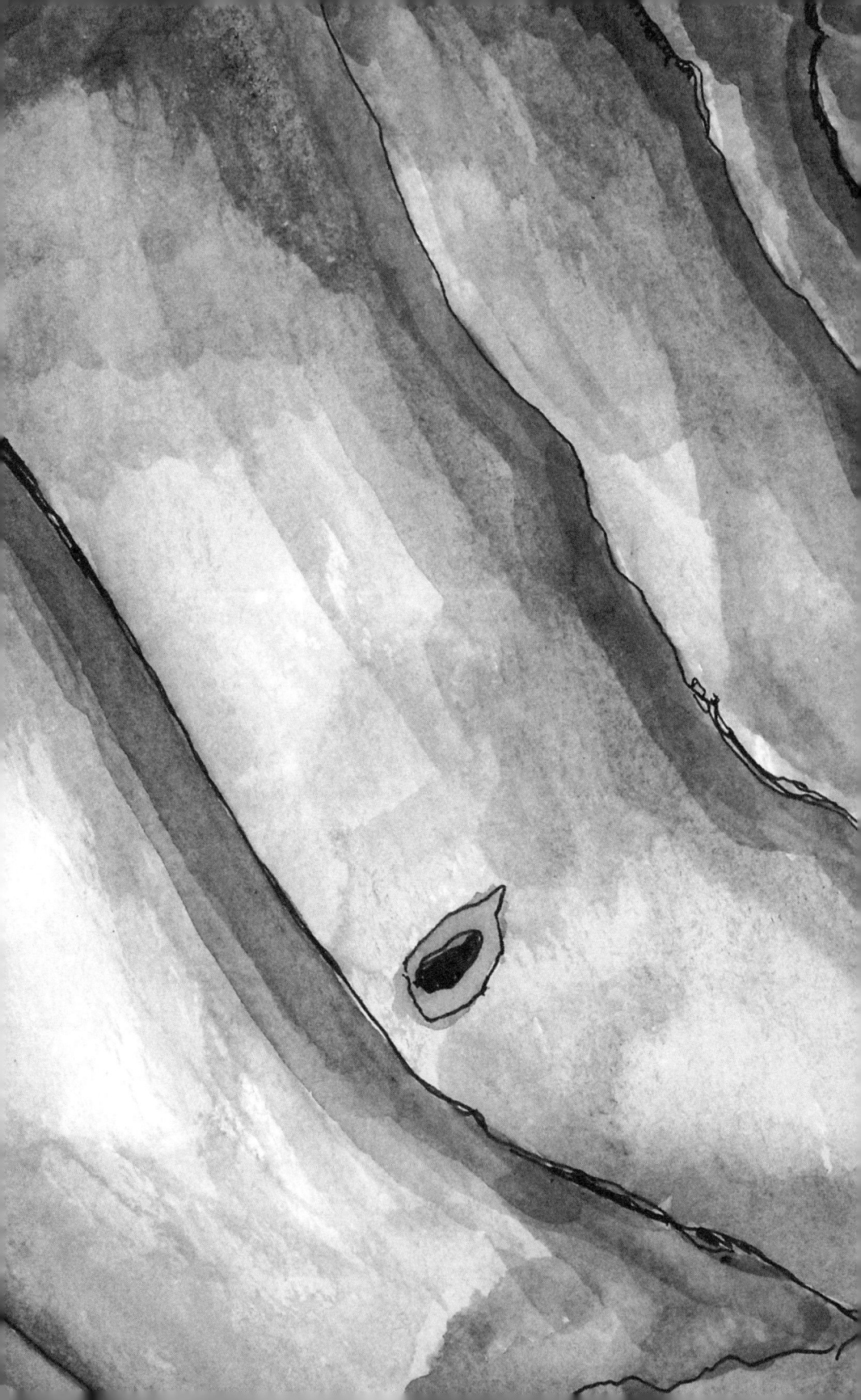

Figuring Things Out

Four billion years ago, life began in hydrothermal vents with an anaerobic, sulfur-eating organism considered to be the oldest-known common ancestor of every creature on Earth. The question is, How do we draw the lines connecting individuals, populations, species, genera, and families from today's oaks to this oldest-known common ancestor?

. One hundred sixty-four years ago, Charles Darwin wrote his theory of descent with modification, putting everybody on the right track on how to go about answering this question. Today, *Oak Origins* presents the story of what has been achieved (with oaks as the focal point) in understanding evolution ever since Darwin's genial idea, and as such, it is not one story but three. The first is the story of evolution, the story of the Tree of Life—specifically, the oak tree of life. The second is the story of how our understanding of this tree of life has evolved into what it is today. The third, which is at once the sum and more than the sum of those two stories, is Andrew Hipp's story.

What is an oak? as an acorn? as a tree? as a species in the genus *Quercus*? as an individual being in a specific population? What are oaks today? What were they 50 million years ago, or 10 million years ago? In what sense do they exist beyond the descriptions and the names that we give them? These are the questions that Andrew intertwines with their answers and his reflections based on his profound understanding of one of the most interesting subjects in the history of *Homo sapiens* trying to figure things out: evolution.

From the first sentence, "It is not hard to answer the question, *What is an oak?*" to the last paragraph, with its "strands of gossamer trailing between the branches," his story is not an easy one to tell. It is not a story that can be simplified without transforming it into something else. It is a story that plunges you into the history of the species concept, the tectonic and climatic transformations of the planet Earth, the ecology of population dynamics, plant sexuality and developmental biology, the plasticity of genotype and phenotype, and ecology as a key factor in maintaining the integrity of species while they benefit from hybridization and introgression, which, in turn, provide the raw material of natural selection and adaptability.

The questions and hypotheses that have oriented research across a variety of disciplines exploring different aspects of oaks over the past several decades, as even just a quick look at the bibliography will show, have resulted in major advances in understanding the oak tree of life, mapping it out and grasping the relationships between species defined as ecological, morphological, and genetic entities. In the introduction the author tells us that he wrote this book because it is a book he wants to read—because it is a book he would have liked to have read when he was working as a naturalist some thirty years ago. I too would have liked to have had such a book when I became interested in the genus *Quercus* twenty years ago. Of course, then, in both cases, it would not have been this book.

Depending on where you live—northern France, southern Spain, northwestern Mexico, southeastern United States, Taiwan, China, or Indonesia—the images the word "oak" evoke will differ greatly, be they of the leaves, of the acorns, or of the habit, which is the overall appearance. If an oak for you is *Q. cerris* (southern and Central Europe), then you might have a hard time recognizing *Q. alnifolia* (endemic to Cyprus) as an oak; if *Q. chrysolepis* (northern Mexico and southwestern United States) is your idea of an oak, then *Q. glauca* (Asia and Eurasia) will surprise you; if *Q. ellipsoidalis* (Canada, northeastern United States) means oak to you, then *Q. minima* (southeastern United States) will raise an eyebrow. The marvelously delicate illustrations by Rachel Davis that are a part of this book will give you a good idea of the outstanding diversity contained in the word "oak."

Yes. It is easy to answer the opening question of *Oak Origins*, *"What is an oak?"* But often, in the field, it is not so easy, even for the well versed, to determine who is who; in other words, what species the individual you are looking at belongs to. If you ever have the opportunity to try, I encourage you to do so for three reasons.

First, this is where it starts. The Tree of Life is constructed starting with individuals, then grouping them into species, and then those into families, and so on, going from right to left, as it were, but its meaning is inscribed from left to right.

Second, it is a great deal of fun, especially if you are in a place like Mexico where the species diversity is the highest of any country and the taxonomy is complicated by hybridization, rapid evolution of species, and our growing understanding.

Third, it will help you understand the passion that has gone (and that goes) into thinking about these things, this passion that Andrew shares with many other individuals, many of whom you will meet as his story unfolds and that are as central to it as the oaks themselves.

Evolution is a fact, with or without *Homo sapiens*. But only *Homo sapiens* know this, can write its story, can enjoy learning about it.

You will find poetry, suspense, and humor in Andrew's science, and in his writing. You can enter his story in many different ways, but once you are in it you will be captivated, and like me, I am sure you will turn the last page with some of your questions answered, some of your answers questioned, and some new ideas perhaps already become questions.

Béatrice Chassé
Former President of the International Oak Society
Editor, *International Oaks*
Cofounder, Arboretum des Pouyouleix (France)

What Is an Oak?

It is not hard to answer the question, *What is an oak?* if we limit ourselves to characteristics. Oaks are woody plants. They are usually trees, though many oaks are shrubs, some reaching only to your ankles. They have alternate leaves and buds clustered at the tips of the branches. Their seeds are encased in acorns, whose caps (or cups, depending on which term you prefer) cover the base of the nut or almost entirely enclose it. Their male flowers are borne in small clusters on catkins, thread-like structures that sway in the breeze. They are wind-pollinated. You can tell whether any plant in the world is an oak based on whether or not it has this combination of characteristics.

But oaks have an existence beyond characteristics. The oak in my front yard is, like you and I, an ecosystem of cells swimming with mitochondria and, like all other green plants, chloroplasts. It hosts communities of fungi, bacteria, and viruses. Each oak is an organism teeming with organisms. Yet the oak in my yard originated from a moment of conception, the joining of the sperm cells inside a pollen grain with the egg cell inside an ovule. The ovule became an acorn. That acorn became this tree. Any time between birth and death, someone can become acquainted with this oak and know it over time, perhaps from childhood until old age. It is an individual.

Oaks are also species, groups of organisms that are similar enough in form, ecology, and genetics that we recognize each species as an entity. Like other tree species, oaks have scientific names that are almost universally accepted. *Quercus ellipsoidalis*, for example, is a species first described by Reverend Ellsworth Jerome Hill in 1899, from a population in southern

Cook County, Illinois. It is commonly called Hill's oak or northern pin oak. The species ranges from southern Ontario to northern Illinois, Indiana, and Ohio. It extends west to Iowa and Minnesota and east to Toronto. It thrives in moderately dry savannas and open woodlands. When I meet a new Hill's oak, an individual tree I have never encountered before, I almost always recognize it as a member of its species by its deeply cut, bristle-tipped leaves and acorn caps that are smooth on the inner surface. It has end buds with silvery hairs on the tips or the endmost half. It lives only in the upper Midwest. Hill's oak as a species is composed of oak trees, saplings, and seedlings that all share a morphology and ecology. It is one of about 425 oak species recognized today. Species and their attributes constitute the diversity of life.

When we talk about the oaks of the world, the entire genus *Quercus*, we are also talking about the Tree of Life. The Tree of Life is a metaphor for the daisy-chained moments of reproduction that connect our children back to their grandparents, humans to chimpanzees, mammals back to all the other animals, and, eventually, all living and once-living organisms to one another. Its many leaves—birds, bacteria, starfish, red algae, roses, grasses, and uncountable others—are outcomes of genealogies that we can trace. The oak tree of life represents a single bough on the larger Tree of Life. The oak tree of life is only a metaphor for the complex history of oaks, but the ancestors of the oaks were real trees. You could have sat beneath one of them to eat your lunch, if you had been around and known where to look 56 million years ago or so. The oak tree of life started as a seedling in their shade.

If someone asks you what an oak is, you will be safe citing characteristics. But we experience oaks as something much larger and more interesting than a collection of attributes. Collecting acorns beneath bur oaks in DuPage County, Illinois, where I live and work, I think of the improbable pollinations that produced them. I think of the migration routes the oaks of our region took as the species spread following the Last Glacial Maximum, some twenty thousand years ago. I find mosses and lichens that have blown from the bur oak canopy to the forest floor in a storm or yellow chanterelles emerging from the leaf litter, and I see the oak entangled in a web of interconnected organisms. I think of distantly related oaks in eastern China or Oaxaca. Oak organisms, oak species, and the oak tree of life offer different but complementary answers to the question, *What is an oak?*

I wrote *Oak Origins* in part because it is a book I would like to have read when I was working as a naturalist in my twenties. I had learned the enjoyment that comes from being able to put names to species. Knowing the names of a handful of woodland wildflowers gave me access to the understory. Knowing the names of prairie plants made it possible to see the diversity growing in a record album–sized patch at my feet. As I learned plant names, the natural world became richer and even more beautiful. But I had only an inkling of the Tree of Life and vague ideas—mostly misunderstandings—about how plants had evolved and were evolving.

This is also a book I would enjoy reading today, as a researcher who spends a lot of his time working on oaks. This is the book I want to be able to pull off the shelf to contextualize the things I see in my research and my walks in the woods. In my work as a plant systematist, I use genomic data, herbarium specimens, and field experimentation to understand plant taxonomy and evolution. My lab group and colleagues and I use these data to reconstruct evolutionary histories—portions of the Tree of Life—as scaffolds on which to build our understanding of biodiversity and natural history. We investigate how distinct species are and how frequently genes move between them. We investigate how the Tree of Life shapes prairies and forests. This work is natural history with new tools and data behind it. Like my work as a naturalist, it depends on the insights of many others who have seen, thought about, worked on, and written about things that are well outside my expertise. Systematics, genomics, population genetics, functional ecology, community ecology, paleobotany, plant anatomy, reproductive biology, and a host of other disciplines cast my observations into relief. *Oak Origins* is a portrait of oaks from several angles, bringing together historic and current research to tell the history of our evolving understanding of these magnificent plants.

I wrote this book not only for myself, however, but also for you, whether you are a naturalist, scientist, teacher, gardener, or simply interested. *Oak Origins* is a collection of stories braided through the forests and other plant communities where oaks thrive. Every organism stands at the intersection of a dense map of pathways. The paths are interactions that connect organisms to each other through ancient and recent history. Enter by way of a tree, and you traverse footpaths through biogeographic and landscape history. Start with a mushroom, and you follow trails of decomposition and nutrients moving from organism to organism. You might end up in any part of the landscape if you only walk for long enough, no matter where you begin. Oaks are one doorway into this web. Because oaks are connected to so many different organisms, they are one of the best places to start.

In studying oaks over the past twenty years and in writing this book,

I have found the world even more surprising and densely interconnected than it looked before. I hope that in reading *Oak Origins*, you do as well.

Conventions in This Book

Oak Names

The ambiguity of common names makes it impractical to avoid using scientific names in this book. For some very widely recognized North American and European species—northern red oak (*Quercus rubra*), California valley oak (*Q. lobata*), or pedunculate oak (*Q. robur*), as just a few examples—I use common names. When I do, I include the scientific name in parentheses the first time I mention the species in any given chapter or section, often abbreviating the oak genus *Quercus* as "*Q.*" Where no common names are in widespread use or where the use of common names is likely to be confusing, I use scientific names only. Where a particular oak species is referred to commonly in a chapter, a paragraph, or even a few adjacent sentences, I endeavor to revert to the common name only unless doing so will create confusion. Scientific names at the genus level and below are italicized, which is standard practice. I provide a cross-reference between common names and scientific names as an appendix.

The Tree of Life

Tree of Life is capitalized when I am referring to the evolutionary history of all organisms; single portions of the Tree of Life are often referred to as their own *tree of life*, uncapitalized. Thus, the *oak tree of life* is not capitalized. I use the term *clade* to distinguish entire subsets of the Tree of Life that have a single evolutionary origin: oaks are a clade, and so are birds. I endeavor to use *lineage* only to refer to a stretch of the Tree of Life that you could trace with your finger without ever backing over your path (making a single line, though sometimes it's a twisting line). A lineage in this usage may represent a single species or a succession of species that evolved along the path between a species and the root of the Tree of Life.

I provide standardized common names for the eight formally recognized sections of oaks, with the scientific name provided in parentheses the first time a section name is used within a chapter (and more often where needed for clarity). Each of the sections and subgenera of oaks is a clade. However, the repeated use of *clade* could become tedious, so I refer to these named clades as *groups*. Common names of the sections and subgenera are capitalized, even if they are not formally recognized. For example, *Red Oak Group* refers to *Quercus* section *Lobatae*, a clade of about 125 species. In the text, I sometimes lop off *group* where the usage is unambiguous: the terms *Red Oak Group* and *Red Oaks* refer interchangeably to *Q*. sect. *Lobatae*. By contrast, in referencing *Q. rubra*, the most widespread Red Oak species of eastern North America, I use the common name *northern red oak* without capitalization. Similarly, I use *Beech Family* to refer to the family Fagaceae, but *American beech* to refer to the species *Fagus grandifolia*. All common and scientific names of oaks and their relatives used in this book are cross-referenced in the appendix.

Geological time

Geochronology (epochs, eras, and ages) follows the June 2023 revision of the International Commission on Stratigraphy (ICS) Chronostratigraphic Chart (Cohen et al. 2013; see also https://stratigraphy.org/, accessed June 30, 2023). I use Early Cretaceous and Late Cretaceous (capitalized) for the ICS Lower Cretaceous and Upper Cretaceous, respectively. My Early, Middle, and Late Miocene, Pliocene, Pleistocene, and Holocene correspond with ICS dates. As no early, late, or middle periods are formally recognized by ICS for the Paleocene, Eocene, and Oligocene, I do not capitalize them, and I use the following correspondences, which follow the recommendations of Head et al. (2017): early and late Oligocene correspond to the Rupelian and Chattian, respectively; early, middle, and late Eocene correspond to the Ypresian, Lutecian plus Bartonian, and Priabonian, respectively; and early, middle, and late Paleocene correspond to the Danian, Selandian, and Thanetian, respectively. I use "mid-Cretaceous" informally in the book to reference the Albian-Cenomanian (ca. 100 million years ago).

Paleoclimate

Timing of climatic transitions and relative temperature since the end of the Cretaceous mostly follow Westerhold et al. (2020, fig. 1); Cretaceous climates follow Scotese et al. (2021). Other sources used appear in the notes section.

1

Flowers and Acorns

Populations Arise and Migrate

Spring in the upper Midwest begins downhill from the oaks. Skunk cabbage blooms in wet soils, generating enough heat to melt the snow around it. Male redwing blackbirds return from the South to mark off aural territories in the cottonwoods and cattails. Woodcocks move northward following the thawing soil and, with it, awakening earthworms. The male woodcocks gyre and twitter overhead, showier than the blackbirds. Roots of eastern white oak (*Quercus alba*) and northern red oak (*Q. rubra*) often get an early start, before the shoots start growing; probably those of other oak species do as well. Silver maples bloom. Then the forest understory becomes visible. Bloodroot flowers, spring beauty and false mermaid emerge. Mayapple shoots huddle like umbrellas with their handles stuck in the soil, leaves wrapped tightly around them, then fan out abruptly when temperatures rise. Yellow-rumped warblers come back to town.

Oak buds open about this time. They swell, and leaf tips appear at their ends. Then the leaves expand to the size of squirrel's ears as male inflorescences spill out from behind the bud scales. The male flowers are initially closed tight like fists, borne in squat, stiff spikes called catkins or aments. The developing catkins elongate and droop, then dangle. They become thread-like and hang from the branch tips like tinsel from a Christmas tree. The female flowers are much less conspicuous, growing individually or a few to a bunch along the stem; look closely, or you may mistake them for vegetative buds. Like the male blackbirds and woodcocks that arrived a few weeks earlier, the male oak inflorescences are showier than the female flow-

Bur oak (*Q. macrocarpa*) leaves and catkins. The catkins are still descending, and the anthers are closed. May 12, 2022, the Morton Arboretum, Illinois.

ers. They have to be: male oak flowers produce clouds of wind-borne pollen to fertilize the thousands to millions of ovules growing in the forest around them, and they will only succeed if they are numerous.

When a catkin reaches its full length, the tiny, petalless flowers attached to it open. Each male flower is little more than a bouquet of stamens. The leaves stop growing as the male flowers open, at least in some species, allowing wind to reach the stamens and resources to flow to the growing flowers. Each oak stamen is tipped with a small sac called the anther. The anther is one to three millimeters long, smaller than a grain of rice, and reminiscent

Eastern white oak (*Q. alba*) stamens, detail. The anthers are splitting open to release pollen. Hipp and Garner 2017008, May 12, 2017, the Morton Arboretum.

of two hotdog buns fused together lengthwise. It is borne at the end of a filament about as thick as a hair.

On a warm, dry day a week or two after the catkins first appear—humidity less than 45% is ideal—the anthers crack open lengthwise and release their pollen. They empty in a few hours or a couple of days. If humidity is higher than about 60%, the pollen gobs up and won't go anywhere. A few days of cool, damp weather will kill off a batch of pollen and you'll find the trails and sidewalks littered with sodden catkins. But with dry weather, the pollen blows on the wind, thousands of grains from every anther, billions from every tree.

The female, or pistillate, flowers have been developing in seclusion since the previous year, each concealed inside a bud, sometimes in solitude or more often with one or a few others. In some species, the pistillate flowers are borne on elongate stalks. Swamp white oak (*Q. bicolor*), for instance, produces pistillate flowers individually or paired on stalks that may be an inch to nearly three inches in length. The flowers of some of the Mexican oak species, particularly in the Red Oak Group, are spread out on long stalks that look like stout catkins, arching but not drooping, but their pistillate flowers are still nowhere near as numerous as the staminate flowers of a catkin.

Three arched stigmas emerge, reddish, pink, or yellowish, from the tips of the pistillate flowers about a week after the first catkins emerged. Stigmas are the landing platforms for pollen. Pollen may land on more than half of the pistillate flowers on a tree each spring. Fifteen or more pollen grains may land on a single stigma. Leaves that had stopped growing as the staminate flowers ripened start growing again once the pollen is all dispersed, expanding in size by as much as 50% in a day or so.

Pollen may travel just a few meters to reach a receptive stigma. In other cases, most pollinations come from trees less than 50 meters away. In some forest stands, the majority of pollinations come from trees more than 100 meters away, even if many suitable trees are closer. That flight is more than 2.9 million times the length of an average oak pollen grain, comparable in scale to a soccer ball flying from Chicago, Illinois, to Nashville, Tennessee. Some pollen grains fertilize ovules 30 to 80 kilometers away, while a very small number may travel 100 kilometers or more. These journeys must be thrilling, as pollen lofts to hundreds of meters above the treetops.

Pollen grains that arrive alive and intact on the surface of a receptive stigma typically germinate within about twenty-four hours. Each germinating pollen grain produces a pollen tube, the role of which is to penetrate the skin of the stigma and convey two sperm cells down to the pistillate flower's ovary. A pollen grain rarely has a stigma to itself: ten or more pollen grains may germinate on the surface of a stigma during the week to ten days during which the stigma is receptive. The styles—the stalks that hold the stigmas—conduct the tangle of pollen tubes downward toward the ovary. The style, however, is not a passive conduit. It is a road with tollbooths. As the pollen tubes tunnel through, interactions between the style and the pollen decide who will pass and who won't. Pollen grains from the same tree on which the pistillate flower is growing are often stopped near the top of the style, or at least slowed down as they grow. If a lot of self-pollen clogs the style, the flower may abort. Oaks rarely if ever produce viable seeds by self-pollination, and when they do, the seeds commonly abort or are low in weight, making them unlikely to survive in the wild. Pollen tubes from species different from the mother tree may be stopped dead in their tracks on the surface of the stigma or forced to branch off to the side so they can't make headway through the style. In other cases, they are slowed to a crawl. Pollen tubes drop out one by one. Pollen that is blocked ends its journey in a state of suspended animation.

In oaks and many of their close relatives, even pollen tubes that complete the trip to the base of the style are compelled to take a pause of weeks to months before they can continue their journey to the ovule. This hiatus is called delayed fertilization. Oaks and their close relatives—the other genera of the Beech Family (Fagaceae), Birch Family (Betulaceae), Sweet Bay Family (Myricaceae), and relatives—and about twenty other plant families exhibit delayed fertilization. In most if not all oaks, the pollen tubes pause at the juncture between the three styles. Delayed fertilization gives time for pollen tubes to catch up with one another, irrespective of when they land on the stigma. This may, in some contexts, level the playing field for pollen of different species. Delayed fertilization may thus encourage cross-pollination between species while reducing the risk of self-fertilization. However, when it is a competition between pollen grains of different species, all starting on the same stigma, pollen from the same species as the mother tree generally wins the race.

The canopy fills in with leaves. Spring wildflowers fruit, and some die back to the ground. More than a year may pass before the oak pollen tubes take up their journey again. Delayed fertilization in many species in the Red Oak, Ring-Cupped Oak, and Golden-Cup Oak Groups (*Quercus* sect. *Lobatae*, sect. *Cyclobalanopsis*, and sect. *Protobalanus*) holds the pollen tubes at bay for a year or more. (If these section names aren't familiar to you already, you can skip ahead to chapter 5 for a summary. For now, it suffices to know that each section is a subtree—a clade—of the oak tree of life.) Numerous species in these three sections exhibit biennial acorn maturation: pollination in one year is followed by fertilization and acorn development the next. In most of the temperate North, pistillate flowers in the Red Oak Group remain dormant through summer, through leaf-fall, through snow, and into the next spring. In California and Mexico, however, some species in the Red Oak Group pro-

Quercus Group *Protobalanus* pollen grain, from near the Eocene-Oligocene Boundary, Florissant Fossil Beds, Colorado. The fossil shows the characteristic oak pollen rugby-ball shape with three furrows along the sides. It also demonstrates that sect. *Protobalanus* was in the Rocky Mountains at the beginning of the Oligocene. Today, the section is centered in California. Illustrated from a scanning electron microscope image provided by Johannes Bouchal.

duce annual acorns. There is variation among trees in some species: in cork oak (*Q. suber*), some trees produce annual acorns, others produce biennial acorns, and the proportion of biennial acorns increases from south to north. In many oaks, including all the White Oak Group (*Quercus* sect. *Quercus*) so far as we know, the pollen tubes are arrested for just a few weeks before they

Female flower longitudinal section, circa 25× magnification, Gambel's oak
(*Q. gambelii*). The black structures at the tip of the flower are the styles. Below them are
two of the six ovules inside a female flower; only one typically develops into a seed.
The thick tissue around the outside is the fruit wall. The entire flower is nested
within the cupule, which will develop into the acorn cap. Illustrated from
Brown and Mogenson (1972, fig. 1).

continue. These species have annual acorns. Leaves in the forest canopy
just have time to expand and shade the understory before the pollen tube
reaches its destination in such species.

Six ovules wait inside each ovary. Often, multiple pollen tubes that germinated together on the stigma make it all the way to the micropyle, the
portal through which the sperm cells reach the ovule. Each pollen tube that
makes it successfully through the micropyle and into an ovule releases two
sperm cells, or gametes. One male gamete merges with the egg cell inside
the ovule to form a zygote. The zygote will grow into an embryo. The other
male gamete merges with two additional cells inside the ovule to form the
endosperm, food for the growing embryo.

Typically, the fertilization and growth of one oak ovule suppresses the
others in that ovary. This is why most acorns enclose only one seed. But once
every 100 to 300 acorns or so, 2 ovules grow all the way to the seed stage,
sharing the acorn together. In some populations, this number may reach
as high as 1 out of every 40 or 50 acorns. Oaks that grow from a two-seeded

acorn compete for resources while they are developing, and they consequently grow more slowly over their first year of life.

The stigmas dry up and darken after fertilization. The stalk at the base of the female flower often thickens. It becomes an umbilical cord, conducting water and photosynthate—sugars formed through photosynthesis inside the trees' leaves—into the acorn. The nutrients feed the embryo as its cells differentiate and multiply. The nutrients also provision the baby plant for its first year of growth. Sugars, water, proteins, and lipids combine to form endosperm, a kind of mother's milk for the seed. The growing embryo absorbs the endosperm into its cotyledons—thick, fleshy, embryonic leaves—as it develops. Within the growing acorn, the cotyledons form a heart shape or a sort of funnel at first, depending on the oak species. The lobes of the heart or the edges of the funnel soak up the endosperm and store it as food that the seedling will use in the future. The cotyledons swell to fill the developing acorn, swaddling the embryonic shoot and root growing inside. They grow to dominate the mass of the acorn.

The cup that surrounds the flower expands to accommodate the nut growing inside. Over a month or two, the acorn ripens.

The innovation of packing a to-go meal into one's cotyledons evolved many times across the flowering plant tree of life, but probably only once in the Beech Family (perhaps around the time the dinosaurs were driven extinct; more on that in chapter 4). Energy-rich, animal-dispersed nuts are part of the success of the oaks and their closest relatives, including the chestnuts (*Castanea*), stone oaks (*Lithocarpus*), Eurasian and American chinkapins (*Castanopsis* and *Chrysolepis*, respectively), and tanoak (*Notholithocarpus densiflorus*). The oaks and these five closely related genera produce large cotyledons that remain hidden near the surface of the soil as the seedlings grow. In many other species of flowering plants, the cotyledons unfurl, thin, green, and leaf-like inside the seed, ready to start photosynthesizing quickly after emergence. Alongside oak seedlings, you may see the strappy cotyledons of maples, the glove-like cotyledons of basswoods, or the leaf-like cotyledons of beeches (*Fagus*). They feed the plant by photosynthesizing. By comparison, oak cotyledons feed the plant on food the mother tree produced by photosynthesis, not by photosynthesizing themselves. They are like the yolk sac attached to the belly of a young turtle. They often remain inside the acorn shell as the seedling grows. They provide food for the growing plant, regardless of whether they are exposed to the light or embedded in soil or leaf litter.

Fagus crenata seedling. The seedling shows broad photosynthetic cotyledons and two young leaves. Illustrated from a photo by Alpsdake (Suzuka Mountains, Japan, May 12, 2017), Wikimedia Commons (Creative Commons license CC-BY-SA-4.0, shared here under the same license).

If you live in the temperate North, it is late summer by the time the acorns are ripe. Afternoon cicadas give way to early-evening crickets and katydids. Aborted acorns have been falling for a week or more. Pull a ripe acorn from a tree or pick a fresh one off the ground and look at it closely. The pointy end of the acorn is where the style and stigmas used to attach. You may still see withered styles at the tip. At the base of the acorn is the cupule, the acorn cap or cup. The cupule embraces the nut inside. It helps get the young fruit through fire and freezing winters, helps protect the growing embryo from hungry mammals and insects.

Cut the nut open with a pair of pruning shears. Sometimes you can skin the acorn by cutting through just the leathery fruit wall, but it's easier to cut the whole thing in half or quarters and pry out the seedling. Look closely at the inside of the shell: if you have a member of the Red Oak (*Q.* sect. *Lobatae*), Golden-Cup Oak (*Q.* sect. *Protobalanus*), Cork Oak (*Q.* sect. *Cerris*), or Holly Oak (*Q.* sect. *Ilex*) Groups, the inner surface of the shell will be hairy. The shell may also be hairy inside if you have a member of the Ring-Cupped Oak Group (*Q.* sect. *Cyclobalanopsis*). Acorn shell inner surfaces in the White Oak Group (*Q.* sect. *Quercus*) are smooth.

Set the shell aside. Everything inside the shell is the embryo. The bulk of the embryo is made up of the cotyledons, provided they haven't been eaten by weevils, fungi, or other critters. In many species, you will be able to separate the cotyledons by hand, like the halves of a sliced bagel. In others, you'll need to cut the cotyledons apart to find the tiny root tip inside. The energy in the cotyledons is good not just for the oaks, but for the entire forest community. Acorns are almost 90% carbohydrates by weight. Acorns are typically about 5%–10% protein by weight, compared to the roughly 25%–30% protein content of black walnut or European beech. Nonetheless, acorns are a staple for jays, turkeys, rabbits, raccoons, opossums, deer, mice and voles, squirrels, and more than 180 different species of birds and mammals. For some, acorns may be the difference between life and death. They comprise 5%–17% of the diet of bobwhites and ruffed grouse. They are 40%–99% of a stellar jay's diet, depending on the season. White-footed mice, deer mice, chipmunks, and white-tailed deer in the eastern United States thrive on acorns. When acorns are bountiful, these animals produce more offspring. The deer and deer mice then provide food—their own blood—for ticks. Among these are the deer ticks, which carry the bacterium that causes Lyme disease. As a consequence, Lyme disease increases predictably two years after a bumper crop of acorns. The chain of oak leaves, acorns, deer, mice, ticks, and Lyme disease bacteria ferries sun-energy through the forest.

Another ecosystem—a smaller one, but rich nonetheless—thrives inside every acorn. Weevils lay their eggs inside acorns developing on the tree. Their offspring grow inside, sometimes decimating the cotyledons. After they have eaten their fill, weevil larvae chew a hole in the acorn shell and squeeze out like a ribbon of toothpaste. They tunnel into the soil to pupate belowground. Acorn weevils emerge as adults the next year to climb up the nearest tree and start the process over again. The same acorn that serves as a nursery for weevils may also hold anywhere from 30 to 90 fungal variants. Many, if not most, of the fungi within an acorn before it drops are inherited from the acorn's mother, either passed directly into the acorn or shared from the foliage that surrounds it. Once it falls to the ground, the same acorn may host from 40 to 250 fungal variants, some from its mother, some from the leaf litter and forest floor around it. Most of these fungi live in the seed coat and acorn shell, but some live within the embryo itself. Some fungi afflict the plant. Others make it stronger, protecting it from other fungi or herbivores as the seedling grows.

Many organisms eat acorns without providing the tree any obvious benefit. But some birds and mammals—particularly jays and rodents—engage in a practice called scatter-hoarding, in which they collect acorns and other nuts and hide them individually to collect later. Scatter-hoarders rely on their wits, memory, and senses to hide and then relocate acorns, pine seeds, and other fruits. In some years, they hide many more acorns than they can retrieve. Acorns wouldn't go far without scatter-hoarders. Scatter-hoarders move nuts into a broad range of environments and hide them in places where the acorns are less likely to dry out or be eaten. Oaks have, as a consequence, evolved to take advantage of scatter-hoarders. When an oak fills the developing acorn with food, it isn't just providing energy for its own offspring. The oak is making a trade-off between enticing seed-dispersers and provisioning its young. The mother tree sacrifices most of the seeds it produces in its long life to seed-eating scatter-hoarders so that a few have a chance to grow.

Wherever you find oaks, you also find scatter-hoarding rodents—squirrels especially, but also mice, rats, agoutis, and others. Most if not all of these rodents can sense which acorns will taste the best or last longest in the pantry. They generally abandon or immediately eat those that are infested with larvae or poorly defended against future insect infestation. Gray squirrels and fox squirrels can detect how sound an acorn is by turning it in their paws, then holding it in their teeth and giving it a quick shake. A weevil-infested acorn is lighter than a sound acorn, and the squirrel discerns its mass just as we would feel the difference in weight between a baseball and a tennis ball by shaking first one and then the other. A squirrel will often cache an acorn that is free of insects. It will often abandon a weevil-infested acorn or eat it right away, getting the goods before the weevils can. In fall, you may find shells of squirrel-devoured acorns piled on fallen logs or tree stumps throughout the forest. Look on the outsides of the shells for scratch marks the squirrels left as they turned the acorns in their paws, then held each acorn as they chewed it open.

North American squirrels are reasonably good taxonomists: they at least know the difference between the White Oak Group and the Red Oak Group. Acorns of most White Oak Group species germinate the year that they fall. Some germinate on the tips of the branches even before falling. The squirrels know this tidbit of oak natural history and tend to eat the White Oak Group acorns more quickly. By contrast, most Red Oak species of the temperate North require a dormant season before they germinate. They fare better in a cache. This generality does not hold for all species. Trails in the upper Midwest are littered with eastern white oak (*Q. alba*) acorns in the weeks after they fall, many with the first few millimeters of the root snaking out through the cracked acorn shell to find the soil. Beside them, acorns of the closely related bur oak (*Q. macrocarpa*) lie dormant, waiting for winter cold to break their dormancy and then spring to wake them up. But move farther south, and bur oak does not have to go through a period of dormancy. The two species are White Oak Group members with different solutions to the problem of getting through an eastern North American winter.

Once it reaches the soil or damp leaf litter, the root tip emerging from the acorn quickly elongates to form a taproot several inches long. It will soon bristle with fine roots and root hairs. Acorns move their nutrients from the cotyledons into the unpalatable root when they germinate, sequestering the food where squirrels cannot easily reach it. Numerous rodent species can recognize an acorn that has broken dormancy by the presence of a shoot or, at least for eastern North American squirrels, erosion of the waxy coat on the acorn's surface. Breakdown of this waxy coating allows moisture to enter the acorn, helping germination along, and allows the odor of the embryo to diffuse out. A squirrel who determines that an acorn has broken dormancy will often nip off the root tip to halt germination, particularly in the quick-growing members of the White Oak Group.

When it has decided to cache an acorn, a squirrel will pound the acorn into the ground, using its incisors as a mallet and throwing its whole body into the job. It will look as though it has abandoned the acorn, but squirrels have a remarkable memory for cache locations, in some cases organizing cached seeds by species and using the locations of fixed landscape objects—a big tree or a boulder, for example—to triangulate the position of their acorn caches. One February, I watched a gray squirrel front-crawl through a two-foot bed of powdery snow. It surface-dove and remained below the snow for several minutes, retrieving an acorn it had cached months earlier. I followed its trail and found a handful of other caches, all surgically excavated. But many acorns escape retrieval and germinate. Acorns that are cached by squirrels survive at higher rates; tucked away underground, they are less likely to be found by hungry mice and insects. Squirrels are accidental gardeners, curating the oaks and shaping the structure of the forest.

Jays are as important to oaks as the rodents are. There are seven or eight acorn-dispersing jay species worldwide. Their close relatives, the rooks, also disperse acorns, carrying several at a time and then caching them singly. Acorn woodpeckers, tufted titmice, white-breasted nuthatches, red-bellied woodpeckers, and other birds move acorns around as well. Acorn woodpecker caches can be huge: you may have read about the woodpeckers who stashed seven hundred pounds of acorns in the wall of a California home in 2023. But among birds, only the jays and their close relatives, the rooks, are scatter-hoarders. Most other birds cache acorns high in a tree or deposit them on the ground's surface, where the acorn is less likely to grow into adulthood if it germinates at all.

Jays have all other acorn dispersers beat for efficiency. Squirrels may move acorns more obviously—a group of squirrels can move eight thousand acorns or more in a couple of days—but jays carry more nuts at a time, and farther. In one study, researchers watched a stand of eleven trees visited by blue jays over the course of a single season. Each jay spent an average of four minutes feeding on and collecting acorns. It pried the cap off, then stuffed the acorn into its gullet. Each jay packed in as many as four acorns before grabbing one more to carry in its bill. Each flew off to a cache anywhere from 100 meters to 1.9 kilometers away, though that's not the limit: rooks have been documented caching acorns 3 or 4 kilometers from the source, blue jays in Wisconsin may cache acorns as much as 4 or 5 kilometers from a mother tree, and European jays can move acorns up to 8 kilometers. The jays disgorged their acorns into piles on the ground and then hid them singly, either covering the seeds with leaf litter or pounding them into the ground with their bills. Few if any acorns were left uncovered. In the end, the jays moved 108,000 out of an estimated 221,000 acorns produced by the trees. They ate another 49,000 acorns while they were collecting and left behind roughly 64,000. Many of the acorns the jays left behind were infested with weevil larvae, a protein source eschewed in favor of unmolested acorn cotyledons.

Jays, like the squirrels, have a huge effect on which trees grow where. They favor viable acorns that are a bit on the small side or at least thin, so they fit easily into their gullet. This may help small-seeded trees disperse farther. North American jays also generally make their caches in relatively open habitats where oak seedlings are likelier to thrive. This same pattern has been observed in Holm oak (*Q. ilex*) acorns, which European jays bury singly as well, almost invariably under leaf litter or a centimeter or so beneath the soil's surface. For this reason, a jay-dispersed acorn stands a better shot at the future. Jays are such good seed-dispersers that they have been utilized to move seed into areas under ecological restoration: by one estimate, a pair of jays may save an oak restoration project $22,000 in planting costs. That cost savings pales in comparison to the work jays did thousands of years ago, reforesting North America as the glaciers receded. But we'll get to that in a few pages.

Oaks strike a balance between giving their dispersers what they want—an energy-rich meal that stores well—and making sure some of their acorns make it to adulthood. Oaks do so in part by using a chemical to control their dispersers' diets: they lace their acorns with tannins. Tannins hamper a variety of digestive enzymes that animals, fungi, and bacteria depend on, making proteins difficult for birds and mammals to digest. They can weaken the kidney and liver. They are toxic to many insects. They are found in oak wood, leaves, fruits, and galls. Tannins are not unique to oaks, but produced by a wide range of trees and other plants. For humans, tannins are mostly beneficial, at least in moderation. Tannins from grape skins, seeds, stems, and oak barrels contribute to the texture of wine. Tannins cure leather. They were the central ingredient of permanent ink used throughout the western world until the twentieth century.

Most important to this story, tannins make acorns bitter and a little toxic. If you have any doubt, crack the shell of a ripe acorn and nibble on the cotyledons. Doing so will probably make you pucker. If so, you can soak, boil, or bury the acorns in mud to leach the tannins away before you try another. Tannins make acorns an imperfect food. It's not impossible for organisms to develop counterresponses to tannins. A Japanese wood mouse gradually introduced to an acorn diet will ramp up production of the salivary enzymes that bind tannins. It will develop colonies of bacteria that break the bonds between tannins and proteins. After a week of acclimation, these mice do just fine on a diet of acorns. But mice that have not been afforded the decency of a breaking-in period rapidly lose weight if they are fed an all-acorn diet.

Ironically, most acorn-specialists cannot live on acorns alone. If you feed a scrub jay only acorns, it loses weight. Gray squirrels lose weight if you feed them only northern red oak acorns. Even acorn woodpeckers lose weight if they have only acorns to eat. Across their range, from northern Oregon to Central America, acorn woodpeckers depend on oaks. Because acorn crops are somewhat unreliable, you typically find the woodpeckers only where there are at least two oak species to provide food. (The most prominent exception appears to be in Colombia, where there is only one native oak species, *Quercus humboldtii*, but there are also acorn woodpeckers. Remarkably, the acorn woodpeckers there appear not to depend on the acorns for food. They do drink sap from the trees and are, for the most part, distributed only where *Q. humboldtii* grows.) Acorn woodpeckers painstakingly drill holes in trees, one at a time, taking about thirty minutes for each hole. In a year, an individual may drill a hundred holes, and those holes serve generations of birds. An average granary tree may have 1,800 or more holes in it and take three or more years to excavate. Some granaries have been documented to have 50,000 holes; these are the cumulative works of decades. Acorn woodpeckers collect acorns individually, fit them into holes point-first, guard their granaries, and move acorns to tighter holes when the acorns dry and shrink and start to fall out. Despite all this work, stored acorns only provide about 6% of an acorn woodpecker's annual energy requirements. Acorn woodpeckers feed their offspring primarily on insects. Yet if their granaries aren't packed full, fewer woodpecker offspring make it to the next generation.

Oaks don't suffuse their embryos evenly in tannins, like cloth soaked in a bucket of dye. Instead, tannins are most concentrated toward the growing tip, near the pointy end of the acorn. Thus, the portion of the acorn that surrounds the growing shoot tip and root tip is particularly bitter, more difficult to digest, and less palatable to fungi. The ends of the cotyledons beneath the acorn cap are lower in tannins and richer in lipids, high-energy fats, and oils. This makes the base of the acorn more attractive to acorn herbivores: it lures them away from the growing oak embryo. Acorn shell thickness in at least some oak species also increases from base to tip, making it more difficult for weevils to tunnel straight into the precious tip of the acorn than into the less-defended base.

Rodents, birds, and weevils as a consequence often begin feeding at the less-defended end of the acorn. Squirrels will hold an acorn in their paws and chew through the base of the acorn, beneath the cap. Jays hammer their way in from the base. Grackles score the acorn around the middle and split the acorn around its waist, losing acorns in the process but gaining easier access to the cotyledon tips. But seed-dispersers don't always follow the rules. The floor of my local forest preserve was littered with northern red oak (*Q. rubra*) acorns in fall 2023. Every red oak acorn I found that had been handled at all by a squirrel was eaten in part, and invariably from the tip—the end that is supposed to be better-defended by tannins—destroying the embryo and leaving a scattering of shells. By January, the squirrels had moved on to the eastern white oaks (*Q. alba*) and were spreading shells and half-eaten cotyledons everywhere.

In a year of plenty, seed predators often abandon the acorns they are working on after eating less than a third of the cotyledon mass. If they have been eating from the acorn base, the young plant may be unharmed. Fortunately, the mother tree packs more into the cotyledons than her offspring will need. Partially eaten acorns are often abandoned yet still able to germinate. The seedling is able to grow with only part of the cotyledon present. Even tiny seed predators often leave plenty of food for the seedling: a germinating acorn may have a hole in the shell from a weevil that has chewed its way out and two or more weevils feeding on the cotyledons as the seedling gets its start in life. One of the largest-seeded oaks in the world, *Q. insignis* of southern Mexican and Central American cloud forests, can lose up to one-third of the cotyledon with no obvious cost to the young oak. The loss of that much cotyledon actually appears to help the acorn germinate.

Acorn defenses are clever, but in a typical year, 70%–90% or more of the acorn crop may be devoured by bears, deer, turkeys, moth caterpillars, sap beetles, midges, and other animals. In a year of low nut production, nearly all the acorns in a forest may be eaten before they can germinate. Those that do remain often fare poorly. I recently heard Laura DeWald, tree improvement specialist at University of Kentucky, speak on the challenges of collecting eastern white oak (*Q. alba*) seeds for propagation. "In a crappy [seed production] year," she said, "don't bother." You'll waste time collecting. Acorns are hard to find in a year of poor acorn production, many of the seedlings die young, and those that survive often don't grow well.

Some oak species manipulate the timing of oak crops within a year to satiate predators: for example, *Q. schottkyana*, a Ring-Cupped Oak from Southwest China, produces more and smaller acorns early in the season, sating the wee-

Quercus sect. *Virentes*: *Quercus minima* leaves and acorn. Illustrated from a photograph provided by Béatrice Chassé.

vils, only to produce bigger acorns later in the year. These late acorns are more likely to make it all the way to germination. Some individual oaks are prodigious acorn producers year after year, regularly outshining their peers. All things being equal, we might expect natural selection to favor these trees and acorn production to ratchet up in the population generation after generation. But if oaks all ticked along, steadily producing acorns or even gradually increasing their pace, squirrel, mouse, and jay populations, along with the creatures who depend on acorns without dispersing them, would swell to take advantage of the abundance. Acorns would be a predictable resource.

Oaks and selected plants scattered across the Tree of Life—from pines and podocarps to grasses, fruit trees, and nut-bearing trees—game the system. They saturate the market at irregular intervals, giving their seed predators more than they can eat in some years and depriving them of food in others. This has the effect of controlling scatter-hoarders' population sizes to ensure higher acorn survival. This phenomenon of coordinated variation in seed production is called masting. Masting involves most of the populations of a given species across an entire region fruiting in synchrony. For many plant species that do not mast, variability in seed production is simply a result of the fact that a plant can store resources until it has enough to invest in fruit. Jack-in-the-pulpit, for example, is an understory herb of eastern North America that lives as a male for years until it has enough food packed away to produce fruits. Then it becomes female for a year, makes a lot of bright-colored, juicy, toxic fruits. It goes back to being male the next year. But this individual behavior doesn't satiate the squirrels and raccoons and other critters that eat jack-in-the-pulpit fruits. One year some plants fruit. The next year, others do. A population of jack-in-the-pulpit plants all doing their own thing produces fruits year after year. Masting species take episodic fruiting a step further: when we have good crops of bur oak acorns at the Morton Arboretum, I know we will find bumper crops of bur oak acorns across the entire Chicago region and much of the upper Midwest.

Masting is determined in part by how individual trees use their resources from year to year, similar to jack-in-the-pulpit. Seed production the year after a mast is almost always reduced: trees will have devoured their resources and need to build them back up before masting again. This only explains low seed production, however. It doesn't explain how masting is coordinated across populations and regions. Oaks synchronize seed production primarily by cueing in on weather. For some populations, flower production varies from year to year. In years of high flower production, trees throughout the region produce more acorns, provided they have enough stored carbohydrates to provision all those acorns with food. In other species or populations, male flowers are produced consistently, but weather influences how effectively pollen moves on the wind. The female flowers in these species simply don't get all the pollen they need in years of low production, whereas they are saturated in years of high production. Masting in these latter species and populations is driven by weather when the flowers are shedding pollen, or after pollination when delayed fertilization is happening and the fruits are beginning to develop.

The environment can, in other words, either encourage pollination or veto acorn production. A well-timed rainy season may make it possible for the oaks of a species to produce an abundance of oak flowers. A rainy spring that extends too long may quash the male flowers. Warm temperatures in the spring can synchronize flowering across the region. A late freeze can kill the pistillate flowers or throw them out of sync with one another. A drought in midsummer can cause fertilization to fail or the female flowers to abort. These drivers overlap and intersect to make masting relatively irregular, something that seed predators cannot easily respond to. Because spring weather is often particularly variable in the Temperate Zone, where oaks evolved and dominate today, the early flowering of oaks helps to synchronize masting.

Insects can influence masting as well. Periodical cicadas of eastern North America spend their long childhoods feeding on tree roots. Beneath the ground, they pierce the roots with their straw-like mouthparts, drinking from the trees' xylem, which carries water and some amino acids but essentially no sugar. Cicadas take a long time to develop on such a low-nutrient diet. The nymphs feed on small rootlets, moving around to fresh ones as the rootlets die. In these first years, the cicadas can reduce tree growth by as much as 30%. After thirteen or seventeen years belowground, periodical cicadas emerge en masse. They molt into adulthood and fly into the treetops, where they wedge their mouthparts into the branches and continue feeding from juices flowing through the xylem. But the cicadas are not there primarily to eat: their calls—sirens blaring each day from morning until early evening, all through summer—are designed to attract mates. The cicadas find each other in the treetops, copulate abdomen-to-abdomen, facing opposite directions, then go on their way.

After mating, the females lay their eggs in the branches. They do a little harm to the tree in the process. In ovipositing—laying their eggs—female cicadas slice the bark with their ovipositors and deposit eggs inside the slit. The effect of these branch cuts on tree growth is relatively minor for a large tree, but detectable: across the landscape, oak trees put on about 4% less growth in years when the periodical cicadas emerge. On a small tree, the effect can be substantial. In that year and the next, the oaks also produce fewer acorns. Unexpectedly, the periodical cicadas seem also to have a positive effect in the following years: two years after cicada emergence, acorn production rises. This may be a consequence of the slug of nutrients from decomposing cicada bodies. Once again, energy from the sun works its way through the ecosystem in surprising ways.

The geographic ranges of oak species track changing climates, fires, human migrations, competition from other species, and changes in how humans manage forests and savannas. You might notice range shifts in your own lifetime. Sessile oak (*Q. petraea*) expanded its range by 20 meters in a single oak generation in one French forest. Holm oak (*Q. ilex*) migrated at rates of 20 to 60 meters per year over the past century through forests along the French Atlantic Coast. At the boundary of species ranges, oaks move even faster: the northern range limit of northern red oak (*Q. rubra*) may be expanding by an average of 500 meters per year in Quebec, and Mongolian oak (*Q. mongolica*)

Germinating acorn of northern red oak (*Q. rubra*) with pine needles. The acorn shell has cracked open, exposing the pale cotyledons inside. The taproot is visible at the base of the acorn, penetrating the soil beneath it. The wing of a sugar maple fruit leans against the side of the acorn. The circular scar at the top of the acorn in the image is actually the base of the acorn. The scar shows where the cap was affixed. July 17, 2022, Fallison Lake, Wisconsin.

appears to be migrating into boreal forests at rates of roughly 1,200 meters per year. Even these rates, however, may not to be high enough for trees to keep up with the current rate of climate change.

To understand how fast oaks and other trees can migrate, scientists often study the past 20,000 years. This period—the late Pleistocene and the entire Holocene—started with the continental glaciers slowly receding from their position at the end of the Last Glacial Maximum to their current holdouts at the Earth's poles and high elevations in the mountains. It has included periods of relatively rapid climate change. Glaciers have been coming and going across the Northern Hemisphere for the past 2.6 million years or so, the duration of the Pleistocene. There have been eight glacial advances and eight glacial recessions in the past 800,000 years alone, driven primarily by regular changes in the circularity of the Earth's orbit around the sun. These changes in Earth's orbit affect how much solar radiation reaches us each year and consequently how warm the Earth is. Each glacial advance was driven by reductions in heat from the sun reaching the far North each summer. Temperatures cooled, and more snow fell in the winter than could melt away in the summer. The accumulating snow crushed and fused to ice under its own weight, packed to more than three kilometers thick. Masses of ice flowed outward, squeezing continent-scale glaciers over the northern half of North America, Europe, and the northernmost reaches of Central Asia (East Asia was left mostly unglaciated). Advancing glaciers scraped away forests and prairies, steppes and savannas. They ground soils to dust. They plowed up oaks and everything else. They carried boulders, gravel, tree trunks, and other debris southward and deposited them far from home. The cooling climate transformed much of the rich, deciduous forests within hundreds of kilometers of the glacial line to boreal forests and spruce woodlands.

Each glacial maximum ended as the Earth's orbit around the sun stretched and flattened, bringing the Earth closer to the sun twice each year. This caused temperatures to rise. The pressure on the ice relaxed. Ice chunks abandoned by the retreating glaciers became buried under rock and soil and then melted to form depressions in the ground, kettles that pooled with glacial runoff. Water coursing beneath the glaciers deposited sediments that formed sinuous eskers. Torrents melting from the glacial margins reshaped rivers. The repeated advance and retreat of ice rearranged a flora that was already pretty modern. We would have recognized most plant species living prior to the Pleistocene glaciation, though the mammoths, mastodons, giant sloths, and steppe lions would have been a shock. The communities, however, were novel. The range of each species expanded, shifted, and receded in its own way.

Expansions and contractions of geographic ranges are recorded for many plant genera in deep stacks of lake-bottom sediment. Tens of thousands of lakes in the pockmarked terrain of the far North, left as the glaciers receded, form a library of plant community history. Pollen grains from oaks, ashes, pines, spruces, grasses, sedges, and a host of other plant species living 20,000 years ago rained on the surface of lakes every year, just as they do today, mostly close to where the plants that shed them were growing. Pollen settled to the bottoms of the lakes and was covered by mud and detritus. The same thing happened year after year, with each year's deposits layered thinly over those of the previous year. The layers of sediment form a record of plants growing in the surrounding area over successive time slices. Paleo-

ecologists combine sediment cores from lakes around the world with information about ancient climates and estimates of genetic variation in contemporary populations. In aggregate, these tell the stories of plant species'
migration and the assembly of plant communities.

At the end of the Last Glacial Maximum, the oaks of Central and western
Europe were mostly taking shelter in the Iberian Peninsula and northern Africa, Italy, and the Balkan Peninsula. The oaks of eastern North America were
spread broadly across the southeastern United States, trailing northward up
the Mississippi River Valley. On both continents, there were holdouts within
a few hundred kilometers of the ice margin or even closer—probably more
in eastern North America, where forest trees may have been present at low
densities far to the north even at the Last Glacial Maximum—but the mass
of oak distribution was restricted to the south. The oaks of California and
Mediterranean Europe were mostly withdrawn into localized refugia. And
the oaks of East Asia, where temperatures had cooled but continental glaciers were absent, were concentrated in mountain populations, with at least
some evergreen species spreading along the southern China coast or into
Southeast Asia.

As the glaciers receded, rivers wound across the continents, scouring
out floodplains. In North America, forests of spruce, aspen, and hardwoods
such as ashes, elms, and oaks spread along the glacial margin. This forest
looked a bit like today's boreal forest, but it was an assemblage of species
that no longer co-occur: cool temperatures near the glaciers intersected with
the seasonality of the upper Midwest to produce a climate that doesn't exist today. Temperatures continued to warm, and the glaciers shrunk farther
northward. Boreal and spruce forests were replaced by maples and temperate pines, then by oaks, prairies, and grassland. In some places, oaks raced
ahead, forming populations nearly at the glacial margin. Oaks may have hit
migration rates as high as 500 meters per year in Europe—perhaps spik-
ing to as high as 1,000 meters per year along the Atlantic coast in the Early
Holocene—and 350 meters per year in eastern North America. Those velocities are estimated from pollen records and may be unrealistic: there may
have been northern populations near the glacial margin that enabled oaks
to recolonize rapidly following the Last Glacial Maximum, particularly in
North America. If so, these rates based on pollen cores are probably overes-
timates. But even if we focus just on the movement of the core of each species
as we understand it, making a more conservative estimate of migration
rates, oak migration reached average rates of at least 100 to 130 meters per
year in eastern North America.

By 10,000 years ago, oaks covered much of their range in eastern North America and were continuing to spread into what is now Canada. Temperatures warmed variously from region to region, and northern environments experienced sustained warm and dry conditions, sometimes lasting hundreds to thousands of years. Fires flashed across hilltops. Prairie and steppe expanded across the center of North America and Eurasia as woodlands and savannas backpedaled. Temperatures cooled to their most recent low in many places at about three hundred years ago, the depths of what we now call the "Little Ice Age." The Northern Hemisphere stayed cool into the nineteenth century, then began to warm again, rapidly.

We don't yet know how different oak species responded uniquely to recession of the glaciers after the Last Glacial Maximum. This kind of detail is not readily interpreted from the pollen record alone, where oaks are usually only identified as far as genus. Genetic data may help us figure out these histories. For now, the story is an aggregate over populations and species.

In 1899, Clement Reid, a largely self-trained geologist and naturalist, published a book entitled *The Origin of the British Flora*. He was struck by how short seed dispersal distances must be, based on what nut trees could do by themselves (which amounts to dropping their seeds on the ground and hoping for the best). It was initially not clear to him how oaks could have traversed the ground needed to repopulate the continents so quickly following glaciation. He noted that "the oak, to gain its present most northerly position in North Britain after being driven out by the cold, probably had to travel fully six hundred miles, and this without external aid would take something like a million years." This conundrum became known as Reid's Paradox. Squirrels alone couldn't explain the rate of tree migration: squirrels move acorns an average of about 10–20 meters from the source tree and rarely as much as 40 meters. With many trees they average 2 meters or less. Reid suspected that birds might do the trick. A few years before publishing his book, Reid documented finding rooks "in the middle of an extensive field, bordered by an oak-copse and scattered trees . . . feeding and passing singly backwards and forwards to the oaks." He found "stabbed and pecked" acorn husks in the field and one he presumed had been stashed by a rook. Might rooks explain the reoaking of the British Isles after the glaciers receded?

Researchers working primarily in Europe and the United Kingdom in the 1920s and 1930s demonstrated that jays and rooks regularly hoarded food, as Reid had suspected. Since the 1950s, there have been numerous studies demonstrating that jays often cache their acorns more than a kilometer away

Germinating acorn of northern red oak with taproot. A few fine roots are evident near the midpoint of the taproot. July 17, 2022, Fallison Lake, Wisconsin.

from the source tree. A rook may cover 4 kilometers at a time. These numbers dwarf the 350 to 500 meters per year needed to explain the migration of oaks as the glaciers shrank northward. But there is a catch: oaks don't reproduce when they are only one year old. Instead, they run through a generation in something closer to twenty years within a closed forest, with some estimates of average length of a generation—from seed of a mother tree to seed of her offspring—running as high as fifty to seventy years. To move a population of oaks northward at 500 meters per year, their acorns need to cover 500 meters times the number of years in a generation each year. At this rate, acorns would need to disperse anywhere from 7 to 10 kilometers per year to keep up with the observed rate of Holocene migration, assuming an average generation time of twenty years.

Jays likely helped the situation along by caching acorns in the open, not under a dense forest canopy. Trees may grow more rapidly in the open sun, with less competition from other oaks. As a consequence, the average age at which a tree produces acorns may be younger for a jay-dispersed seed. But at the height of their movement, oaks may have raced along faster than even jays could take them. One simulation of oak migration history in Europe suggests that the gradual diffusion of acorns toward their modern ranges was punctuated by occasional long-distance dispersals of 20 to 60 kilometers or more. A single tornado or flock of migrating geese might move millions of sedge or birch seeds in one event. These seeds, however, are dispersed by migratory birds or directly by the wind. What could account for long-distance dispersal in the oaks, with their relatively heavy acorns?

One answer in North America may have been the passenger pigeon. Passenger pigeons reached their height of recorded population sizes in the mid-1800s, when one flock in Kentucky was estimated at more than 2 billion birds. It darkened the sky for five hours as it passed by. Passenger pigeons filled eastern North American forests through much of the nineteenth century, foraging in treetops and on the forest floor. They could pack thirty acorns or beechnuts in their gullet and carry them for miles until the nuts worked their way down to the gizzard. The gizzard is a grinding machine that birds need because they don't have teeth; it does the work our molars would do. Acorns that reached a passenger pigeon's gizzard were destroyed. But a pigeon often spit up acorns in exchange for better food. And any bird that died in flight would fall to the ground, where acorns might have a chance to grow from the fallen corpse.

Any single passenger pigeon was probably not likely to disperse many oaks in this way. But with so many passenger pigeons traveling across the country, the rare success stories probably contributed to the long-range dispersal of acorns. Of course, we will never know for sure. Passenger pigeons had naturally fluctuating population sizes and appear to have been on a downward slide by 150 years ago, but humans dealt the final blow. We logged the forests they depended on and hunted them relentlessly, beating the birds with long sticks as they roosted, knocking them to the ground to feed our pigs. Passenger pigeons cycled acorn energy through North American forests for more than a million years. But in less than 20,000 years, humans drove them extinct. The last passenger pigeon died in captivity in 1914.

Humans probably aided oak migration during some periods of our history. Populations of modern humans migrated from Africa into the Middle East and western Europe around 60,000 years ago. Our close relatives—the ancestors of the Neanderthals—had moved into Europe half a million years earlier. Humans encountered a cooler world in Europe as they moved northward. They encountered oaks at the same time. Prehistoric sites in the Middle East and Europe hold evidence of humans grinding and roasting acorns. Our myths show that acorn-eating was a part of early human life. Humans regularly stored acorns in pits in the ground. (Unlike squirrels, humans caching acorns in the ground are hoarders rather than scatter-hoarders.) As people migrated, taking acorns with them, it would have been hard to avoid losing a few acorns along the way and thus planting oak trees along the migration route. In Europe, the inferred rate of human migration very closely matched the rate of oak migration at approximately 40,000 and 18,000 years ago. In some areas of eastern North America, masting trees such as oaks, hickories, and walnuts were clustered around Native American villages, where they may have been managed for food production.

Humans have lived in North America since before the end of the Last Glacial Maximum. We continued eating and moving acorns around the landscape well into the modern era, sometimes even across the ocean. North America's northern red oak (*Q. rubra*) was introduced to Europe in the seventeenth century, probably from the northern portions of its natural range. It thrives now in more than twenty countries across the Atlantic. It has become a weed, replacing native species and reproducing naturally, having escaped from the insects that it encounters at home. Insects have trailed along with some oaks: human-planted stands of turkey oak, *Q. cerris*, across northern Europe over the past 500 years have brought with them at least seven gall-forming insects of the wasp genus *Andricus*, the life stages of which alternate between members of two oak sections: the White Oak and the Cork Oak Groups, *Quercus* sect. *Quercus* and *Q.* sect. *Cerris*, respectively. In summer 2023, I visited a site in Boise, Idaho, where pedunculate oak (*Q. robur*) acorns appear to be reproducing along the river. It is clearly coming from trees planted in the parks and around town. Perhaps in a million years, pedunculate oak and northern red oak will be thought of as native to both North America and Europe.

When we look at a range map for any oak species, we are looking at the outcomes of countless individual processes following one after the other: a tree that outcompeted a hundred others, growing from an acorn that survived after a mouse nibbled off the ends of its cotyledons, after a jay had moved the acorn from a mother tree two kilometers away, after a warm dry spring and a rainy summer, on soils left behind by now-vanished ice, after a lean year, when almost every acorn in the population had been devoured. A species is a web of individuals, knit together by reproduction and migration history.

After their travels across seasons and landscapes, possibly after being cached, nibbled on, infested with weevils, or ingested and then disgorged, a few acorns germinate. The fruit wall and seed coat become permeable to water either immediately or after a period of dormancy. The seedling swells. The shell cracks. Then the root emerges. In some Eurasian species—for example, the "Kerrii" group of the East Asian Ring-Cupped Oak Group (section *Cyclobalanopsis*) and *Q. alnifolia* and *Q. aucheri* of the Holly Oak Group (section *Ilex*)—the root emerges from the base of the acorn, at the scar left by the acorn cap. In most species, the root emerges from the acorn's pointed tip. Nutrients from the cotyledons pour into the developing root, where they are less palatable. The root elongates into a taproot, which can in some species extend to ten inches in the first season, or thickens into an underground tuber, as with at least some of the Southern Live Oaks and Ring-Cupped Oaks.

Then the shoot develops. The stalks at the base of the cotyledons stretch out during development, pushing the cotyledons away from the growing plant. If a squirrel or mouse chews the cotyledons off at this point, the seedling will likely be just fine, little if at all affected by the loss of its mostly exhausted food source. In its first full growing season, the seedling will produce photosynthesizing leaves. The seedling has plenty of hard work ahead, but it now has a shot at shaping the future. Only about 2% of pistillate flowers make it all the way to becoming acorns that germinate, ignoring all the acorns that are devoured by squirrels and jays and insects. Of those acorns, perhaps 1% grow to their second or third year as seedlings. Every oak seedling stands at the end of a long trail of improbabilities.

Imagine that you could draw a one-generation map of reproduction for all the oaks of one species. First, draw an arrow from each mother tree to each of its living offspring. The mother tree may be dead, so don't worry if the base of an arrow hangs in empty space or hovers over a stump. Next, draw an arrow from each father tree to all the mother trees it pollinated. This arrow represents the pollen grains that, against the odds, hitched a ride from a catkin to a pistillate flower, then fertilized the ovule that grew into the seedling that grew into a tree. Do this for each of the thousands or millions of oaks in the species you've chosen. Most trees will be a mother to many and a father to many others. Individuals that stand 25 to 100 meters away from each other in real life will have lots of pollen-flow arrows between them on your map. Pollen arrows will tend to point in the dominant direction of the wind in some areas of your map, though only imperfectly. Seed arrows will point wherever the squirrels and jays carried the acorns.

If you step back from your map, you won't see the individual arrows. You'll just see the density of genes moving across the landscape. Clusters of individuals within a single forest will form a thick fog. A fine web will reach out to clusters 100 meters away. A few spidery vectors will span long distances, tracing the route a jay flew before disgorging its acorns or the flight of a pollen grain that traveled between forest stands on the wind and touched down, improbably, on a receptive stigma. At some point, stands of a single species will be so far from each other that they will be connected only via intermediate stands of oaks, stopping-off points along the leapfrog game that genes play across continents.

The clouds of arrows on your map are populations, groups of individuals who regularly interbreed. Some populations are nearly discrete, living within a single forest. Other populations are much more diffuse and difficult to define. The degree of connectedness varies even within a single species: individual stands of pedunculate oak (*Q. robur*) near the southern edge of the species' range, for example, rarely exchange pollen with stands just four to six kilometers away, while other isolated stands of the same species mate with trees more than eighty kilometers away. It is often unclear where one population ends and another begins.

Imagine how your drawing of the web of reproduction and migration in oak populations would look if you drew it anew every 100 years. Draw one map for each century, going back 1–2 million years. Stack up your drawings and assemble them into a flip-book. As you turn the pages, you'll see populations squeeze together to the south as glaciers expand and spread back northward as the cold recedes. Grasslands will pool across the midcontinents, severing oak populations from each other. Connections will flicker as the climate fluctuates. Some populations will be only rarely connected to the others, perhaps brought into the fold by an exceptional storm carrying a cloud of pollen or an errant jay with a gullet full of acorns. Others will be at the heart of a dense web. Populations will arise as acorns travel off into new territories and became established, connected to their ancestors by pollen movement and the stray acorn. Other populations will wink out of existence. Your flip-book will end with the geographic distribution of the species as it looks today.

Now draw the same set of maps, but this time use two very closely related species instead of one. In eastern North America, you might choose *Q. ellipsoidalis* and *Q. coccinea*, Hill's oak and scarlet oak. In East Asia, you might choose *Q. aliena* and *Q. serrata* or *Q. acutissima* and *Q. variabilis*. In western Mexico, you could use *Q. scytophylla* and *Q. sideroxyla*. Start your flip-book again from the millennia before your two species separated from each other. You won't see two webs. Instead, you'll have one, representing the population that gave rise to the two species you know today. Populations will come into and out of connection with one another in the first pages, forming one shifting ancestral species. As you continue flipping the pages, your web of individual populations will begin to separate into two. They might separate in geographic space, or they might separate in habitat or even flowering time. The separation will likely be gradual. The connections between these separating populations will thin like fibers between your fingers as you stretch a cotton ball to the point of breaking. At the end, you'll see two species standing where previously there was only one. Your species will meander across the map, meeting at times but still largely separate from each other.

You will be challenged to say at what moment your species split into two. If you go page by page, is there a clean break point between two centuries? Probably not. For most organisms—certainly for oaks—there is no single moment at which two populations cross the line and become separate species. And after they separate from each other, closely related species still have semipermeable boundaries. Genes slip between them, in some cases for a very long time. Species are a kind of population that has found its own place in the world, distinct from others. Individuals within the species reproduce with each other more than they do with other species, but the boundaries are porous, at least initially. And for every population that becomes a species, there are unnumbered populations that might have become species but never will.

Pollen grains shed by a bur oak on the edge of a prairie knoll in northeastern Illinois or a forest in southeastern Missouri convey alleles across the landscape. Most land on the surface of a leaf or a trunk or on the ground, where they will die. They may land on a red oak stigma, where they don't stand a chance. They may bump into a mass of big bluestem in the surrounding prairie. The vast majority will never make it to a related oak; this is the inefficiency of wind pollination. But a small number will stumble across receptive stigmas on a nearby bur oak, swamp white oak, swamp chestnut oak, overcup oak, eastern white oak, post oak, or other close relative. These pollen grains will, if they arrive alive, begin (or at least try) to grow a tube toward the ovule, then halt for a few weeks in anticipation. Most will be shut down by the mother tree before they can reach the ovule. Very, very few will pollinate ovules. Of the resulting acorns, the majority will be eaten by fungi or animals. A small number will become seedlings, and of those seedlings, a small number will become trees.

Every stamen packed with pollen grains, every oak draped with catkins and dotted with pistillate flowers, represents a universe of possible futures. These potential lives fall away in droves every spring as billions or tens of billions of pollen grains from every tree slam against the sides of cars, lodge

Northern red oak seedling. The acorn is still attached to the base of the plant by the cotyledonary petioles. These are visible as strap-like extensions of the acorn, wrapping around the base of the plant. July 17, 2022, Fallison Lake, Wisconsin.

in the bark of maple trees, land on trillium petals, are lofted to 3,000 meters in the air, or fall to the ground undispersed in a wet spring. Every seedling is the outcome of a long chain of good fortune. The words *population*, *species*, and *Tree of Life* stand in for successions of individual couplings that extend back millions of years. All of oak diversity traces back to flowers and acorns, these moments of reproduction.

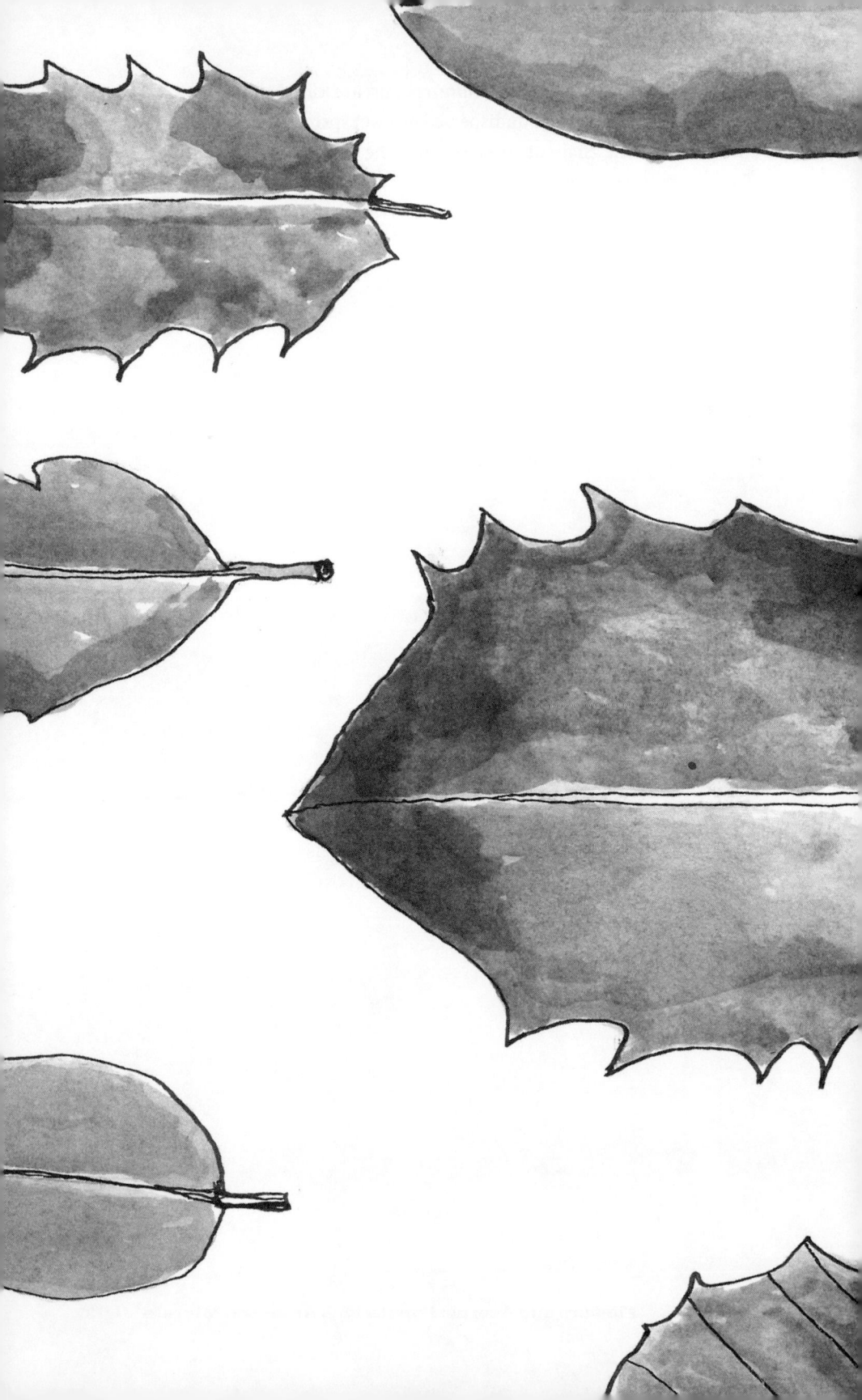

2

Variation

Populations Evolve

Find an oak close to your home. Make a collection of leaves in early summer, after they are fully expanded but before they are battered by storms and chewed by insects. Use a ladder, if you can, to get some that are high in the crown. Choose some that are lower as well. Pick a few leaves growing close to the trunk, where they are shaded and protected from the strongest winds. Make sure to get others that are fully exposed to the sun. Find a long branch on your tree, and collect a set of leaves from the base of the branch to its tip. As you collect, note the position of each leaf on the tree.

Lay your leaves out on the ground or press them in a phone book for later. When you have a free hour, try sorting your leaves by size or by shape. You can line them up from smallest to largest or group them by how deep the sinuses are between their lobes. However you sort them, you'll find variation. Some of the differences you find will be associated in predictable ways with where on the tree you collected each leaf. Leaves from the edge of the crown are likely to be thicker and smaller. They developed with relatively high exposure to sun and wind, which put a premium on being tough. In many species, the more exposed leaves will also be more deeply lobed, which helps keep the leaves cool. Leaves from the interior of the crown are likely to be thinner and larger in area. They were protected from the elements and could throw their resources into capturing sunlight. If your tree should be cut down in the future, watch to see if it produces new shoots from the base. Collect a few leaves from the stump sprouts. They will probably have lumpy outlines. They will resemble the leaves of the adult tree they sprouted from but will be less deeply lobed, if lobed at all, and they may be as much as two or three times the size of the leaves on the adult shoots.

The variation you find among the leaves on your tree is called plasticity. It is the variation that a single individual exhibits over time or, if it is cloned and grown in different places, over space. Plasticity gives individual trees the ability to weather environmental changes. When the root of an eastern white oak (*Quercus alba*) grows into dry soils, it tunes its chemical protections to maximize collaborations with symbiotic fungi. If a Holm oak (*Q. ilex*) is stressed, it grows fine roots that are better able to gather resources from the soil. All organisms are plastic to a certain degree, but plasticity is especially important to plants, and even more so to trees. Trees can sense light, soil nutrients, hormones, water availability, temperature, and other attributes of their growing environment, but they are rooted in place and cannot go indoors when the world treats them roughly. They cannot put on a coat when a frost is forecast. A bur oak (*Q. macrocarpa*) in the forest where I work has been weathering the elements for more than 250 years. There is a 464-year-old eastern white oak in Buena Vista, Virginia. A sessile oak (*Q. petraea*) growing at high elevation in southern Italy was radiocarbon-dated to more than 870 years old. Plasticity allows old oaks like these to respond to variable springs by flowering at the right time for the year, and to warmer or drier summers by growing tougher, more deeply lobed leaves that are more resistant to drought and better able to cool themselves in a breeze.

An isolated clone of Palmer's oak (*Q. palmeri*) in the Jarupa Mountains of Riverside, California forms a thicket about 25 meters across at its widest. It is a single individual genetically, composed of dozens of sprouts from a root system that has survived millennia of burning. It is hypothesized to be more than 13,000 years old. Plasticity has allowed this clone to persist and spread as global temperatures rose by 4°C and a city grew up around it. Today, it grows at a hotter, drier, lower-elevation site than the rest of its species. It was left behind as populations migrated upslope in response to climates warming through the Holocene. Palmer's oak is one of many oak species that resprout in response to fire, browsing, or avalanches. Other species explore new territory via underground runners, or rhizomes. Clones of *Q. havardii* and *Q. hinckleyi* in the arid southwestern United States may form thickets 30 meters across. Mature trees of Pyrenean oak (*Q. pyrenaica*) and Holm oak (*Q. ilex*) also spread by rhizomes, as do several other oaks of dry habitats. Plasticity allows oaks to adapt to the environments they have crawled into and to survive in place as the world changes around them. Every spring, these clones shed pollen grains holding gene copies that were selected by the

Canyon live oak (*Q. chrysolepis*), leaves from a single tree. Illustrated from a photo provided by Béatrice Chassé.

environments in which they first established. Ancient trees and long-lived, shrubby clones are genetic time capsules.

Plasticity is only one source of natural variation. It is generally not heritable: a tree that has become stunted by the elements will not pass that life-form down to its offspring. Genetic variation, by contrast, refers to the diversity of gene copies that different organisms receive from their parents. Genetic variation is heritable. Every population of oaks contains substantial genetic variation, which shapes all aspects of form and natural history. Genetic variation becomes particularly obvious in species with wide geographic ranges. Bur oak spans about 2,000 kilometers of latitude from its northernmost populations in Manitoba to its southernmost populations near Houston, Texas. It ranges about 2,400 kilometers in longitude from the Black Hills at its northwestern limits to scattered populations in New Hampshire at the east edge of its range. Its leaves vary in size from smaller than my hand to larger than a dinner plate. Gabe Ribicoff, an undergraduate researcher in our lab, returned from autumn fieldwork in 2022 with a set of three bur oak acorn caps collected in Minnesota, northern Illinois, and Missouri, which were nested within each other like a set of bowls. The largest one nearly filled his palm; the smallest was about the diameter of a U.S. penny. This range in size is well beyond the variation we find on a single bur oak tree. It is also larger than we would find if we were to plant any of those trees into a wide range of environments.

Ribicoff's nested acorn caps reflect the combined effects of plastic variation, as trees responded to the environment in which they grow, and the heritable, genetic variation that the oaks from which he collected the acorns inherited from their ancestors. Plastic variation enables individuals or populations to survive in uncertain environments. Heritable variation enables populations and species to adapt and evolve. When you put plastic and heritable variation together, you can begin to figure out how a species will respond to what the environment throws at it. Evolutionary biologist Theodosius Dobzhansky wrote, "Variation is the fountainhead of evolution." The variation among trees of a single species within one population is the starting point for the next generation.

Modern biodiversity science traces back to *The Origin of Species*, published by Charles Darwin in 1859. Darwin had spent five years traveling the world as a naturalist on the HMS *Beagle*, and then twenty years after that working over his ideas, before publishing the book. He considered his travels and work on the *Beagle* "by far the most important event in my life." He had documented landforms, geology, plants, and animals all along the way. In his travels along the coast of South America, Darwin collected fossils of giant ground sloths, an extinct horse, a relative of modern elephants, and a "rhinoceros-sized rodent." He marveled at the rainforests of Brazil. He studied bands of oyster shells embedded in bedrock that had been uplifted over vast periods of time. He correctly inferred that coral atolls were ancient coral reefs that had formed around volcanoes. He collected more than thirty finch specimens in the Galápagos Islands; the finches were so diverse in form and life histories that he did not even realize at first that his nine species, all new to science, would turn out to represent four closely related genera of a single bird subfamily. By the age of twenty-seven, Darwin had seen and thought about more wild landscapes and biological diversity than most of us see in a lifetime.

You might imagine that with a background like this, Darwin would have set the opening chapter of *The Origin of Species* in some exotic place, probably in South America. Instead, the first chapter of Darwin's book is rooted in gardens and pigeon lofts. Darwin's five years on the *Beagle* were a turning point in his thinking about the origins and evolution of diversity, but they were only part of a life packed full of experiments and observations. As a student, he had favored oyster-trawling and collecting specimens in Edinburgh's tidal pools to his medical studies. After failing in medicine, he studied to be a clergyman but directed much of his time and attention to collecting beetles and plants. After spending five years on the HMS *Beagle*, Darwin raised a family. He took careful notes on his firstborn's behaviors as an infant. He followed "humble-bees" around his garden to see which flowers they were pollinating. When he could not keep up with the bees on his own, he recruited a few of his children—he had ten, of whom seven lived to adulthood—to stand around the yard calling out "Here is a bee!" as one flew past, so they could map each bee's route. The year before he died, Darwin published a book on how earthworms create and churn soil. He lived as though, in Annie Dillard's words, his "least journey into the world [were] a field trip, a series of happy recognitions."

Darwin knitted his observations on flowers, birds, fossils, landscapes, and all the natural world into his theory of "descent with modification through variation and natural selection," which he presented in *The Origin of Species*. Darwin's theory of biological evolution explains the diversity of life on Earth by marrying the Tree of Life—*descent*—with the evolution of form and ecology—*modification*. It explains evolution as an outcome of *natural selection*, which tends to favor the fittest combinations of organismal attributes in every generation. Variation gives evolution the raw material from which to select organisms fitted to their environment. And the organisms closest to us—garden plants, domestic dogs, or pigeons—provide us the clearest opportunities to study variation and evolution.

The Origin of Species opens with a simple observation: cultivated plants and domestic animals are often more variable than their closest wild relatives. All of today's domestic dog breeds are descended from one or two ancestral wolf populations, initially domesticated more than 15,000 years ago in Europe and more than 12,500 years ago in East Asia. But compare a dachshund with a Chihuahua, poodle, bulldog, collie, huskie, great Dane, terrier, foxhound, or any of hundreds of other named breeds. The variation among these breeds far exceeds the variation we find in the wolf species from which they are descended. Pigeons are similar: carriers, tumblers and rollers, runts, barbs, pouters and croppers, wattles and homers, and fantails are all so distinct that no ornithologist encountering them in the wild would place these pigeons in the same species. Yet all domestic pigeons are descendants of a single species, most likely from the Mediterranean, tracing back 3,000–5,000 years or perhaps longer ago. One of Darwin's insights was that the wide range of variation we observe today must be due to human selection, "our domestic production having been raised under conditions of life not so uniform as, and somewhat different from, those to which the parent species had been exposed under nature."

Domesticated animals and plants show us that great biological variation arises gradually, as humans decide what they want from the many varieties they breed. Look back at the cut flowers or gooseberries your parents used to buy, Darwin writes: in the span of a human generation, they have become bigger and better just from the effects of humans selectively breeding the ones they like best. Animal breeders study the variants in merino sheep "like a picture by a connoisseur . . . so that the very best may ultimately be selected for breeding." Plant breeders walk the fields, pulling the "rogues" and slackers that they do not want. They select heritable variants—variation that can be passed down from one generation to the next—and in so doing, they nudge each generation down a road of modest, barely perceptible changes toward sometimes-fantastic ends. A pigeon with slightly longer tail-feathers became the starting point of a lineage that led to a fantail. The variation we find today in pigeons, dogs, and wheat arose gradually through this succession of conscious decisions, each selection inching toward a new variety.

The products of human selection are bite-sized examples, microcosms of diversity. And because we can put approximate times on how long humans have been breeding dogs and pigeons or cultivating wheat and blueberries, these examples show us how rapidly selection can work. If so much variation can arise in thousands of years, how much more could arise in 100 times that much time, or 10,000?

In his second chapter, Darwin turns his attention to natural selection, by which evolution operates on populations without human intervention. About thirty years before the first edition of *The Origin of Species*, Charles Lyell published his hugely influential *Principles of Geology*. He argued that the same processes at play in the world today had shaped the world of the past, and that Earth must be much more than 6,000 years old, the age generally supposed at the time Lyell was writing. The exact age of the Earth was uncertain; by the time Darwin wrote his final (1872) edition, the Earth's crust was believed to have hardened anywhere from 20 million to 400 million years ago. While the older end of this estimate is about eleven times younger than we now know the Earth to be, it still allowed a long time for evolution to occur. It would give natural selection time to work wonders that far surpassed what humans had done with gooseberries.

Darwin's second chapter has no flagship organism, nothing to rival the paragraphs lavished on pigeons in his first chapter. But who should show up on page eight, gobbling up more ink than brachiopods, birds, or butterflies? Oaks. Darwin relays botanist Alphonse de Candolle's observation of more than a dozen attributes that can vary on a single oak tree, "sometimes according to age or development, sometimes without any assignable reason." Darwin knew that oaks are paragons of plasticity. They also exhibit high heritable variation within and among populations. Darwin writes that the European pedunculate oak (*Q. robur*) was so closely studied that twenty-eight varieties had been recognized based on variation in morphology. Yet most of that variation could be boiled down to three subspecies that accord closely with three species we recognize today: pedunculate oak, sessile oak, and pubescent oak (*Q. robur*, *Q. petraea*, and *Q. pubescens*). Filter out the "comparatively rare" individuals that are intermediate in form, and the remaining groups of trees would be about as distinct from each other as any oak species. Should these three entities be recognized as species, subspecies, or varieties? For Darwin, this decision was subjective. "I look at the term species as one arbitrarily given, for the sake of convenience, to a set of individuals closely resembling each other." What was clear to Darwin is that the species we recognize today—whether finches on the Galápagos Islands or oaks spread across western Europe—arose gradually under natural selection, over the course of tens of thousands to millions of years, just as domestic pigeon breeds arose gradually under human selection.

Natural selection only works when variation is accompanied by two conditions. First, some portion of the variation among organisms has to be heritable. We can easily see evidence of heritable variation: the members of a family resemble each other more, on average, than they do members of other families. Likewise, the members of a species or a genus tend to resemble each other more than they do members of other species or genera. Second, variation must influence one's chances of getting offspring into the next generation. In simple terms, reproductive success—how effective a given organism is at getting its offspring into the next generation—is called fitness. When variation among organisms is heritable and influences fitness, organisms of a particular form, with a particular set of attributes, will tend to increase in number in some environments and decrease in others. Natural selection leads to organisms who can see farther or run faster, tolerate drought, fight off viruses, or perceive a bird flitting at the edge of their peripheral vision. Natural selection favors different combinations of traits in different environments: a polar bear is better suited to life in the Arctic than a black bear would be. Little by little, natural selection tunes populations of organisms to their environments through differences in individuals' probabilities of survival and reproduction.

Darwin's theory of descent with modification through variation and natural selection has been probed, prodded, used as a springboard, and refined over the more than 160 years that have passed since the first edition of *The Origin of Species*. The theory is central to explaining the magnificent diversity of living organisms, all with attributes that appear to be tailor-made: eyes seemingly crafted by someone who knew animals would need to see; leaves custom-built for life in the desert, the forest understory, or a bog. It starts with variation that any of us can see: forms of flowers and leaves, bones and shells; anatomy; behavior; fossils; pedigrees of domesticated animals and cultivated plants; and geographic distributions. Descent with modification is the starting point for almost all research in evolutionary biology. It is our starting point as well for understanding the diversity of oaks.

The Earth bakes in one place and is blanketed with snow in another. Tundra grades southward to boreal forest, then to temperate forests. Air laden with moisture blows in off the ocean and rides up the west face of the mountains, depositing rain as it goes. It slides down to the east, blowing over arid grasslands spread across the middle of the continent. Natural selection adjusts populations as they cross these gradients, weeding out unfit individuals much as a plant breeder pulls out the rogues. Species morph in response. A plant population in the driest portions of the range of its species is rarely the same in physiology or morphology as a plant population in the richest soils. At some point, though, species reach the limits of their adaptive potential. They drop out because environments have changed too much and competition is too stiff, or because they haven't had the chance to migrate any further.

Darwin argued that common oaks exhibit high variation in part because they have large geographic ranges. Natural selection twists the dials on genetic and trait variation as a species migrates across environments. But genetic variation becomes increasingly difficult to distinguish from plastic variation as species become geographically and ecologically widespread. This is the problem of distinguishing between "nature" and "nurture." Teasing apart the relative importance of heritable and plastic variation helps us to predict which trees will survive from year to year and what the evolutionary trajectory of the forests will be. In eastern white oak (*Q. alba*), trees inherit the growth rates of their mothers. This is heritable variation that has the potential to affect fitness. In California valley oak (*Q. lobata*), individuals that leaf out farther from the population average have smaller acorn crops and lower growth rates. This is also heritable variation that affects fitness. That variation—including trees that are less fit because they grow more slowly in today's environments—will be part of what enables the species to adapt to earlier springs in the coming decades. Variation in fitness itself is essential to the evolutionary success of the population and the species.

Common gardens are one of our best tools for investigating how heritable variation affects species. Common gardens are experiments in which plants from multiple populations or species are grown together in a single environment. These experiments attempt to eliminate the differences in soil, climate, daylength, competition, and other aspects of the environment that account for much of the plastic variation we see across species' ranges. In so doing, a well-designed common garden highlights heritable variation so we can measure it. Imagine two massive oaks standing a hundred meters apart from each other in a natural forest. They differ in the straightness or crookedness of their trunks, the shape of their crowns, how much early wood they put on each spring in comparison to the late wood of summer, or any of a variety of other traits. Some of this variation is due to plasticity (one is standing at the bottom of the hill, one at the top); some is heritable. It is very hard to know in the natural community how much of the overall variation is due to which cause. But if you collect acorns from each tree and grow them out randomly in a common garden, you can measure how much of the variation you observed in the field persists across an even environment: in simple terms, heritable traits make the offspring of a single mother tree similar to one another and differentiate between offspring of different mother trees. There have been many oak common gardens and provenance trials—common gardens that focus on what trees will grow best in a particular environment—and in aggregate, they show that the traits enumerated in our imaginary pair of oaks are heritable in at least some species and under some ecological contexts. And these traits are also all highly variable within

Quercus sect. ***Cerris***: ***Quercus cerris*** **leaf**. Tree in cultivation, preserved as a specimen in
the Morton Arboretum Herbarium, Accession 70700.

populations. This combination of variability within populations and heritability gives natural selection a lot to work with.

Common gardens are often planted in a design called a reciprocal transplant experiment. A reciprocal transplant experiment involves at least two populations. Seeds from each population are collected and planted into both their home environment—near where they were collected—and the environment of the other population or populations in the experiment. Differences in how each population performs at the different sites allow us to estimate how much of its variation is plastic. Differences among the populations planted into a single garden provide an estimate of heritable variation. Reciprocal transplant experiments allow us to investigate whether populations are locally adapted to the particular environment in which they grow. A reciprocal transplant experiment by Victoria Sork and colleagues showed that northern red oak (*Q. rubra*) can be locally adapted to different microsites that lie within shouting distance of one another. In a hilly forest outside St. Louis, the researchers planted acorns collected in each of three different microsites of the forest back into all three microsites. Each microsite differed in the direction the slope faced, which controls temperature and moisture of the microsite. They replicated their experiment using acorns from six mother trees from each microsite. By tracking seedlings over the first year of life, Sork and colleagues found that seedlings growing in the microsite where they were born were chewed less by insects than seedlings relocated from just 100–200 meters away. This suggests that northern red oak populations, at least at this site, are so locally adapted to herbivore resistance that they suffer if they are moved less than a city block away from their home.

Such fine-scale oak variation within a forest may be rare; we don't have enough studies to know. But we do know that oaks are often adapted to their particular site or region, generally separated by tens or scores of kilometers. One of the most massive oak common garden experiments was established in the 1980s in Europe. The experiment included more than 150,000 sessile oaks (*Q. petraea*) reared from 116 source populations across the range of the species. The oaks were planted into twenty-three common gardens in six European countries. Each garden, as a consequence, grew in a climate similar to that of some source populations but different from that of other sources. This makes the study like a huge reciprocal transplant experiment. The authors of one of the studies from this continent-sized experiment calculated the total precipitation available to plants at each source site and each garden. Plants growing in gardens with water availability similar to their home environment tended to grow the most rapidly and survive the longest. This is evidence that different populations of sessile oak are adapted to the climate in which they and their immediate ancestors evolved. Other experiments have used similar analyses of transfer distance, the ecological difference between a common garden and the sites where seed was gathered, to predict how well trees will grow. Smaller transfer distances generally result in trees that grow more rapidly, survive longer, or otherwise perform better. However, there is high variation in this response, indicating that local adaptation among populations is tempered and complemented by genetic variability within populations. Many oaks appear to be adapted to their climates, but ample variation within oak populations gives natural selection a handhold by which to guide species' evolution.

Even without reciprocal transplant experiments, common gardens show that natural selection shapes patterns of trait variation among populations. In California valley oak (*Q. lobata*) and sessile oak (*Q. petraea*), buds open earlier for trees grown from populations that evolved in warmer climates. This makes sense: if you are an oak from a warmer population, where late-spring frost is uncommon, you are probably better off opening your buds early to put on more growth each season. Cork oak (*Q. suber*) seedlings from the southern areas of their range produce deeper roots, an apparent adaptation to the greater drought stress they face. This finding also aligns with our expectations.

But some studies are not nearly as straightforward. A common garden study of cork oak showed that populations from drier sites tended to have relatively thick, tough leaves that were more resistant to drought and herbivores. Leaves from plants collected in more mesic sites invested less in resistance to drought and herbivores. So far, so good: we generally expect plants from more stressful areas to protect themselves more from insects and other stresses. However, plants from drier sites in this experiment had higher photosynthetic rates for their size than plants from more mesic sites. They grew faster and won the race when grown in drier soils and climates. This is surprising: all things being equal, you might expect that investing in thick leaves would come at a cost to the plant, resulting in slower growth rates. You can't throw all your resources into growing well-protected leaves and also into growing rapidly. The expected trade-off between growth rate and leaf defenses revealed itself, though, under more mesic conditions: with a little water, dry-site populations lost their growth advantage. The benefits of stress-tolerance balanced out against the benefits of growing quickly when resources were less limiting. Natural selection can thus tilt the scales in one direction or the other—toward growing rapidly or toward tolerating challenging environments—depending on the environmental conditions of the population.

One of the most widespread oaks of the Americas, *Q. oleoides*, thrives in forests and savannas from Tamaulipas in eastern Mexico to Costa Rica, a span of roughly 2,000 kilometers from northwest to southeast. It is a primarily bottomland species that climbs to about 800 meters in Central American mountains. At higher elevations, *Q. oleoides* sees an average of 3 meters of rain in a year. At lower elevations, it makes do with about 50 centimeters. Different *Q. oleoides* mother trees produce offspring that differ greatly in performance, and the species has relatively short-distance pollen movement. These two conditions, combined with the strong difference in selective pressure between the lowlands and the highlands, might be expected to drive local adaptation. Despite this fact, reciprocal transplant experiments between the two environments give no indication that the populations of the dry lowlands fare better in drier climates, nor that the populations of the wetter uplands have an advantage in wetter climates.

Does this mean that there is no adaptation to climate in the species? Far from it: *Q. oleoides* populations from wet climates have thicker leaves that resist dehydration, and they can accumulate amino acids, carbohydrates, cations, and other solutes in their cells to draw water in, even when soils are relatively dry. In this way, the wet-climate populations are able to keep photosynthesizing right through the long dry season of the seasonally dry tropics. Populations from drier climates have evolved a different strategy. They drop their leaves more quickly when dry conditions set in, which allows them to avoid rather than resist drought. This tropical oak has evolved to split the difference between drought-tolerance in the wettest environments and drought-avoidance in the driest. Stated another way, plasticity itself has evolved within *Q. oleoides*: different populations respond to drought in different ways, modifying how they grow based on what the environment throws at them.

An experimental study of the Mediterranean Kermes oak (*Q. coccifera*) also provides evidence for the evolution of plasticity. Some populations of the species grow in forests, where light varies from shaded to dappled to full sun. Others grow in dry shrubby communities (garrigues) or exposed rock outcrops, where they find little or no shade. Acorns from these different populations were grown together in a greenhouse experiment with varying levels of controlled light, ranging from 100% full sunlight down to 20% sunlight, approximately the amount they would face under full shade in natural forests. Plants grown from acorns collected in the forests, where seedlings have to make do with more variable light, were able to shift their growth rate and photosynthetic activity more in response to changes in experimental light environment. This suggests that plasticity of forest populations may have evolved in response to the variable light environment on the forest floor.

In some cases, plastic variation seems to run counter to adaptative variation. Wild populations of sessile oak growing at low elevations in the Pyrenees produce lots of relatively large acorns. Wild populations growing at high elevations, where the climate is colder, produce fewer and smaller acorns. Acorns from those populations show the opposite trend in a common garden: the high-elevation populations produce more and larger acorns than low-elevation populations. Natural selection in this case appears to favor trees that pour resources into seeds even in the short growing season of high elevations. It may be that the harsh environments and shorter growing seasons of high elevations, combined with the need to produce acorns, select for high fruit production. The result in the field is lower overall fruit production at high elevations than at low elevations; but the common gardens studies suggest this disparity would be even greater without overcompensation by natural selection.

Plasticity and local adaptation—including the evolution of plasticity itself—work hand in hand to tune oak populations to their environments. Many leaf traits that vary in sessile oaks as you climb the Pyrenees—leaf area, leaf mass per area, the rate of gas exchange through the pores in the leaf—do not correlate with the elevation of origin when grown in a common garden. This tells us that much of the variation in natural populations is plastic. And in a study of flowering time in California blue oak (*Q. douglasii*) planted in replicated common gardens near the northwestern and northeastern edge of the species' range, the climate of the source population explained only 16% of the variation in flowering time. The particular garden in which the tree was growing and the year in which observations were made explained 68% of the variation in flowering time. This points to high plasticity in flowering time, which may give oaks a little breathing room as climates change.

As species cross continents and climates, the differences among populations increase. The heritable components of these differences illustrate natural selection at work, beginning with offspring from a single individual and working its way to species. The variation among populations, including local adaptation, begins with evolution within populations. Evolution of populations is, in turn, a potential step toward the evolution of new species.

The things we can observe and measure on an organism are called its phenotype. The collection of gene copies that make individuals distinct from one another are collectively called its genotype. Every oak parent contributes one copy of all its genes to its offspring. Each copy of each gene is called an allele. Alleles join forces when the sperm cells in a pollen grain fertilize an egg cell inside a pistillate flower. As a consequence, every cell within an acorn, seedling, sapling, or grown-up oak contains two alleles for each gene, one from each of its parents. We describe such an organism with two copies of the genome as diploid. Humans are diploid, as are oaks.

Heritable traits are passed along in pollen and egg cells through alleles. Every allele holds the instructions to create a protein or a set of proteins, depending on how the allele is translated inside the cell. Proteins build leaves, roots, trunks, and all the other organs that make an individual. Proteins serve as signals and messengers between cells within a plant and among trees within a forest. They defend against insects or lure insects. Some proteins are involved in translating DNA into other proteins. Each individual's genotype holds the complete set of instructions that build the organism and then shape that organism to its environment.

Except for genetic clones, no two oaks have exactly the same genotype. This is the case for two reasons. First, mutations occasionally arise in the long, intricate chain of 750–800 million nucleotides that make up the oak genome. Nucleotides are the building blocks of DNA. Each nucleotide is made up of a base—A, C, T, or G, abbreviations for the "letters" of the genomic code—along with a sugar molecule and a cluster of phosphates that together make the long side rails of the DNA molecule. Every time a cell divides, the entire genome is copied, nucleotide by nucleotide, into the new cell. DNA duplication is an efficient process, but small copying errors slip in. Nucleotides can be duplicated, deleted, or mistranscribed. Errors are rare: a study of a 234-year-old pedunculate oak estimated an average of 30–40 single-nucleotide mutations per generation across the entire genome. Some minor proportion of single-nucleotide mutations may cause changes in the phenotype. A single nucleotide change can turn an inland brown mouse of Florida into a light-colored mouse of the coastal dunes. A point mutation can increase muscle mass in sheep or induce human eye disorders. Many point mutations working together influence your height and the texture of

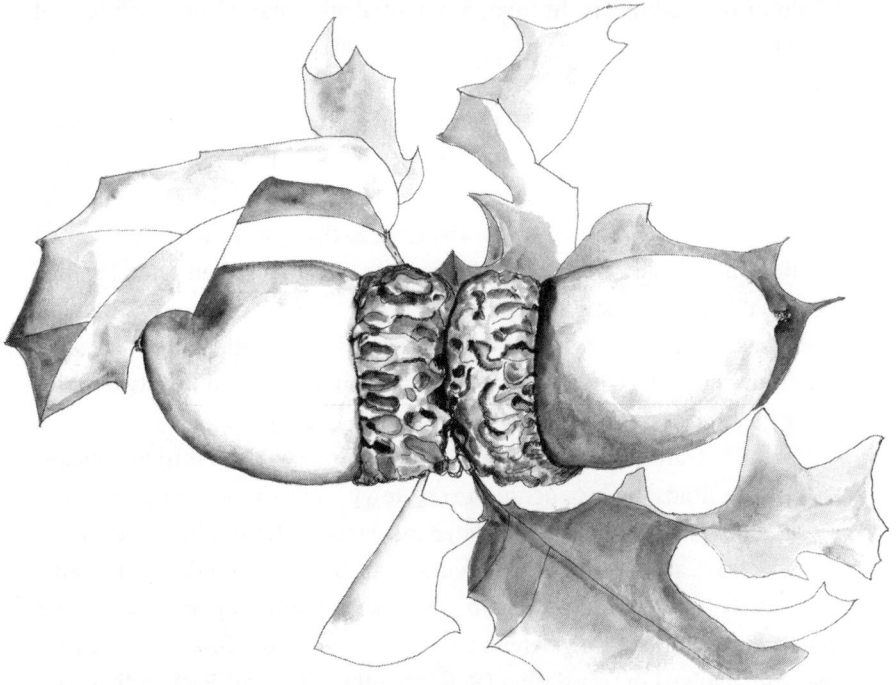

Quercus sect. *Quercus*: *Quercus pungens* **leaves and acorn**. Tree in cultivation, UC Davis Shields Oak Grove, October 2018.

your hair. The vast majority of DNA mutations, however, land in areas of the genome that do not affect the phenotype at all. These have no effect on the proteins that a gene creates or the conditions under which a gene is turned on or off. They are invisible unless we look at an individual's genome.

The second source of genotypic variation is far more common: every individual carries a unique combination of alleles from its parents. Alleles are mixed anew every time a pollen grain or egg cell is created. The offspring of a couple can range from short to tall, have curly hair or straight hair. The variation among children generally has nothing to do with novel mutations that arose within one child and not the others: it comes from their particular environment in combination with the unique set of gene copies—alleles—dealt out when the egg and sperm that eventually became them joined forces. Genes and regions of the genome are shuffled every time an egg cell or pollen grain is created. Sex is the wellspring of variation in oaks, as it is in humans and all sexual organisms.

Measuring the phenotypes of numerous individuals in a common garden allows us to quantify the genetic variation in traits that make organisms fit. Characterizing the genotypes of those same individuals can help us figure out what genes and genome regions help organisms survive and thrive. Oak powdery mildew provides a fascinating example. Powdery mildews are fungi whose mycelia form a thin, whitish coat over the surface of leaves in a variety of plant species. More than fifty powdery mildew species are known in oaks. In the early twentieth century, one powdery mildew species—*Erysiphe alphitoides*—arrived in Europe, perhaps from East Asia, and took the continent by storm, spreading rapidly among pedunculate and sessile oaks (*Q. robur* and *Q. petraea*, respectively). First sighted in Paris and a handful of locations in western Europe in the summer of 1907, *E. alphitoides* was widespread in France and numerous other European countries a year later. By 1909 it had spread to Russia and Turkey. By 1912 it was in Brazil. It has subsequently been shown to shift onto more distantly related oak species as well as eucalyptus and mango. *Erysiphe alphitoides* is not lethal to most trees, but it saps their energy by draining off sugars produced in the trees by photosynthesis and using it for its own purposes. Infected leaves tend to live shorter lives. Infected oaks grow more slowly than they would otherwise, putting on less wood and less height each growing season.

Some pedunculate oaks turned out to be relatively resistant to *E. alphitoides*, despite the fact that the tree and the pathogen had not evolved together. Genotyping the offspring of a cross between two pedunculate oaks planted into two common gardens, a group of French researchers led by Marie-Laure Desprez-Loustau and Jérôme Bartholomé found twenty-six regions of the genome that each explained anywhere from 4% to 39% of the total variation in powdery mildew infection on the trees. Within two of these genome regions, the researchers found more than 350 genes that might be involved in resistance to powdery mildew infection, including many genes involved in immune system responses. Few if any of these genes are likely to have arisen within European oaks in response to powdery mildew, for the simple reason that the fungus only invaded in the early twentieth century. Instead, the collection of alleles that help pedunculate oak respond to *E. alphitoides* evolved to fight other diseases and stresses and were ready when a new pathogen arrived.

The powdery mildew research is just one of many common garden studies to show that most traits vary among populations and that most traits are shaped by a number of genes, each of small effect. In oaks alone, the same approaches used to study the genes underlying resistance to powdery mildew have shown that growth rates, seed production, leaf size and shape, the number of stomata on a leaf, the seasonal timing of flowering and leaf bud opening, water-use efficiency, and responses to flooding are all shaped not by single genes, but by handfuls or entire crews of genes working in concert. The differentiation of populations and species involves the gradual accrual of mutations and recombined alleles. Much of this variation is shaped by natural selection.

But the variation that helps oaks survive and thrive has evolved in natural landscapes, not in gardens created by researchers. Studying the evolution of genes under the effects of natural drought, fungal pathogens, or other selective pressures might seem straightforward: simply tally up what alleles are present in different oak populations growing in different environments, and we should be able to tie the genotype to the phenotype. There is a problem with this approach, however: many of the genetic differences we find among populations are not due to selection at all, but to differences that accrue over time by processes unrelated to natural selection on the genes themselves. The genotypes that form a population are shaped in large part by the caprices of natural populations: random migrations, establishment of acorns, wind carrying pollen in sometimes unpredictable directions, unfortunate deaths. These processes create genetic variation even when selective environments do not differ. Common gardens get around the conflation of natural selection and more random population processes by remixing genetic variation through artificial crosses and manipulating the environment to directly test for adaptation. We don't have that option in wild populations.

Fortunately, we can take advantage of the fact that a good deal of genetic variation is barely visible to unrelenting, long-sighted natural selection. Neutral or nearly neutral genetic variation provides what researchers refer to as a "null" expectation. That's a dreary phrase that describes the world as it might look if a particular hypothesis were not true. One standard null expectation is that allele frequencies in different populations have evolved to be different from one another in the absence of natural selection. It is the counterpoint to the much more interesting hypothesis that natural selection has driven the evolution of at least some genes in the populations. Alleles that are strongly associated with environment should stand out against the null expectation. Compared to the null expectation of no adaptation, even small numbers of genes can be informative. One study using just thirty-six molecular markers revealed a region of the genome that was highly divergent between two hybridizing species, northern red oak and Hill's oak (*Q. rubra* and *Q. ellipsoidalis*). The genome region was associated with flowering time in a separate oak common garden study and has been linked to growth rate in other flowering plants, and thus it may be important in limiting gene flow and maintaining ecological differences between species. More recent studies using thousands to tens of thousands of genomic markers at a time, or in some cases whole genomes, have identified dozens of genes correlated with temperature, precipitation, or soil pH in a variety of oak species. Some genes have evolved similarly in different lineages: a set of genes under selection in the California valley oak also appear to be under relatively strong selection in the distantly related netleaf oak (*Q. rugosa*), for example. Seven genes are all associated with the same climatic predictors in three different European white oak species. Common patterns of phenotypic variation in different species, in other words, may be driven by genes that respond similarly to natural selection across the continents. These studies offer a glimpse of natural selection's effects on genes themselves.

Natural selection is most effective when there is high genetic variation within populations. Most trees have high within-population variation. This is not because trees exhibit higher rates of molecular evolution than the wildflowers growing beneath them. To the contrary: pick two woody plants whose common ancestor lived 5 million years ago and two herbaceous plants that are equally distantly related, and the genetic similarity of the woody plants will tend to be greater than that of the herbaceous plants. However, woody plants tend to have relatively high diversity of alleles within populations compared to the differences in allele frequencies between populations. In other words, a woody plant population gives natural selection lots to play around with. This is especially true of species that rarely self-pollinate, have large geographic ranges, and disperse their seeds through the wind or hungry animals. Many trees have some of these attributes. Oaks have all three.

Some oaks also grow in communities with exceptionally high intraspecific competition, which provides an additional opportunity for very strong natural selection. I've seen no better demonstration of the effects of competition among oak seedlings than in the French national forests located west of Paris. I visited a couple of these forests with Antoine Kremer, Alexis Ducousso, Laura Truffaut, and several others in 2014. Kremer, Ducousso, and their colleagues have been studying oaks at the Institut National de la Recherche Agronomique (INRA, now INRAE) in Pierroton, France, since the late 1980s. Early in his career, Kremer and colleagues established a set of common gardens and experimental plots in a few forests managed by the French National Forest Office. At the edge of some of these forests, where the light penetrates from the road, oak seedlings grow as thick as ground cover. Oaks crowd around your feet, overlapping, practically nipping at your ankles. They are racing to get ahead of each other.

Walk 150 feet, and you can see the effects of this selection. Sessile oaks 360 years old or older grow in these forests. One of the most famous is the Oxford Oak, which measures 43 meters tall and 22 cubic meters in volume. It is a particularly striking tree, but there are many equally impressive oaks in these forests. These old trees are the legacy of Jean-Baptiste Colbert, France's first minister of state under Louis XIV. Colbert established a network of French national forests in the late seventeenth century as a source of lumber for shipbuilding. Today, the French navy no longer depends heavily on big wooden ships. The trees have an important function nonetheless: they provide staves for wine barrels. Some exceptional old trees sell for more than 30,000 euros. Even this cost is a bargain—about 100 euros per year of growth—when you consider the intense work of natural selection over the first decade of each tree's life, weeding lawns of oak seedlings down to a comparative handful of trees.

These forests, like even-aged managed oak forests across much of Europe, are initiated by first cutting the trees to a density of about 100 per hectare. This allows the mature trees to breed freely. The germinating acorns form a dense ground cover. A hectare may support more than 100,000 seedlings. After a decade of open pollination and free reproduction, the remaining adult trees are cut. Now, only the seedlings remain. For the next decade, the seedlings fight for water, light, and soil nutrients. One seedling is weakened by insect herbivory, then overtopped by a neighbor with better-defended leaves. Another gains an edge when a late-spring freeze wilts its neighbors. Competition weeds out as many as 98% of the seedlings by the end of ten years, leaving only the strongest trees to survive. After that, foresters become more actively involved, cutting saplings and then small trees. They thin the forest to make way for the straightest, fastest-growing trees. This is a period of artificial selection for trees that are most useful to people. Its effects, however, pale in comparison to the natural selection that operated on the trees in their first decade or two of life.

Colbert's forests are an inadvertent experiment in natural selection. His oaks have high genetic diversity and initiated their long journey to the twenty-first century under strong competition. They also happen to have been planted at a turning point in the climate. Colbert was appointed to his position during the Little Ice Age, a cold period that gripped the Northern Hemisphere from the early fourteenth to the mid-nineteenth centuries. Temperatures dropped by an average of half a degree Celsius worldwide, which hardly seems worth mentioning. But bitterly cold winters and droughty summers hide behind that average. Rivers and lakes froze. Birds are reputed to have dropped from the sky in flight, and the king of France's beard to have frozen one night while he was sleeping. The Ming dynasty fell as crops failed and famine spread. Droughts contributed to the Great Fire of London in 1666 and slowed tree growth. Europe pivoted from feudalism to a market economy as depressed grain production made long-distance trade more profitable.

All the while, freezing winters across Europe killed oaks and other trees that were not equipped for frost. The countless deaths and victories that played out in carpets of oak seedlings three hundred years ago are recorded in the genotypes of Colbert's old oaks. To read this history, Dounia Saleh, Kremer, and colleagues sequenced the genomes of even-aged sessile oak stands in each of three French national forests: Bercé, Tronçais, and Réno-Valdieu. They selected four oak stands from each forest, one from each of four founding periods: 1670 to 1705, the depths of the Little Ice Age; 1830 to 1855, as Europe was leaving the Little Ice Age; and 1957 to 1961 and then 2001 to 2013, both during postindustrial global warming. This makes a total of twelve forest stands, three replicates from each of four time periods. Saleh and Kremer's team collected leaves and extracted DNA from them, pooled the DNA from each forest stand, and sequenced genomic data from each DNA pool. They used these data to estimate how many copies of each allele were present in each forest stand. They then asked how the number of copies they found of different alleles in each forest stand—the allele frequencies of each forest stand—correlated with one another between time periods and between locations. Alleles that are favored by the environment should increase over time, due to natural selection. And if different populations have undergone the same kind of selection, we would expect similar alleles to increase in frequency in those populations and a different set to decrease. By looking at the genes of today's trees in different populations, Saleh and colleagues peered back in time at the first decade or so of each tree's life to ask how the cold conditions of the Little Ice Age or the warming conditions coming out of it shaped the genetics of the oak forest.

First, Saleh and colleagues compared data between time periods within each forest. They found the alleles that became more common in any particular forest between any two time periods also became more common in that same forest between the next two time periods. This suggests that local soil conditions, insect populations, microclimate, or other environmental factors unique to a given site over time are shaping the populations. More exciting, however, was what they found in comparing different forests: alleles that became more common in Bercé or Tronçais also became more common in Réno-Valdieu when they compared the same interval between time periods. This means that between any two time periods, natural selection favored a particular set of alleles across the landscape, even in forests more than seventy kilometers apart. This is evolution in action, but not simple local adaptation (which was also at work). Rather, climatic conditions across the region as Europe was emerging from the Little Ice Age selected for the alleles we find today. This was further supported when Saleh and colleagues identified which genes were increasing in concert across forests. Many were involved in responding to extreme temperature and drought or fighting off fungi such as powdery mildew and root pathogens. A follow-up study showed that selection for alleles also shaped growth, with all trees across sites changing in growth rate in similar ways over time. Natural selection by climate most likely caused populations to evolve in parallel across time periods and across the landscape.

Each pollination is a roll of the dice. Every population is the outcome of billions upon billions of rolls over years of drought and fire, drenching rains, insect infestations, disease outbreaks. A seedling dies, and with it dies the unique combination of alleles that suited it to a possible future. Another seedling lives and is prolific, giving birth to homebodies or travelers, all adapted to other possible futures. With births, deaths, and migrations, alleles are born, go extinct, and recombine with alleles from other genes. At every step, genomes persist or die off. The genotypes left in a population are our record of the history of individuals that make that population. They are also our window into how the population will fare a century from now. The uncertainty of tomorrow's world makes genetic diversity a population's key to success.

A population or species is more than just a map of who's sleeping with whom and where their offspring make their livelihoods. It is a biodiversity engine, testing new genetic variants in the real world. Its offspring bring this variation into the next generation as combinations of traits that will get tested over and over again in environment after environment, perhaps for centuries.

In the 20,000 years since the Last Glacial Maximum, bur oaks have expanded to their current distribution in eastern North America, where they range from southern Canada to near the Gulf of Mexico in Texas. In that same time, pedunculate oak has extended its range northward to within 6 degrees of the Arctic Circle, migrating from refugia in the peninsulas of Iberia, Italy, and the Balkans. But oaks do not evolve or migrate instantaneously. They suffer a delay. Sessile oak populations near the southern edge of their range appear to be maladapted: they grow better if they are transplanted northward. California valley oak trees grow fastest not at the temperature where their parents evolved, but in cooler temperatures. Oaks may be evolving more slowly than our climate is changing.

The delayed responses to climate change that at least some oaks exhibit will likely become even more pronounced in the next hundred years. Alleles of sawtooth oak (*Q. acutissima*) in eastern China are strongly correlated with climate. Associating these genotypes with future climates under two different climate change scenarios suggests that numerous populations are hundreds of kilometers from where they will be best adapted in 2070. Similarly, associating the alleles present in European oak populations today with their current and predicted future environments suggests that adaptation alone will not be enough for oaks to keep up with climate change. These studies do not consider all the possible compensatory effects of plasticity, nor the capacity for plants to adapt rapidly via evolution at gene regulatory regions. They raise the question, nonetheless, of whether the combination of evolution at the population level, gene flow via pollen, and migration via acorns will enable oaks to keep up with climate change.

Every species and population exhibits heritable variation. Every individual organism is, to some degree, plastic. Which form of variation matters more in trees? This all depends on context. Charles Darwin's son Leonard, considering in 1913 the question of whether nature or nurture mattered more to human variation, asked whether the question might not be as illogical as asking whether the width or length of a rectangle contributed more to its total area. Both matter. The ability of oaks to acclimate rapidly to cooling temperatures, to flower at different times depending on what the season throws at them, and to grow tougher or more deeply lobed leaves depending on the conditions of the year may help them to survive in place as climates

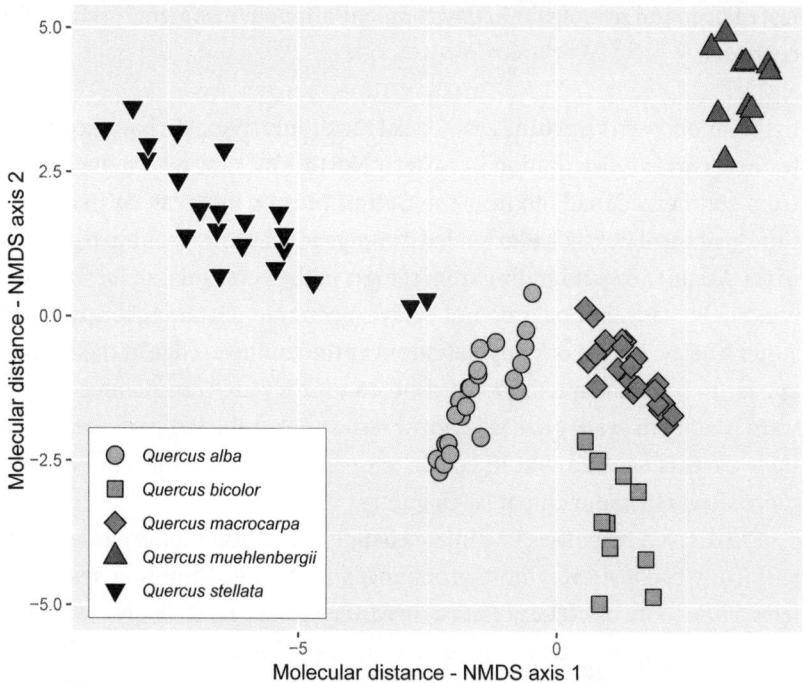

Clouds of molecular (DNA) variation, eastern North American White Oak Group
(*Q*. sect. *Quercus*). Genetic distances were estimated using 75 single-nucleotide DNA
variants, which were visualized using nonmetric multidimensional scaling (NMDS).
Data and methods from Hipp et al. (2019).

change, at least temporarily. This plasticity is especially important for trees
that can live for centuries. But plasticity is not enough on its own to allow
a plant to survive everywhere. If it were, we might inhabit a world clothed
in a single, hypervariable species. Heritable variation is needed if plants are
to evolve. It allows trees to pass novel adaptations down to their offspring
within populations. Lineages of populations and species bear the legacies of
natural selection long past and are molded by natural selection acting today.

Close your eyes and say the words "northern red oak," "pedunculate oak,"
"California valley oak," "*Quercus rugosa*," or the name of any other species
you choose. What do you have in mind? Perhaps you will see an individual tree. Perhaps you have looked closely at your species in many places,
and sheaves of images scatter across your field of view. But the individual
trees in your mind don't resolve into a species until you imagine a second
species. Variation in one species distinguishes it from all others. Species
cluster in different areas of ecological, morphological, and genetic space.

They may overlap a bit in your mind, and in the real world: there is some hybridization, and some individuals are genetically, morphologically, and ecologically ambiguous. But heritable variation separates them into distinct clouds—species—as reproduction, migration, and natural selection connect the evolutionary trajectories of some populations and individuals while they separate others. Whether the oak species you imagined are sisters or distant relatives, the resulting clouds of variation represent what we mean when we say the names of species.

3

Species and Their Hybrids

The swamp white oak in front of our house is hemmed in by pavement. A species of floodplain forests and seasonally flooded swamps, it is healthy in spite of the abuses taken by its soil, a packed clay that has been driven over, mixed with gravel in places, and capped with concrete or asphalt in others. Like almost all swamp white oaks, our tree has leaves that are shallowly lobed. Its bark curls off in places. Its acorns are borne on long stalks. Its leaves are two-toned, dark green above and whitened with velvety hairs beneath, giving the species its scientific name: *Quercus bicolor*. Donald Peattie, in *A Natural History of Trees of Eastern and Central North America*, writes that swamp white oak "proclaims its unmistakable identity at a glance and at a distance," even when we're driving down the highway, as the wind flips the leaves to yield flashes of light. Across the range of the species—from its northern edge in Central Wisconsin to its southern edge in Kentucky, and from Central Missouri in the west to a few trees straggling out to eastern Massachusetts—this combination of characters makes swamp white oak easy to recognize. You might confuse it with a few other species, but it is recognizable even in populations that are new to you.

A mile west of my house, the suburban residue of a mixed oak forest has been growing in place for thousands of years. The forest is thick with sugar maple, northern red oak (*Q. rubra*), eastern white oak (*Q. alba*), and bur oak (*Q. macrocarpa*). The latter two are closely related to the swamp white oak in our front yard. Bur oak is marked by acorns with shaggy-margined caps, frilled at the edges as though they'd received bowl cuts and then had their hair curled all around the edges. The fruits can be quite large, almost palm-sized in the southernmost populations in Texas, giving it its scientific name (*macrocarpa* means "large-fruited"). It has deeply furrowed bark and branches that are often corky, clothed to resist the fires that blew across the landscape regularly before Europeans took the area. It is the icon of Midwestern oak savannas and prairies. Eastern white oak is generally easy to recognize as well by its strongly lobed leaves, the lobes rounded at the tips, the undersurfaces hairless. Its bark is thin and gray, obvious at a distance as you scan the forest. Step close, and the bark resolves into a landscape of cracked plates, often dotted with tiny fungal cups of *Aleurodiscus oakesii*, fresh and open or dried, curled in on themselves. The acorns grow on short stalks. They begin germinating in fall along the trails of upland forests across eastern North America. Eastern white oak is a majestic and dominant tree of many of the forests I walk, the "king of kings" in eastern North American deciduous forests.

Swamp white oak, bur oak, and eastern white oak are species I learned as a naturalist and experience almost every day on my walks. They are also species I study as a researcher because they are genetically, ecologically, and morphologically distinct from each other despite the fact that they hybridize with each other. You can find swamp white oak and bur oak playing together in bottomland forests of eastern Missouri and southern Illinois. But as you head northward toward Wisconsin and Minnesota, bur oak climbs uphill into the savannas. If you instead go southwest into Kansas, Arkansas, Oklahoma and Texas, swamp white oak drops out altogether. You can also find eastern white oak and bur oak in the same mesic forests together. But get to a place in the wood where there's a little variation in topography, and you'll generally find eastern white oak ranges farther upslope and bur oak farther downslope. These oaks are what we often call "good species," entities that we can recognize in nature and that are generally distinct from each other across their range, even when they grow together at the same site.

Oaks have the greatest number of species and highest biomass of any major tree genus surveyed in Mexico or the United States. Their species include dominants of forests across the Northern Hemisphere. Iconic oaks characterize a wide range of forests: *Q. insignis* and *Q. skinneri* in the cloud forests of Mexico and Central America, *Q. humboldtii* nearly to the equator in the montane forests of the northern Andes, California valley oak (*Q. lobata*) in the foothills of the California coast ranges, northern red oak (*Q. rubra*) across the hardwood forests of eastern North America, sessile and pedunculate oak (*Q. petraea* and *Q. robur*) across Europe and into western Asia, *Q. glauca* and

Q. kerrii in the East Asian evergreen broad-leaved forests. Humans have, for tens of thousands of years, depended on recognizing different oak species so they could use the most suitable trees for lumber, shipbuilding, charcoal, and acorn-eating. Our understanding of the world's estimated 425 oak species is essential to our ability to manage forests, make conservation decisions, and decide what trees to plant if we have the chance to do so.

Yet oaks have, for a long time, challenged us to say what we mean by "species." Their variation can be baffling, making it difficult at times to separate species based on appearances alone. This problem is generally surmountable with careful study. Variation within species consequently undermines our ability to identify oak species without undermining the existence of oak species. The bigger problem for our understanding of oak species is that most oaks can hybridize with numerous other species. When they do, their hybrids are generally fertile. The eastern white oaks and bur oaks of our area would most likely produce offspring together if they were cross-pollinated. The swamp white oak in my front yard could bear acorns sired by my neighbor's bur oak, and those acorns would probably grow up into healthy seedlings given half a chance. Numerous studies of genetic diversity in co-occurring oak species show that closely related oak species often swap genes in the wild. The high diversity of oak species living together in a region makes the number of possible crosses dizzying. Oaks have earned—perhaps justly—a reputation for promiscuity.

Despite all this, we continue to use oak species names. I will make a statement at supper like, "I saw a great bur oak this morning" as casually as I might say, "Our old friend Steve called last weekend." What do I think the name "bur oak" refers to? Is it something as real as "human," "great blue whale," or "white lady's slipper orchid"? If oaks cross so easily, do those names point to something with biological meaning—and if so, what?

Approximately 600,000 years ago, a population of our distant relatives migrated from Africa into Europe. Within about 200,000 years, they diverged into the genetically distinct lineages we know now as the Neanderthals and the Denisovans. The Neanderthals were the more geographically widespread of the two. They evolved large brains, broad skulls, and protruding faces. They were relatively stocky and well adapted to northern climates. They were social; they made tools; they buried their dead. They thrived and spread across Ice Age Eurasia.

Meanwhile, the lineage that would lead to modern humans evolved in Africa. This set of populations diversified genetically and spread across the continent. By 230,000 years ago, our ancestors had the taller stature, rounder cranium, and flatter face that set modern humans apart from Neanderthals. (We often call our ancestors "anatomically modern humans," but throughout, I'll use "modern humans" to refer to *Homo sapiens*, our species, whether I am talking about today's humans or yesterday's.) By 100,000 years ago, modern humans had spread across Africa and into the margins of southwestern Asia. The Neanderthals and the Denisovan lineage remained mostly separate from modern humans for hundreds of thousands of years, becoming sufficiently distinct genetically and morphologically that many scientists, if not most, recognize them as different species: *Homo neanderthalensis* for the Neanderthals, *H. sapiens* for us.

The ancestors of modern humans who spread across the Northern Hemisphere left Africa around 60,000 years ago. They encountered Neanderthals and Denisovans in Europe and Asia. It's hard to imagine how these distant relatives viewed each other, but we know they interbred at least intermittently. Several modern human fossils have been found whose ancestors had mated with Neanderthals or Denisovans. Sequenced genomes from some of these fossils record at least three episodes of gene flow from Denisovans or Neanderthals into modern humans over the roughly 20,000 years when they lived in the same regions. One fossil has been found from an individual whose mother was a Neanderthal and father was a Denisovan.

By 40,000 years ago or so, the modern human lineage was the only human species left. The Neanderthals and Denisovans were gone, but not altogether. Bouts of interbreeding, followed by tens of thousands of years of natural selection, left as much as 3% of the modern human genome carrying Neanderthal and Denisovan genes. Some Neanderthal alleles worked their way back into modern human populations in northern and Central African populations as *Homo sapiens* continued expanding. Today, all or nearly all humans carry some Neanderthal alleles, and many, especially populations centered on Oceania, share a small percent of their genome with the Denisovans.

Should we consider *Homo sapiens* a species distinct from *Homo neanderthalensis*? It depends on what you mean by "species." In high school biology class, many of us learned about species not from the examples of oaks or humans, but from that of horses and donkeys. Most of us learned that horses and donkeys are viewed as different species because when they are mated with one another, the offspring are often sterile mules or hinnies (depending on which species is the mother and which is the father). We were taught that their offspring are dead ends. I grew up, as you may have, assuming that each species is marked off from all others by the fact that species can't interbreed, at least not in nature.

The idea that species are defined by who can share genes with whom became widely known in the early twentieth century as the biological species concept. What most of us learned about species in our biology classes, however, was overly simplistic. Hybridization is not uncommon between many species we recognize. Aristotle noted more than 2,300 years ago that even a mule will occasionally give birth, and there is genomic evidence of hybridization between zebras and asses despite the fact that those lineages have been separated by nearly 2 million years. Nonetheless, the notion that different species can't interbreed is old. When Charles Darwin wrote his chapter on hybridization for *The Origin of Species* in 1859, he addressed the "view commonly entertained by naturalists . . . that species, when intercrossed, have been specially endowed with sterility, in order to prevent their confusion." He acknowledged that the view "certainly seems at first highly probable, for species living together could hardly have been kept distinct had they been capable of freely crossing." A popular belief in Darwin's time held that while different species were not crossable, the varieties within a species were. If you had enough information on hybridization rates, you should be able to identify the limits to all the world's species.

Yet many hybrids in *Fuschia*, *Petunia*, *Rhododendron*, and a host of other genera were known to produce seed with little problem. Joseph Gottlieb Kölreuter, the German botanist of Linnaeus's time who first documented plant barriers to self-fertilization, made numerous successful hybrids between well-established species. And Carl Friedrich Gärtner, a well-known botanist who strongly influenced Charles Darwin, conducted experiments showing that many forms that "the best botanists rank as varieties" nonetheless failed to produce offspring. In other words, experiments already showed that our notions of species did not accord strictly with the failure or success of hybridization experiments. Populations and varieties were, in Darwin's view, just steps along the path to becoming what we somewhat arbitrarily name as species. There was, for Darwin, no clean distinction between species that cannot cross with one another and varieties that interbreed freely. Instead, varieties and species formed a continuum, with the completely reproductively isolated populations on the one end and those that are completely interfertile on the other.

What made a good species, for Darwin, was that it remained distinct and recognizable even when it grew naturally with other close relatives for many generations of potential interbreeding. Today, when we talk about species, we think something very similar: species are populations that have become different enough that even when they continue to grow together over time, we can recognize them as distinct from one another. One may outcompete the other. The two may interbreed, but their hybrids will be outnumbered by offspring that we can generally classify as one species or the other. Pull out the intermediates, and good species remain. What made modern humans and Neanderthals distinct species was the fact that they were ecologically, morphologically, and genomically distinct from each other, despite the fact that they swapped genes.

Many potential hybrid combinations are never found in the wild, even for oak species that grow together regularly, but more than two hundred oak hybrid combinations have been given formal names. It may well be that all 145 species of the White Oak Group (*Quercus* section *Quercus*) are capable of hybridizing with one another if you dust the pistillate flowers of one with pollen from the other. The same is true of the Red Oak Group (*Q.* sect. *Lobatae*), which numbers an estimated 125 species worldwide. Imagine all the possible combinations: for the White Oak Group alone, there are thousands of potential crosses.

If we take the average generation time in bur oak and eastern white oak as fifty years, then these two species have had perhaps two hundred generations together in the woods of our town since oaks recolonized following the recession of the last glaciers. Given all that time and the fact that they can interbreed, it is a wonder that we can recognize them as distinct species. The California Red Oaks may have them beat: fossils that look like today's *Q. wislizeni* and *Q. agrifolia* appear together at sites going back perhaps as far as the Middle Miocene (approximately 16–12 million years ago), and back to the Pliocene with high confidence (2.6–5.3 million years ago). These two species hybridize in the wild today, yet they are still distinct morphologically and genetically. The Remington Hill Flora in the Sierra Nevadas is situated at the boundary between the Miocene and Pliocene. One of the common species in that flora is *Q. ×morehus*, a hybrid between *Q. kelloggii* and *Q. wislizeni*. The same hybrid occurs in California today, suggesting that these two may have been hybridizing for more than 5 million years while remaining morphologically and genetically distinct. Moreover, oaks can hybridize in the wild with species that have been separated for tens of millions of years: the California scrub White Oaks exchange genes with Engelmann oak (*Q. engelmannii*), despite the fact that their most recent common ancestor lived more than 30 million years ago. Humans, by contrast, no longer hybridize with chimps or bonobos. Yet the most recent common ancestor of humans and chimpanzees lived just 7–8 million years ago.

Prior to the twentieth century, botanists identified hybrids via the gradual accumulation of damning evidence. If you found an individual that was intermediate in morphology between two established species, and if the two parental species could reasonably have mated to produce this individual, then perhaps you'd found a hybrid. Finding hybrids this way required that the parents be sufficiently different from one another that intermediates would be distinctive. Hybrids disappeared into the shadows when parent species were similar in appearance. Botanist George Engelmann wrote that hybrids had been noticed more frequently among the species of the Red Oak Group (particularly in eastern North America) in part because the species were easier to distinguish from one another than species of the White Oak Group. Because the method of studying hybrids at the time was observation and argumentation, without a clear set of quantitative expectations for the data, there was a great deal of disagreement among botanists. Were hybrid individuals in natural conditions oddballs with no future, members of hybrid populations, or possibly the progenitors of novel hybrid species?

This kind of detective work is still a first step to studying hybrids today, especially when we are making initial observations in the field or museum. I wrote the first draft of this chapter fresh from a trip to a local forest preserve where I found Hill's oak (*Q. ellipsoidalis*), black oak (*Q. velutina*), and northern red oak (*Q. rubra*) growing together. All three of these oaks can interbreed. Most of the individuals I found were easily assigned to a species based on the hairs on the inner surface of the acorn cap and the end buds, shapes and texture of acorn cap scales, leaf shape, form of the tree, and texture of the bark. During this visit, I also identified two plants that I suspect are hybrids between black oak and Hill's oak, based on the acorn caps, end buds, and leaves. They occurred only in the area where I found black oak and Hill's oak growing together, and they were intermediate between the two species in several characters. The scientific study of hybrids does not stop at this kind of reasoning: I returned to the population later in the summer to gather leaf tissue for molecular study, when collecting permits came through.

Merely observational study of hybrids began to give way to experimental crosses and numerical analysis in the twentieth century. Daniel MacDougal, director of laboratories at the New York Botanical Garden, published a paper in 1907 in which he described an experimental investigation of offspring from the "Bartram oak." The original Bartram oak was discovered along the Schuylkill River near Philadelphia in the mid-eighteenth century by the

North American botanist and plant explorer John Bartram. He noticed it as a tree sporting leaves that ranged from entirely unlobed, like those of willow oak (*Q. phellos*), to distinctly lobed with bristle tips, like northern red oak or the local black oak (*Q. velutina*). In 1802, the French botanist François André Michaux paid the tree a visit with Bartram's son William. Michaux agreed that this oak growing along the river was a novelty. He named it as a species, *Quercus heterophylla*, in his 1812 volume of *Histoire des arbres forestiers de l'Amérique septentrionale*. Bartram's oak was cut down in 1842, but acorns from it were collected before it was felled, and these were planted on the property. Over the course of the nineteenth century, the original Bartram oak and its offspring were visited by several American botanists. Some thought that the Bartram oak and its offspring were hybrids. Others found the idea of hybridization ludicrous.

MacDougal and two colleagues collected seventy-five acorns from one of the Bartram oak offspring in 1905 and grew them out. MacDougal found that the seedlings ranged almost continuously from leaves typical of willow oak to leaves nearly typical of northern red oak. Darwin had reported this same kind of result from second-generation hybrid crosses: crosses between hybrids can show the full range of variation between the parent species. The authors concluded that Bartram's oak was not simply a cryptic oak that looked intermediate between two species. Its descendants varied as one would expect of the hybrid offspring from willow oak and northern red oak. The case of the Bartram oak was solved by experimentation. "*Quercus heterophylla*," MacDougal wrote, is really "a medley of oak trees which possibly includes the first generation of a cross between *Q. rubra* [northern red oak] and *Q. phellos* [willow oak], secondary hybrids with either parent, as well as successive generations in which various combinations of ancestral qualities may appear." A return to this question with genomic tools in 2020 would show that MacDougal's inference from experimentation was correct.

Plant biologists working in the 1930s and 1940s—particularly G. Ledyard Stebbins and Edgar Anderson—enriched our understanding of species by documenting that alleles can move between species even when hybrids are rare. Anderson laid the groundwork for studying hybrids in the field. He had been hired as the Missouri Botanical Garden's plant geneticist in 1922 and conducted research aimed at understanding species. Through detailed field studies of natural plant populations, Anderson showed that it is not the obvious hybrids that matter most to evolution. Rather, it is only when alleles from one parent species slip through these hybrid individuals and disappear into the genetic background of the other parent species that hybridization has a lasting effect on the evolution of species.

In a foundational study, Anderson and a colleague demonstrated that alleles from Ohio spiderwort (*Tradescantia ohiensis*) influenced the form of both prairie and long-bract spiderwort (*T. occidentalis* and *T. bracteata*, respectively) when Ohio spiderwort grew with either of them. But where Ohio spiderwort grew with zigzag spiderwort (*T. subaspera*), Ohio spiderwort phenotype was influenced by gene flow from the latter. What Anderson was describing were not the Bartram's oaks of spiderworts, but the result of initial hybrids crossing with one of the parents, then their offspring crossing repeatedly back to the parent species over the course of several generations. This process is called backcrossing. The first hybrid between two spiderwort species combined alleles of the two species, and alleles of one parental species were progressively reduced in frequency with each backcross to the other parental species. After a few generations, organisms looked almost exactly like one of the parental species—typically the one most dominant at the site or in the habitat where plants were sampled—but with some alleles from the other species incorporated into their genome and subtly influencing their phenotype. The initial hybrid might be long gone. Anderson's work thus showed that alleles move between species, even when you don't find obvious hybrids in the population. He termed this movement of genes *introgressive hybridization* or simply *introgression*. The process can carry genes between spiderwort species, between oak species, or from Neanderthals and Denisovans into *Homo sapiens*.

Anderson never turned his attention to oaks. But in 1947, plant geneticist G. Ledyard Stebbins did. Stebbins, two colleagues, and a group of botany students from Columbia University collected field material from a site in the New Jersey Pine Barrens where blackjack oak (*Q. marilandica*) and bear oak (*Q. ilicifolia*) grew together, as well as a site where only blackjack oak grew. They also measured numerous herbarium specimens of the two species to understand what the standard range of variation was for each. They measured leaf, bud, twig, and acorn characteristics of each plant. They found that at the site where the two species grew together, about 9% of plants appeared to be first-generation hybrids between the parent species. Another 20% were blackjack oak with some bear oak genes. Two individuals were bear oak with a little blackjack oak. By contrast, in the site where only blackjack oak grew, as well as in the herbarium samples, there was little if any evidence of gene flow between the species. Gene flow appeared to be moving traits between the two species when they were growing together, making the species more similar at that site than at other sites across the range. Despite this fact, the two species remained distinct. The species had been growing together for hundreds of generations, swapping genes the whole time. "Almost everywhere two or more related species of oaks occur sympatrically, and occasionally hybridize," Stebbins and his colleagues wrote. "Rarely, however, does this hybridization affect the essential integrity of the species."

A year later, Ernest J. Palmer, a botanist at Harvard University's Arnold Arboretum, wrote a review of oak hybrids in the United States. He noted that co-occurring White Oak species can generally hybridize, as can co-occurring Red Oak species, but the hybrids rarely dominate. "A keen observer with a general knowledge of the native species is likely to find hybrids by careful searching in any region where compatible species grow in close proximity; but he will probably encounter hundreds or even millions of trees of the different species for every hybrid." A few years later, Cornelius Muller of University of California, Santa Barbara argued that many reports of hybrids were based on a misunderstanding of the morphological variation within species. Muller had spent roughly twenty years studying the oaks of Texas and found that hybrids were almost all correlated with ecology. Where *Q. grisea* of igneous (volcanic) soils and *Q. mohriana* of limestones met on abutting rock outcrops, their distributions cut off abruptly. The hybrids were limited to the seam between bedrock types. Elsewhere, where dolomitic soils were equally suitable for the two species, "free hybridization results from this broad intermingling." He pointed out that many potential crosses are never seen in the wild. Ecological differences between the parents, Muller's data showed, plays a role in keeping oaks separate. There is often no good habitat for the offspring of ecologically divergent oak species.

Then, in 1961, John Tucker—who established the renowned Shields Oak Grove at the University of California, Davis, wrote influential articles on oak hybridization and taxonomy, and conducted field and herbarium taxonomic studies of oaks from 1940 until months before his death in 2008—published the first of five articles on hybridization involving Gambel's oak (*Q. gambelii*) and six other species of the southwestern United States, in which he teased apart contributions of each species to the hybrid mess that people were simply calling "*Quercus undulata*." His work showed that complex patterns of gene flow can masquerade as—and cause—variation within species. Shortly thereafter, Tucker and his collaborator Jack Maze described a population from northeastern New Mexico that included individuals very like eastern North American bur oak (*Q. macrocarpa*), others that looked like Gambel's oak of the southern Rocky Mountains, and others that were intermediate. The closest bur oak population in New Mexico lay about 400 kilometers east of their populations. Maze went on to publish descriptions of a similar population from the Black Hills. Both populations appeared to be remnants of Pleistocene migrations across the country, leaving morphological evidence of introgression thousands of years past the time when the two species might have grown together.

Thus researchers working in the 1940s through the 1960s showed that oak hybrids weren't just novelties. Introgressed oak populations had the potential to show us how gene flow affects the evolutionary trajectory of species, and how species remain distinct even in the face of hybridization. Based in large part on his failed experiences trying to create artificial hybrids that would mirror the variation in the Bartram oak, L. Gale had written in 1856, "If the oaks could intermix in their native forests and the resulting hybrids continue to stock the ground, what inextricable confusion must follow. Instead of 75 species of oaks on the North American continent, as now enumerated, we should have millions. But arguments and fact both go to prove the permanency of species." Stebbins, Muller, Tucker, Maze, and others demonstrated that Gale's concern that all hell would break loose *if the oaks could intermix in their native forests* was not borne out: in fact, as he had observed, oak species were well distinguished from one another. Even with hundreds of generations of hybridization and introgressive gene flow, the things we call oak species maintain their ecological and morphological distinctions.

The conflict between evidence for hybridization and gene flow on one hand and species coherence on the other came to a head with a trio of papers published in 1975 and 1976. The first was by James Hardin, curator of the North Carolina State University Herbarium from 1959 through 1996. In a review of eastern North American white oaks, Hardin presented an illustration that has become iconic: a ring of sixteen white oak species whose names were interconnected by a network of lines. Each connection between species indicated a hybrid combination known from the wild. Some species were only minimally connected: dwarf live oak (*Q. minima*) was known to hybridize with only two other species, and Oglethorpe oak (*Q. oglethorpensis*) was, at that time, not known to hybridize with any other species (though it was subsequently shown to hybridize with at least one other species). Other species were strongly connected: post oak (*Q. stellata*) and eastern white oak were known to hybridize with eleven other species each. Hardin argued that

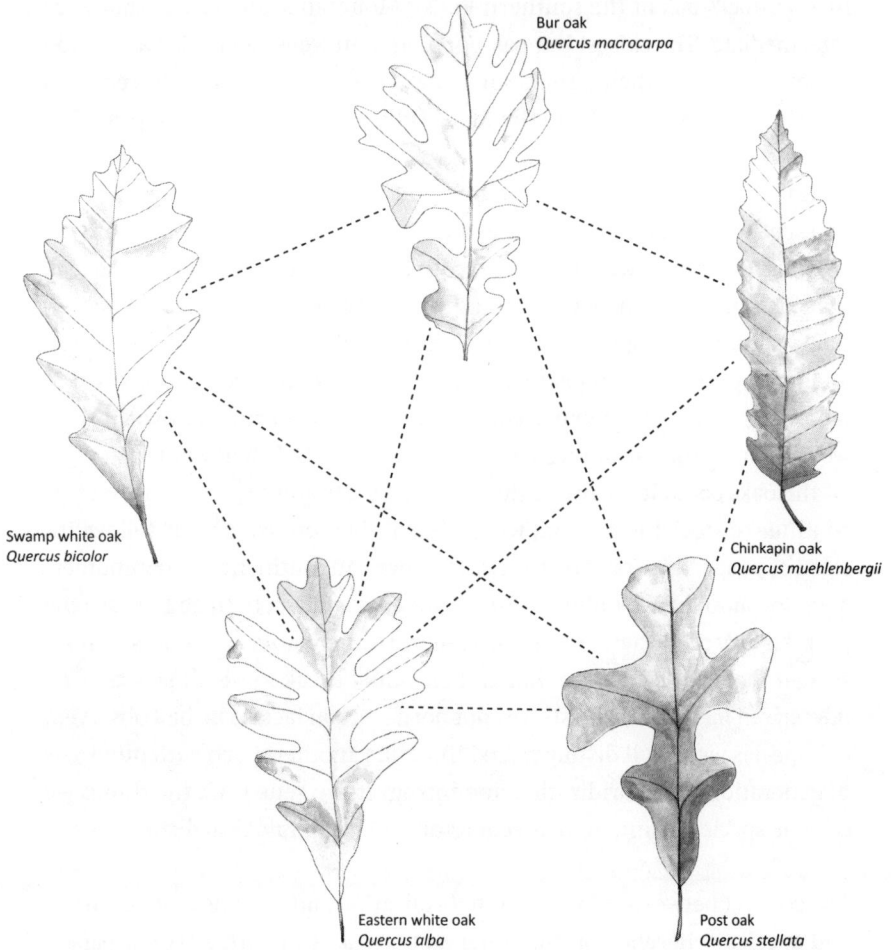

Five interbreeding eastern North American White Oak Section (*Q*. sect. *Quercus*) species. Lines between the species represent reproductive interconnections documented by Hardin (1975). An additional line is added between *Q. muehlenbergii* and *Q. stellata* indicating hybridization between those species as documented in our research (Hipp et al. 2019).

while most variation in the eastern white oak was "intrinsic," due to variation within the species, "a significant, although relatively minor, component of the variation is due to hybridization and localized introgression with eleven other white oaks of the subgenus *Quercus* (*Lepidobalanus*) [the White Oak Group] which are sympatric with *Q. alba* in various parts of its range."

The eastern North American White Oak species together formed a *syn-*

gameon, a term coined by Johannes Lotsy in 1917 to describe a group of reproductively interconnected species that are able to exchange genes. The term was picked up by Verne Grant, Hardin, and others, and oaks seemed to fit the description nicely: the species could be differentiated from one another by taxonomists and naturalists, yet many groups of oaks were interconnected by gene flow to one degree or another.

The same year that Hardin published his review of hybridization in the White Oaks, William Burger, a plant systematist at the Field Museum, wrote an article arguing that we couldn't use reproductive isolation to define oak species. Instead, each species was a population of oak individuals adapted to an environmental niche. Each species' niche, according to Burger, "differs slightly" from the niche of all close relatives, but still shares "the broader evolutionary advances of these same close relations." Oak species, in other words, are ecological and evolutionary entities that find their distinct place in the world ecologically, but that still interbreed opportunistically. They solve their own problems and then share those solutions with one another. Oaks, in William Burger's view, were having their cake and eating it, too.

Leigh Van Valen, an evolutionary biologist at University of Chicago, who was working just down the road from Burger's office at the Field Museum, refined Burger's ideas in 1976. Van Valen argued that our persistent focus on gene flow and reproduction gets in the way of understanding species. What is most important to the integrity of species, he argued, is not gene flow, but the effects of natural selection. He suggested that instead of focusing on interbreeding and the movement of genes between populations, we should be asking how natural selection molds populations into groupings that are ecologically and usually morphologically disjunct—distinct from one another—in other words, the things we call species. Defining species was, for Van Valen, primarily a problem of defining their niche. Oak species, in Van Valen's view, are groups of individuals related *genetically* by either inheritance of genes or by gene flow, and *ecologically* through occupying an adaptive zone that can be differentiated from that of other species. The adaptive zone is defined by climate, by soils, by antagonistic and mutualistic interactions with other organisms, by fire and storm and disturbance; in short, by everything that selects for phenotypes. "It may well be that *Quercus macrocarpa* [bur oak] in Quebec exchanges many more genes with local *Q. bicolor* [swamp white oak] than it does with *Q. macrocarpa* in Texas," Van Valen wrote, but that doesn't make all the oaks of Quebec one species and all the oaks of Texas another. In Van Valen's view, species are maintained "for the most part ecologically, not reproductively." In this view, it is ecological niche and ancestry, more than gene flow, that make a species.

Hardin, Burger, and Van Valen were writing a decade before oak molecular data would become available. Through the 1970s, oak genes had been viewed only through their effects on form, physiology, and ecology. Without known pedigrees or DNA data, phenotypes provide, at best, a murky view of the genome. If we could see the genes, we could look more directly at the patterns of genetic relatedness among oak populations and the boundaries of oak species. Genes should give us a picture of how oak species persisted through long histories of interbreeding.

One indirect way of viewing genes is to look at the proteins they create. In the 1950s, researchers developed tools for deciphering the chemical composition of proteins. Every protein is composed of a chain of amino acids, and each amino acid is encoded by DNA. Proteins are thus just a few steps from the genome. Enzymes, which catalyze chemical reactions inside organisms, are a particularly useful type of protein for understanding species. They vary, and different variants are referred to as isozymes. Many isozyme variants can be detected and distinguished from each other in the laboratory. Some are ecologically important. An isozyme variant that influences the formation of cell walls, for example, might have a strong effect on how oak roots interact with the fungi they depend on, and oaks that interact differently with mycorrhizal fungi might have different versions of that enzyme. Another variant might help an oak adapt to the heavy clay soils of river floodplains of Missouri or the rocky soils of the Sierra Madre Occidental in western Mexico. Most isozyme variants probably have little if any ecological significance, but they still reflect genetic differences between individuals.

The first oak isozyme studies, conducted in the 1980s, did not distinguish closely related species. Instead, isozymes for a given set of genes tended to be diverse and shared between species. There were exceptions. Isozymes showed significant differences between the White Oak and Red Oak Groups. And a few of the more genetically distinctive species within each lineage jumped out: pin oak (*Q. palustris*), for example, is distantly related to the other eastern North American species of the Red Oak Group, and isozymes distinguished it clearly. Among close relatives, however, isozymes generally couldn't separate one species from another, even when the morphology and ecology were clear. This suggested that perhaps there was too much hybridization between related oak species for molecules to distinguish them. Alternatively, perhaps oak species carried so many ancient alleles that isozymes were bound to overlap with their relatives. Perhaps both effects were at play.

The first paper to study oak evolution using DNA sequence data was published in 1991, just four years after the first oak isozyme study. The researchers, Alan Whittemore and Barbara Schaal, working at Washington University in St. Louis, revisited the eastern North American White Oak Group that Hardin, Burger, and Van Valen had studied fifteen years earlier. Whittemore and Schaal relied primarily on DNA from the chloroplast. Chloroplasts live inside green plants and have their own genomes. The chloroplast is inherited exclusively from the mother tree in oaks and most other flowering plants, with rare exceptions. (We know of no exceptions in oaks.) This means that oak chloroplasts move around the landscape inside acorns. Acorns tend not to travel as far as oak pollen does. Oak chloroplast DNA consequently reflects more local gene movement histories than nuclear DNA, which is carried by both parents and roams as far as the wind will take it.

Whittemore and Schaal genotyped 122 eastern white oak, bur oak, post oak, swamp chestnut oak, and southern live oak trees (*Q. alba*, *Q. macrocarpa*, *Q. stellata*, *Q. michauxii*, and *Q. virginiana*, respectively). These five species are all distinct ecologically and in the shapes and sizes of their leaves, their acorns, and their overall growth form. They are, in other words, what most botanists, naturalists, horticulturalists, gardeners, ecologists, and evolutionary biologists would consider "good species." They also hybridize. Whittemore and Schaal collected leaf samples of all five species from sites ranging from Minnesota and Wisconsin to Missouri, Kansas, and Texas. Two or three of the species grew together at most of the collection sites. Whittemore ground the leaf tissue and subjected it to an analysis that cut the chloroplast genome at short, specific sequences scattered across the genome. Comparing lengths of the resulting genomic scraps between individuals provided an estimate of the genetic similarity between individuals. If the biological species concept were a reliable description of oak species—if different species rarely exchanged alleles—the chloroplast should tell which individual belonged to which species.

But Whittemore and Schaal found no chloroplast genotypes to be typical of bur oak, nor of any of the other White Oak species they investigated. Instead, the most common chloroplast genotype found at each site was shared among species at that site. They found, for example, that the post oaks and southern live oaks of Travis County, Texas, had the same chloroplast genotype; the other post oaks all had a different chloroplast genotype. The eastern white oaks and bur oaks of Green County, Wisconsin, shared a chloroplast genotype, to the exclusion of all the other eastern white oaks and bur oaks across their range. It seemed the chloroplast moved across species boundaries as easily as squirrels scampering between trees. In fact, the probability of a chloroplast genotype being shared among species at a single site was about six times higher than the probability of a chloroplast being shared between different populations of a single species. Whittemore and Schaal were detecting the history of interbreeding between species, followed by repeated backcrossing of hybrids by pollination from one parent. The study seemed to demonstrate, in other words, that Leigh Van Valen was right when he said, "It may well be that *Quercus macrocarpa* in Quebec exchanges many more genes with local *Q. bicolor* than it does with *Q. macrocarpa* in Texas."

This finding was supported by chloroplast studies published two years later on the White Oaks of Europe, similarly showing that the chloroplast does not tell us what oak species we are looking at, but rather where the oak grows and what other species it grows with. Within a few years, high-resolution maps of chloroplast genotypes across western Europe showed a consistent pattern: the chloroplast was shared freely among most White Oak species, but it defined geographic regions well. In fact, the French researchers who had worked on genetic diversity of the western European oaks for over a decade argued that *Q. robur* and *Q. petraea* had chased the glaciers northward by leapfrogging. Forests of pedunculate oak (*Q. robur*), they argued, were established by seed spread by jays, and sessile oak (*Q. petraea*) spread north in their wake by flooding the *Q. robur* populations with pollen. Founding populations of pedunculate oak were thus gradually transformed into sessile oak populations through gene flow. It seemed clear that at least from the chloroplast's standpoint, it could not be gene flow alone that defined oak species.

Whittemore and Schaal had, importantly, sampled a second genome alongside the chloroplast: the nuclear genome, which includes the genes that encode almost all of the ecologically and morphologically significant characteristics of species. The nuclear genome is woven together into chromosomes and is biparentally inherited: Mom provides a set of gene copies or alleles, and Dad provides the other set. Thus, if an acorn from one species is fertilized by pollen from another species, the nuclear genome will carry alleles from both parents, one set from each parental species. By contrast, only the mother tree's chloroplast is carried into the hybrid offspring of a cross between two species; the father tree contributes no chloroplast. This makes the chloroplast a poor marker for identifying species for two reasons. First, the population of reproductive individuals that can pass on a chloroplast is half as large as the population that passes along a nuclear genome copy, and that half the population carries only one copy of the chloroplast genome each. As a consequence, any chloroplast variants that enter a population by gene flow have a greater chance of coming to dominate a population by simple luck of the draw. Second, acorns do not travel as far as pollen. When an oak is near the edge of its range, it will tend to be pollinated by other species that are more common locally. But pollen, carrying the nuclear genome, can migrate many kilometers on the wind. The movement of genes within a species can counteract the effects of gene flow between species more easily for the nuclear genome than for the chloroplast. The shorter distance that chloroplasts travel on average thus makes them more susceptible to recording hybridization history.

Using the same methods they had used to study the chloroplast genome, Whittemore and Schaal gleaned a small but important nugget of information from the nuclear genome: a length variant in the nuclear ribosomal DNA repeat region. They observed very little variation in the nuclear genome using this method, but they did find that it followed taxonomy instead of geography. Their markers separated eastern white oak, bur oak, and swamp chestnut oak as one group from post oak and southern live oak as another. This separation of species marks one branch of the oak tree of life as we would know it twenty-five years later. Their work thus suggested that while the chloroplast genome tracked the history of gene flow between species, the nuclear genome might be more faithful to the species we know from the herbarium and the field. This was a harbinger of the next wave of important findings.

A few years later, isozyme data demonstrated that while two closely related European oaks, *Q. robur* and *Q. petraea*, shared most of the same alleles, the relative frequencies of those alleles in each population clustered not by geography, but by species. The two species appeared, in other words, to maintain different collections of gene copies even when the populations were living together, and even when they could exchange genes. Then in 2000, a note published in *Nature* showed a similar result, but using the proportion of shared nuclear microsatellite DNA variants. Microsatellite DNA is composed of repeated nucleotides—AAAAAAA, GTCGTCGTCGTC, or ACACACAC, for example—that mutate particularly quickly during DNA duplication. Because they evolve rapidly, a collection of microsatellites from numerous chromosomes have the potential to provide more information about species differences than DNA sequences from a single gene, or from a handful of isozyme markers. Once again, populations of the two species clustered not by geography, but by morphological and ecological identity. These data suggested that while hybridization was common enough to move chloroplasts between species, it was infrequent enough that genetic distinctions between species were often stamped into the nuclear genome. Even with hybridization, oaks were beginning to look like real species, genetically as well as morphologically and ecologically.

Thus, in the course of about thirty years, oak species had turned the corner from ethereal to tangible. In the 1970s, we could see there were oak hybrids and oak species, but we had little idea of what the genes might tell. Then, in the 1980s and 1990s, seeing that isozymes and chloroplast genomes were shared between good species, it appeared that gene flow was rampant. Finally, research in the late 1990s and early 2000s showed that the nuclear genome could distinguish hybridizing species. The genetic reality of species was heading toward a resolution: oak species were real, oak species did exchange alleles through hybridization and backcrossing, and both kinds of histories—the history of speciation and the history of gene flow between species—were encoded in their genomes. Every individual oak was turning out to hold a collection of oak evolutionary histories. With molecular data, we could begin to tease those histories apart.

Over the first quarter of the twenty-first century, study after study showed that oak species remain distinct from one another but also share alleles through hybridization and introgression—that is, through an initial cross between two species, and then subsequent generations of pollination by one parent species or the other. A sample of studies that use nuclear markers—because they allow us to estimate how much of an individual's genome is attributable to introgression—and a commonly used population assignment method shows that an average of about 3% of genotyped individuals are likely to be first-generation hybrids; these are the conduits for introgressive gene flow. If we treat the threshold for a "pure" species as being 90% of their genomes coming from only one species, these studies find an average of about 80% of individuals as nearly "pure," and the remaining 20% as either hybrids or backcrosses. There's a huge range of variation, however, with some studies finding as few as 1% or 2% hybrids and introgressed individuals and others finding more than 60%.

You can view this in two different ways. On one hand, *so much gene flow!* But on the other hand, *so many good oak species even with all that gene flow!*

Both reactions are justified; every individual encompasses multitudes. Closely related oaks exchange genes. Yet gene flow, selection, and evolutionary history also define the oak species we recognize in the field by their form and niche. Many of the genes making up the local black oak (*Q. velutina*) share a more recent ancestor with Hill's oak (*Q. ellipsoidalis*) growing in the same forest than they do with black oak growing in Wisconsin, downstate Illinois, or New York. But this does not undermine the fact that *Q. velutina* and *Q. ellipsoidalis* are what most of us would consider "good species": they are recognizable entities with distinct life histories, ecologies, and collections of alleles, separate from the other species in their neighborhoods.

Swamp white oaks, bur oaks, and eastern white oaks live within a few meters of each other, separated by differences in soil moisture and texture, fire frequency, the insects that feed on them, and the fungi that colonize their roots and leaves. Yet they grow in the same forest. What forms their boundaries and keeps them from merging? The stigma and style are the first gatekeepers, favoring the pollen of one species, slowing down the pollen tubes of others. Genetic differences between species are gatekeepers, reducing the viability of pollen produced by hybrids, the rate at which their seeds germinate, and even how fast hybrid offspring grow. After that, ecology holds the keys. Frequently burned prairies favor alleles that code for the exceptionally thick bark of bur oak. Inundated soils of river floodplains favor alleles that allow the roots of swamp white oak and bottomland populations of bur oak to thrive. Cretaceous limestones favor *Q. mohriana*, whereas igneous substrates favor *Q. grisea*. Natural selection is the final arbiter of which alleles define or move between species.

A classic textbook about speciation by evolutionary biologists Jerry Coyne and H. Allen Orr adopts a more relaxed version of the biological species concept than most of us learned in high school. Coyne and Orr allow for a little hybridization between species: "In our view, distinct species are characterized by substantial but not necessarily complete reproductive isolation." In the first chapter of the book they address several challenges associated with the biological species concept. Oaks, viewed by most as a "worst-case scenario" for the biological species concept, are among these challenges. But Coyne and Orr point out several studies providing evidence for species coherence, as well as a study showing partial postzygotic mating barriers between *Q. gambelii* and *Q. grisea*. They show that there are limits to gene flow between species, which Muller, Tucker, Stebbins, and Hardin all would have agreed with. If there weren't, we wouldn't consistently see genomic, morphological, and ecological coherence in oaks. Oaks, they conclude, may not present such a challenge after all.

The fact remains that you have to accommodate quite a bit of gene flow to accept oak species. The amount of backcrossing we find in oaks is impressive: in the studies I mentioned in the previous section, nearly 20% of genotyped individuals appear to be hybrids or backcrosses to one parent or the other. Are the oak species involved not "good species"? I doubt it: hybridizing species pairs such as *Q. engelmannii* and *Q. berberidifolia* in California or *Q. virginiana* and *Q. geminata* in the southeastern United States are well characterized ecologically and morphologically as well as genetically. So far as we can tell, almost every oak within pollination distance of a close relative is the potential ancestor of a genetically mixed lineage. They nonetheless retain their distinctiveness even as they grow together for hundreds of generations, and thousands of years.

We still have a lot of basic research to do before we fully understand how oak species come into existence and persist. We know, however, that when we say "oak species," we are talking about groups of individual oaks that look similar to one another, behave similarly to one another, and are genetically similar to one another, but are dissimilar to other species. We know, too, that oak species are separated by a combination of intrinsic barriers to reproduction and natural selection acting on alleles that determine a species' niche. Each effect may be weak on its own: the relative importance of ecology and physiological barriers to reproduction is part of what we have yet to understand. But the aggregate effects are the discontinuities we observe between species. Imagine genetic, morphological, and ecological variation as each defining a kind of space. There are gaps between species. There may be hybrids—in oaks, there are plenty of them—and those hybrids fall within the gaps between species. But good species generally remain distinguishable even after they have mingled with other species for a long time. The White Oaks of Hardin's interbreeding ring are, as Hardin expected, good species genomically, just as they are ecologically and morphologically.

This doesn't mean that taxonomic decisions are always clear. Anyone looking at the data on Hill's oak (*Q. ellipsoidalis*) and scarlet oak (*Q. coccinea*) can see two populations that are separated genetically. These two are allopatric now, or mostly so, yet they are morphologically very difficult to tell apart. Place one in front of me without telling me where it's from, and I may have about a 70% chance of telling you the species correctly. But genotype it or tell me what county it's from, and I'll be right almost every time. So: should we consider Hill's oak and scarlet oak separate species? Separate varieties? Allopatric sister populations are often taxonomically ambiguous. Other species, like the pedunculate and sessile oaks of Europe, are distinct genetically but began swapping genes late in life, perhaps as recently as the current interglacial period. All species have traveled along a passage from their origins as distinctive populations to species. And their distinctiveness may change over time.

When Darwin was writing, the British flora alone had at least 182 named entities that were considered by some to be varieties but by others to be species. This makes it sound as though we will never agree on what species there are in the world, but the situation is not as bad as all that. If you live in a place that has at least two or three common oaks, ask your naturalist friends what oak species they encounter in a walk through the neighborhood. They will probably be able to name more than one, and with high confidence. Then ask someone else with a comparable amount of experience, and you'll hear common answers. In fact, the ninety-one oak species recognized in *Flora of North America* in 1997 are almost all agreed on by botanists today.

Even in the most complex groups, taxonomic study generally leads toward consensus. The Mexican oak species, for example, are particularly hard to understand, with the number of recognized species fluctuating over the course of a few decades. But disagreements mostly converge on agreement with additional research. We know from detailed molecular and morphological work published in the past few years that the widespread *Q. laeta* can be distinguished into at least two genomically and morphologically distinct species, and ongoing work suggests that there may be more. The species *Q. conzattii* and *Q. urbanii*, both with isolated populations separated by the Trans-Mexican Volcanic Belt, are each separable into two species based on genetic data. These taxonomic changes are the result of new field collections, morphological study, and molecular data, and they are generally accepted by the community of plant taxonomists. There are lifetimes of work remaining to be done to understand the taxonomy, ecology, and genetics of the world's oak species. In general, though, when a group of taxonomists has carefully studied a species using genomic data or morphology, we gain a clearer understanding of how distinct that species is compared to close relatives. We wouldn't experience this if the boundaries between species were arbitrary.

The histories of alleles knit species together: trace them backward in time, and you will discover the boundaries of the species that contain them.

Quercus sect. *Lobatae*: *Quercus ellipsoidalis* leaf. Natural population, Taltree Arboretum, Indiana, October 5, 2005. Illustrated from photo of a live specimen vouchered at the Morton Arboretum Herbarium, Accession 173398.

The histories of species themselves braid together and branch to make the Tree of Life. Look closely, though, and you'll find the branches of the Tree riddled with interconnections. Those are produced by hybridization and backcrossing, introgression from one species to another. The hybridizations that connect oak species and branches of the oak tree of life do not mean there are no species and no Tree of Life. Oaks' strength—as exemplified by the longevity of the eastern white oaks and the bur oaks in my local forest preserve—derives from their persistence over the course of tens of millions of years, partly despite, and perhaps partly due to, hybridization and introgression.

4

Origins

Fagaceae

The world in the mid-Cretaceous, about 100 million years ago, was hot and growing hotter. Atmospheric carbon concentrations were three to six times as high as modern levels and global temperatures were 10° to 15°C warmer than they are today. There were no massive glaciers in the north. An inland sea separated eastern North America from the west. A warm ocean cradled the remnants of the northern supercontinent, Laurasia, which was made up of what today are North America, Europe, and most of Asia. Cycads thrived in western Greenland; today, they are restricted to the tropics and some temperate ecosystems of the Southern Hemisphere. Ginkgoes grew on Alaska's North Slope. Mammals migrated between Canada and western Europe on a northern land bridge crossing the Atlantic Ocean, but they were dwarfed by the dinosaurs. Our distant ancestors may have been scurrying through the understory.

Flowering plants—angiosperms—were changing the world. Beetles, snakes, and spiders nested in angiosperm leaf litter, burrowed through wood, copulated in the canopies. Moths, butterflies, bees, gnats, and beetles shuttled pollen from flower to flower. Ants flourished in angiosperm-rich leaf litter and treetops, feeding on insects. Amid this burgeoning cacophony, the ancestors of the birches, walnuts, sweetbays, and a few other families, including the Beech Family, Fagaceae, were an unassuming population of flowering trees, with flowers probably buzzing with insects. That population would ultimately give rise to twelve hundred of today's tree and shrub species. Among those species are the oaks.

Nothofagus alpina **leaves and fruits**, showing the distinctive spiny-backed cupules. Tree in cultivation. Illustrated from a photo provided by Koen Camelbeke.

No botanist exploring the Cretaceous forest could have predicted that the offspring of these trees would one day rule forests across the Northern Hemisphere. Like all branches on the Tree of Life, the distant ancestors of the oaks—the Beech Order, Fagales—started as a species with a little luck and a lot of potential.

Yggdrasil, the Norse Tree of Life, stands at the center of the world. One of its roots dips into the well of wisdom and intelligence. One extends into the home of the gods. One drinks from the spring that gives birth to all waters. Its canopy drips honeydew and knits together the heavens and the earth. The Tree of Life in Genesis stands in the Garden of Eden, bestowing eternal life on anyone who eats from it. There are "trees of life" all around the world: the Caribbean and Central to South American lignum vitae (*Guaiacum sanctum*), which has a range of medicinal and practical uses (including providing the neck of Pete Seeger's banjo); arbor vitae, the white cedar (*Thuja occidentalis*), rich in vitamin C and central to the health of American Indians and European settlers when they arrived; and the baobabs (*Adansonia* spp.), which shelter, clothe, feed, and provide water to entire communities in African savannas.

Charles Darwin had a different vision of the Tree of Life. He included only one figure in *The Origin of Species*, a trifold illustration nestled about three-quarters of the way through chapter 4. The figure looks like two twiggy shrubs—almost like a pair of *Ephedra*—rooted at the base of the page. Interspersed with the shrubs are nine vertical lines. At the very bottom of the page, each line or shrub originates at a letter, A through L. Each letter stands for one species. The lines and shrubs represent genealogies of organisms giving birth to more organisms, populations giving rise to populations. Fifteen horizontal marks traced across the page delimit time intervals from an unspecified past at the bottom to an unspecified later time, perhaps the present day, at the top. All but one of the vertical lines stop before they reach the top of the page. These represent populations that died off. One line reaches the top of the page without branching, representing a population that survived to the present day without producing any new species. The two shrubs flanking it are populations that branched into numerous populations as they engendered generation after generation. Most of the populations they gave rise to died off, but a few reach the present at the top of the page. They have differentiated from their distant ancestors.

If each horizontal mark indicates 1,000 generations and we estimate average oak generation times at about 50 years, then Darwin's Tree of Life might represent a single species. If each marks off 85,000 generations, then one shrub might be the oaks, *Quercus*, and the other the stone oaks, *Lithocarpus*. If each represents a million generations, then Darwin's twiggy shrubs might be rooted on the other side of the Cambrian explosion. Darwin's figure is vague by design. Whatever the span of history, the lines in his illustration represent lineages differentiating from each other by the processes of reproduction, dispersal and migration, and natural selection.

The Tree of Life that Darwin illustrated is a metaphor for the sprawling family tree that connects all the organisms that have ever lived on Earth. If we could see across all of space and time at once, we could observe and draw the innumerable matings, births, and deaths that make the Tree of Life. We could record individuals breeding within or migrating between populations. These individuals would link up by ancestry to form a lineage that extends through time. Each lineage would stretch out to form a branch of the Tree of Life. Lineages would evolve and change gradually over time. We would see speciation events, the gradual or sometimes abrupt cleaving of one species

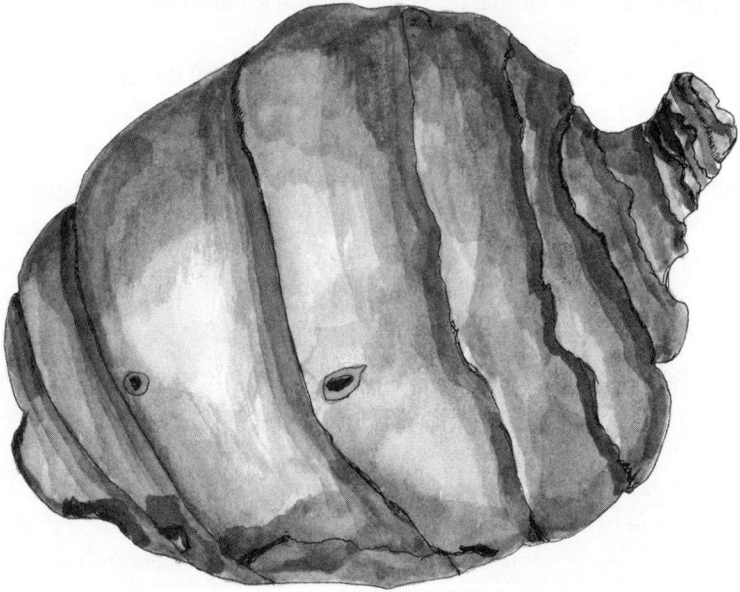

Lithocarpus keningauensis, **enclosed-receptacle (stone-type) fruit, external.**
The cupule wraps all the way around the fruit in this species and others in the genus.
The base of the acorn points to the right and has a visible stalk. Illustrated from a
photograph provided by Paul Manos.

into two. We could draw these events as forks in the tree. We would watch species go extinct, drawing each one as an attenuation of a branch toward its tip. These lineages would narrow and end when the last individuals of their species died.

As we sketched in the details, our Tree of Life would bristle with the stubs of extinct species, like Darwin's, only these stubs would be more numerous by far than the species known from all the world's fossil beds. It would ramify into the shrubs and witch's brooms that constitute genera or species complexes. Thick gray cobwebs in the crotches between branches would stand in for individuals mating between young species. Lines crossing between more distant lineages would record infidelities between species long separated from each other. We could run our fingers back to where all lineages connect, at the headwaters of the Tree of Life. That point of convergence marks the most recent common ancestor of all of life, the anaerobic denizens of

Lithocarpus keningauensis, **enclosed-receptacle (stone-type) fruit, internal**.
Inside of a stone-type fruit after it has been cut open from end to end using a saw.
The seed inside is protected by two strong layers. The thick outside layer is the cupule,
which has grown all the way around the nut. The next layer in is formed by the receptacle, tissue that grows to surround the seed in the enclosed-receptacle *Lithocarpus*
species. The chamber in the middle holds the embryo. The stalk of the fruit points
downward and to the right. Illustrated from a photograph provided by Paul Manos.

hydrothermal vents who gave rise to all living things on Earth. Our portrait
would be a direct observation, not an inference.

Because we are not omniscient, we infer the Tree of Life as best we can
from what we can observe. We document impressions of organisms that
lived in the past, etched into the earth as fossils. We measure bones and
leaves. We study the composition of proteins, tally up chromosomes, piece
together genomes from scraps of DNA. We do this for many organisms. We
use our knowledge of Earth history and molecular evolution to construct
statistical models that describe how the differences and similarities between
organisms might have evolved. We iterate the models in computers. We ob-

tain from our labors explanations of how evolutionary history may have proceeded. Each explanation is a hypothesis about the shape, dimensions, and timing of the Tree of Life. This understanding of the Tree is a shadow cast by real evolutionary histories.

Inferring the Tree of Life is a central goal for many taxonomists, evolutionary biologists, and biodiversity scientists. The science of the Tree of Life is called phylogenetics. Phylogenetic biologists usually don't study the whole Tree of Life at once. Instead, they investigate individual clades. A clade is a group of species that includes all the descendants of a single ancestor, no more and no fewer, plus all the branches connecting them to each other. If the Tree of Life were a real tree, a clade would be a bough or limb that you could prune off, including all the leaves growing from it, with exactly one cut. The group "birds," for example, is a clade: birds arose from a single dinosaur species in the Late Cretaceous. You could cut birds from the Tree of Life with a single snip. "Dinosaurs" are also a clade, but only if by "dinosaurs" we include all the usual suspects—*Brontosaurus*, *Tyrannosaurus*, *Triceratops*—along with the birds. If we say "dinosaurs" without including the birds in our definition, we are excluding all the living organisms within the clade. In other words, we would have to make two cuts in the Tree of Life to get just the nonavian dinosaurs: one at the base of the dinosaur clade, and one more to remove the birds. Oaks form a clade, maples form another, and pines and hawthorns each form their own clades. But "trees" do not form a clade: lots of unrelated plant families have given rise to trees over evolutionary time, and some groups of trees have given rise to herbaceous lineages. Some clades number in the thousands or millions of species or even more: eukaryotes, bacteria, archaea, beetles. Others comprise hundreds, tens, or just two or three species: humans, the genus *Homo*, are a clade of four to a dozen species, of which only *Homo sapiens* has survived to today.

Darwin wrote that the Tree of Life "fills with its dead and broken branches the crust of the earth, and covers the surface with its branching and beautiful ramifications." Every clade on the Tree of Life—grasses, pines, birds, whales, mosses, Enterobacteriaceae—plays a unique role in the ecosystem. The extinction of any species or clade leaves a hole in the interactions that make the natural world. Inferring the shape and timing of the oak phylogeny—the oak tree of life—is thus central to understanding and conserving oak biodiversity.

I did not know the difference between a pine and a spruce the year before I took my first plant taxonomy course. One thing I did know is that I enjoyed being in the woods. I suspected that taking botany classes would be a vehicle to eventually getting a job outdoors, so I enrolled in introductory botany and then a course called "Vascular Flora of Wisconsin." I quickly found I enjoyed knowing the names and personalities of species. Learning to recognize species resolved the wash of green that was previously "the woods" into a rich and familiar assemblage. Scientific names for the common plants of my region—*Asarum canadense, Ostrya virginiana, Andropogon gerardii, Acer saccharum*—were signposts leading into the natural world.

Once I had learned enough species, a genus would begin to make sense: trilliums, birches, sedges, roses, and the many other plant genera pieced together to form knowable groups. Over the course of a spring, I came to appreciate the way in which genera bundle together to form families, sometimes for obvious reasons, other times only subtly. Roses, raspberries, strawberries, peaches, and their relatives made the Rose Family, Rosaceae. Oaks, beeches, and chestnuts, together with a group of genera I had not seen yet, made the Beech Family, Fagaceae. I was gradually learning the Tree of Life from the tips inward by learning to identify trees, weeds, and wildflowers.

It would take a few more years before I appreciated that families aggregate to form larger clades of the Tree of Life. The taxonomy we use today is shorthand for our understanding of phylogeny. It wasn't always. There are many useful but artificial systems of classification. Artificial classifications are concocted for our use and generally rely on just a few characters to make categorizing easy. Linnaeus's sexual system of plant identification—published in 1753 in his encyclopedic *Systema Naturae*—is an artificial system, and perhaps the most famous. It divides plants up based on the number and arrangement of their stamens and pistils. Linnaeus's system was designed to facilitate plant identification (meanwhile, in parallel, Linnaeus worked on a natural system of classification that he never achieved). The system groups many unrelated families together: dioecious species—species with unisexual flowers on different plants—form a single group in his system, tossing together the very distantly related ginkgo (a gymnosperm), asparagus (a monocot, more closely related to lilies than to anyone else in this group), and poplar (a Eudicot, a relative of the willows). Linnaeus's sexual system of classification was stimulating to read, scandalizing to some, and easy to understand. Plant enthusiasts from northern Europe to Quebec and Suriname used his work to collect and identify plant specimens.

Artificial but practical plant classifications have a long history. Theophrastus, in the fourth century BC, classified plants by life-form: tree, shrub, undershrub, and herb. Alfred Kinsey—who conducted years of research on oak galls before he became famous for his work on human sexuality—and the eminent botanist Merritt Lyndon Fernald wrote a guide to edible plants that included such delicious categories as "Nibbles and Relishes," "Rennets," and "Masticatories and Chewing Gums." A wildflower book organized by flower color can get you into the prairie to identify plants even if it doesn't show us how plants are related.

By contrast, natural classifications use lots of information about each organism to organize species into groups that reflect fundamental differences, not superficial or merely practical similarities. Natural classifications since Darwin reflect our growing understanding of the Tree of Life. Taxonomists choose which branches of the Tree of Life to attach names to. In other words, taxonomists almost universally endeavor to name clades. Humans have named clades across the Tree of Life: every living species that has been named by scientists since Linnaeus is assigned—rightly or wrongly, precisely or not—a position on the Tree of Life. Placing a species into a genus, a genus into a family, and a family into an order implies a claim about the shape of the Tree of Life. This is because species correspond to the tips of the Tree of Life. They are the leaves.

The business of phylogenetics is like crawling among the branches of the Tree of Life to see how and where they connect to one another. You can think of clade names as though they were written on these branches. A genus name is written on a branch of the Tree of Life that gives rise to species, usually more than one species. A larger branch that sprouts genera—even if it only has a single genus, like Ginkgoaceae, the Ginkgo Family—bears a family name. In the Linnaean classification system, which is the one most widely used today, names have ranks that tell you their relative depth in the Tree of Life. A large branch bearing families is ranked as an order, and one that sprouts orders is ranked as a class. Each rank tells you what other ranks are contained within it. Knowing that the oak genus *Quercus* is within the Beech Family, Fagaceae, tells you that the first Fagaceae species must be at least as old as the first *Quercus* species. If the Tree of Life were a real tree, about three-quarters of the wood and leaves would be bacteria, all growing from a single branch nearly as thick as the trunk itself. Oaks would be a spray of about 425 leaves near the edge of the canopy. You would probably need an orchard ladder to reach them, once you found them.

"If you do not know the names of things," Linnaeus wrote, "the knowledge of them is lost too." Learning oaks, mosses, wildflowers, birds, butterflies, or any other set of species is more than just learning names. Familiarity is a road to understanding; affection often follows closely behind. Because

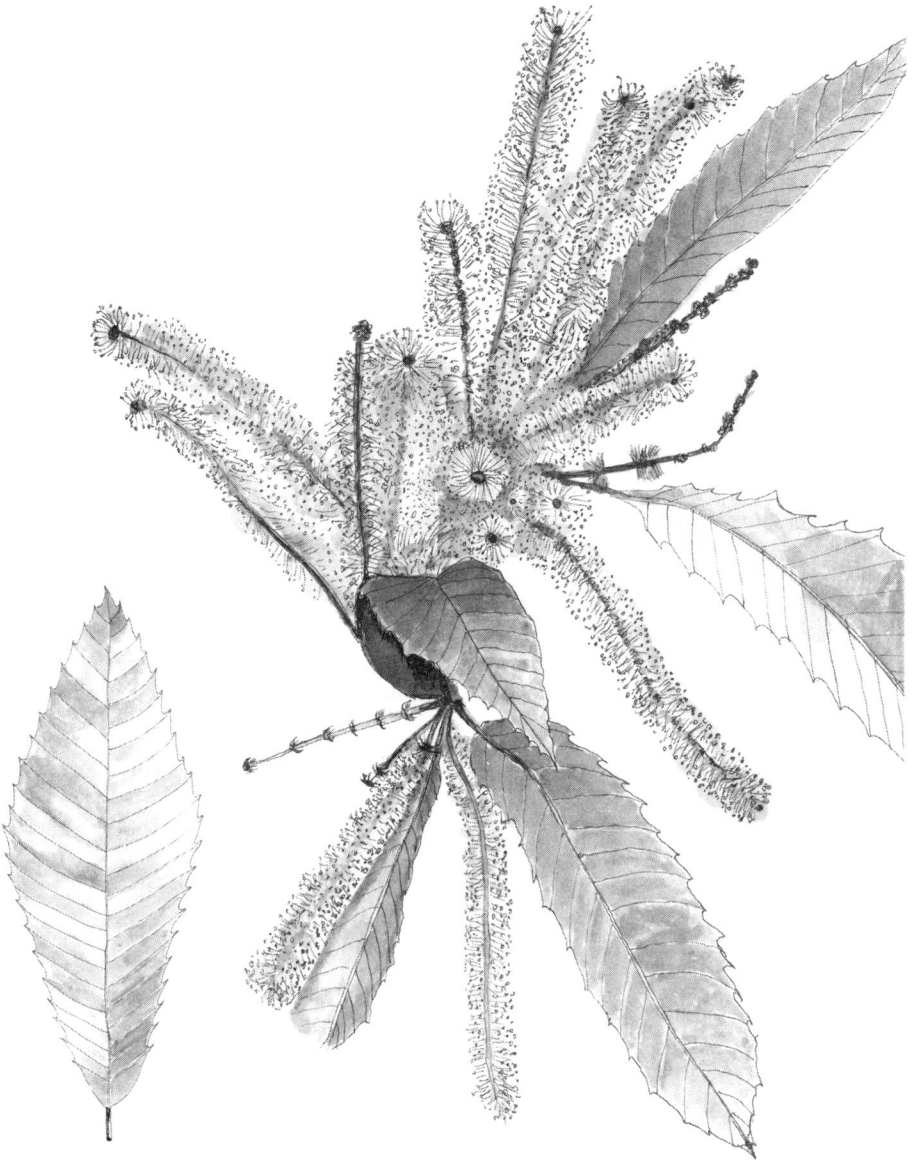

American chestnut (*Castanea dentata*). The inflorescence and the individual leaf are from the same tree. The stout, bottle-brush-like, insect-pollinated catkins contrast with the lax, wind-pollinated catkins of *Quercus* (cf. the illustration on p. 8). Tree in cultivation, Georg-August-Universität Göttingen, June 2023.

of the correspondence between taxonomy and the Tree of Life, taxonomic ranks and their names serve as an introduction to the history of life on Earth.

Until the late twentieth century, members of the Beech Family, Fagaceae, were classified as belonging to a group comprising most of the plant families that have aments. Aments—also called catkins, as you may remember from chapter 1—are inflorescences of tiny, typically male flowers borne together on an elongated spike or dangling stalk. Their flowers are usually unisexual, wind-pollinated, and sessile, meaning that the flowers themselves don't have individual stalks. Their petals are generally highly reduced, so they don't get in the way of the wind. Put these traits together with the fact that most ament-bearing plants are trees, and it may seem natural to classify the ament-producing species together in a single group.

In Adolph Engler's 1892 *Syllabus der Pflanzenfamilien*, a natural classification that endeavored to group families based on evolutionary relationships rather than mere similarity of form, the Fagaceae and the Birch Family (Betulaceae) were placed together in an order called the Fagales. They were wedged between the Walnut (Juglandaceae), Sweetbay (Myricaceae), and Willow (Salicaceae) Families on one side, and the Elm (Ulmaceae), Mulberry (Moraceae), and Nettle (Urticaceae) Families on the other. These families all bear their flowers in aments and were, in several classification systems, collectively called the "Amentiferae" or "Amentaceae." Engler considered aments to be primitive, their tiny flowers a step along the road toward showy flowers. The first herbarium I worked in was, like many of the world's herbaria, still organized according to Engler's system, with the Amentiferae housed in cabinets near the front of the room.

The traditional Amentiferae are not a clade, however, just as trees are not a clade. And aments are not a primitive solution to the problem of moving pollen around: instead, they represent a cluster of traits that have evolved in multiple lineages independently. When the environment selects for several traits simultaneously, species sometimes evolve similar suites of attributes irrespective of their true relations. You can see this in the trilliums and jack-in-the-pulpit, for example. Broad leaves with net-like venation are selected for in forests, where they help plants capture sun in shaded environments. Trillium and jack-in-the-pulpit leaves are consequently similar in shape, thickness, and venation, despite the fact that they are not closely related. Similarly, the vultures of the Old World and the Americas have converged on the practice of circling overhead to look for dead animals. They have a similar body shape, and they have both evolved featherless heads, perhaps to help them regulate their temperature or perhaps so that bacteria have fewer places to hide and fester as the birds devour rotting carrion. But the old-world vultures are related to storks, while the American vultures are more closely related to hawks and eagles. Multiple traits in the unrelated groups have converged in concert, giving the organisms a false appearance of relatedness based on similarity in form and ecology.

By the early 1970s, it was clear that aments represent an amalgam of traits that have evolved multiple times to facilitate pollination. The ament-bearing families had independent evolutionary histories. They were not a single clade. But it wasn't clear how these families were related to one another. My textbook for plant systematics was the second edition of Gleason and Cronquist's *Manual of Vascular Plants of Northeastern United States and Adjacent Canada* (1991). The book was new and about the size and heft of a Bible. I used it in the field regularly for fifteen years and still keep it handy when I'm writing. It represents a transition between the old flowering plant classifications and the new classifications that would arise just a few years later. Cronquist placed the Fagaceae and Betulaceae together into the Fagales, just as Engler had. He placed the Fagales into a larger grouping called the Hamamelidae, with such trees as the Walnuts (Juglandaceae), Elms (Ulmaceae), Mulberries (Moraceae), Sycamores (Platanaceae), and Witch-hazels (Hamamelidaceae). Cronquist and others working with morphological and anatomical data through the early 1990s had dropped some distantly related ament-bearing families (such as the Willow Family) and brought in others without aments. But the higher grouping they recognized still assembled families of plants that are not actually each other's closest relatives. Witch-hazels and sycamores, mulberries and elms, walnuts and sweetbays, oaks and birches were all sharing a home where only a subset of them belonged. When I first opened my copy of Gleason and Cronquist and read the "Synoptical Arrangement of the Orders and Families of Hamamelidae as Represented in Our Flora," I was reading a taxonomy that would soon be out of date.

In 1993, Mark Chase, a researcher at Kew Gardens, led a group of more than forty plant scientists in creating the first molecular outline of the seed plant phylogeny. The study used DNA data from the chloroplasts of species representing more than 250 plant families to infer a highly inclusive phylogenetic tree. Their project included cycads, pines, and other gymnosperms as well as a careful sampling of flowering plant families. It included only three of the families we now know to be the closest relatives of oaks, but it sufficed to demonstrate—among many other findings—that the Beech

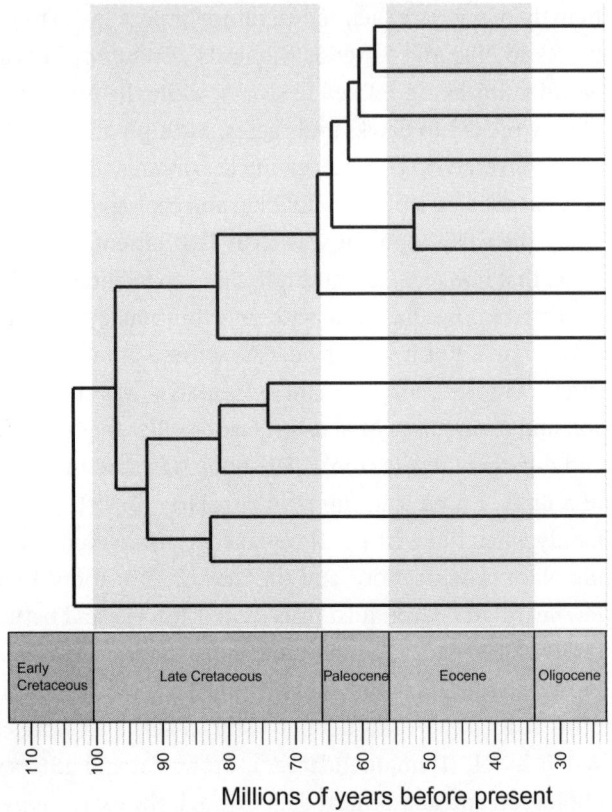

Fagales tree of life. This phylogenetic tree includes one representative from each family of the Fagales and one from each genus of Fagaceae. The horizontal axis represents time measured in millions of years. Horizontal "branches" of the tree represent populations and species evolving over time. Points where the branches split into two represent hypothesized speciation events in the past. The relationships on this tree are vknown with high confidence. The timing of speciation events is more uncertain and will likely change as more analyses are conducted with more fossil and genomic data. Phylogenetic tree estimated from molecular data for 465 genes, provided by Elliot Gardner and Kasey Pham and used with permission; and time-calibrated using molecular and fossil calibrations from several publications (Zhou et al. 2022; Grimm and Renner 2013; Yang et al. 2023; Hipp et al. 2020).

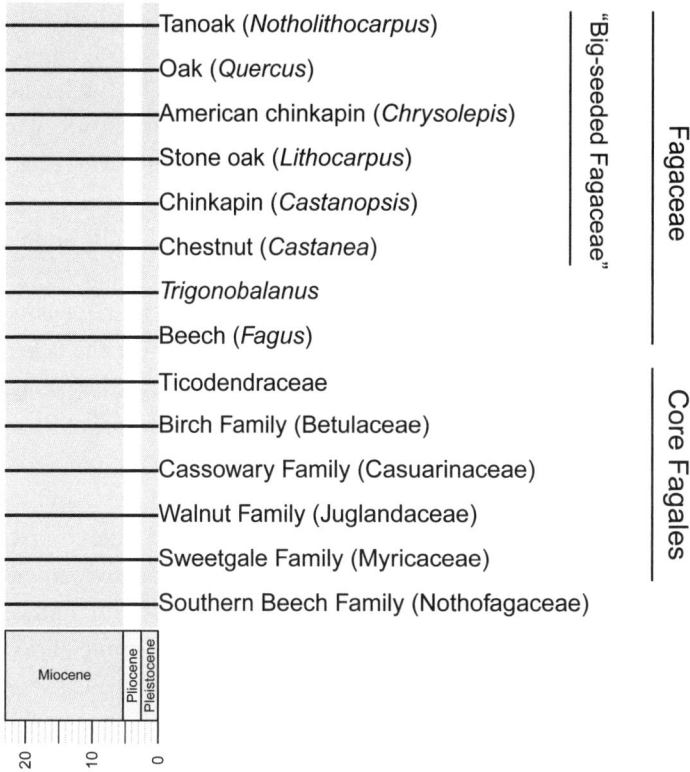

————————Tanoak (*Notholithocarpus*)		
————————Oak (*Quercus*)		
————————American chinkapin (*Chrysolepis*)		
————————Stone oak (*Lithocarpus*)	"Big-seeded Fagaceae"	Fagaceae
————————Chinkapin (*Castanopsis*)		
————————Chestnut (*Castanea*)		
————————*Trigonobalanus*		
————————Beech (*Fagus*)		
————————Ticodendraceae		
————————Birch Family (Betulaceae)		Core Fagales
————————Cassowary Family (Casuarinaceae)		
————————Walnut Family (Juglandaceae)		
————————Sweetgale Family (Myricaceae)		
————————Southern Beech Family (Nothofagaceae)		

Miocene | Pliocene | Pleistocene

20   10   0

Family (Fagaceae), Sweetbay Family (Myricaceae), and Cassowary or She-oak Family (Casuarinaceae) were closely related to each other and only distantly related to mulberries, elms, witch-hazels, and sycamores. The study formed the foundation of our understanding of seed plant evolutionary history for many years, and it is still the starting point for our current plant classification. Mention "Chase et al. 1993" to almost any plant systematist, and they'll know exactly what you're talking about.

Mention the name Paul Manos to any oak systematist, and you'll get a similar reaction. Manos has been a leader in our understanding of oak evolutionary history for decades, and some of his earlier work laid the foundations for our understanding of the relationships among the Beech Family and its relations. A paper Manos published the same year as Chase used DNA data from plant chloroplasts to investigate relationships among the Beech Family and its closest relatives. Manos and his colleagues included the families of Cronquist's Hamamelidae and other potential relatives of the Fagaceae.

Manos demonstrated that the families related to the oaks included, in addition to the Sweetbay and Cassowary Families, the Southern Beech Family (Nothofagaceae, literally the "bastard beeches"), Walnut Family (Juglandaceae), and Birch Family (Betulaceae). Manos and Kelly Steele would go on to publish a paper in 1997 that completed the sampling of families we recognize today as the closest relatives of the Fagaceae.

This molecular work did not give us a name for the Beech Family and its relatives. There are no rules about which branches of the Tree of Life taxonomists should name. Taxonomists nonetheless find it useful to name entities that we can recognize. The members of the clade containing the Beech Family and the other families included in Manos's study fit together nicely. All are woody plants with flowers that are mostly unisexual. Many of them are wind-pollinated. Most bear male flowers in aments or scaly, cone-like inflorescences. Most produce nuts. The clade has a lot of variety, and its composition as we know it today only became clear when taxonomists peeked under the hood at the molecules. Fagaceae and their relatives nonetheless were similar enough in overall appearance to make a useful named clade, or taxon. It made sense for these reasons that in 1998, a group of taxonomists organizing themselves as the Angiosperm Phylogeny Group (APG) formally recognized this lineage as the order Fagales, the name initially used by Engler in 1892 to bundle together just the Beech Family with the Birch Family.

Taxonomists had, at last, found the proper home for the oaks and their relatives. Fagales is the largest portion of the Tree of Life that includes oaks and also comprises only trees and shrubs. Prune the Fagales one branch closer to the current day, and you would lop off the Southern Beeches (Nothofagaceae), an iconic tree family of much of the South Temperate Zone. Prune the Tree of Life one branch deeper in time, and you would add in a clade that includes the Begonia Family (Begoniaceae), the Squash Family (Cucurbitaceae), and several others. You would no longer have a woody clade whose families and genera all had somewhat similar fruits and flowers. You would end up with a subtree of the Tree of Life, but one that lacked the ecological and morphological coherence of Fagales.

Fagales is not the most taxonomically diverse flowering plant family: its approximately 1,200 species look like a small classroom in comparison with the Bean Family (19,500 species), Sunflower Family (25,000), Orchid Family (26,000), and a few other super-diverse flowering plant clades. But the Fagales are ecologically diverse and dominant in a wide range of habitats around the globe. Fagales is a point of entry into the evolutionary and natural history of broad-leaved forests.

Follow the Fagales tree of life back to its root, and you'll find fossils scattered along the way. There are Fagales impressed in coal, rock, or amber dating back to 90 million years ago. But we can't read evolutionary history straight from the fossil record. This is partly because many of evolution's transitions are rare in the fossil record, if they made it into the record at all. Additionally, the processes that create fossils do not preserve all species equally, and they do not preserve all parts of an organism well. When I read Steve Brusatte's *Rise and Reign of the Mammals*, the fossil discoveries come together like a movie. This is mostly due to Brusatte's knowledge and story-telling skill, but it's also partly because mammal teeth, leg bones, skulls, and other skeletal bits preserve well. You can almost see lineages evolving over time. By contrast, we have relatively few fossils of leaves and acorns for crucial periods of oak evolutionary history. There are regions of the globe and time periods when fossilization was less efficient. Many other fossils were beaten up, abraded, or crushed as they were buried and gradually encased in stone.

The fossils we find in good shape are often challenging to hang on the Tree of Life. Fossils are almost invariably fragmentary: isolated branch tips with a few flowers on them, pollen grains from catkins we can no longer find, a few leaves. This limits the number of characteristics we can use to estimate their relationships. The characters fossils present are also often difficult to interpret. Some fossilized "flower petals" have turned out to be the scales of conifer cones.

Even if fossils were known perfectly, tracing the evolution of morphology and anatomy would be more complicated than tracing the evolutionary history of DNA. The form of an organism evolves along complex pathways due to adaptation, genetic drift, and constraints on trait evolution. The evolution of DNA sequences is also complicated: there are mutations, rearrangements, duplications and deletions, patterns of adaptation and convergence, and correlated patterns of evolution among nucleotides due to the structure of DNA. But most of these complications are easier to model statistically than the complexities of an entire organism. DNA also provides more data points, more information about speciation and gene flow history, than any set of morphological traits we could ever hope to measure and count. The oak genome contains millions of variable nucleotide positions. Morphological studies generally quantify at most hundreds of traits.

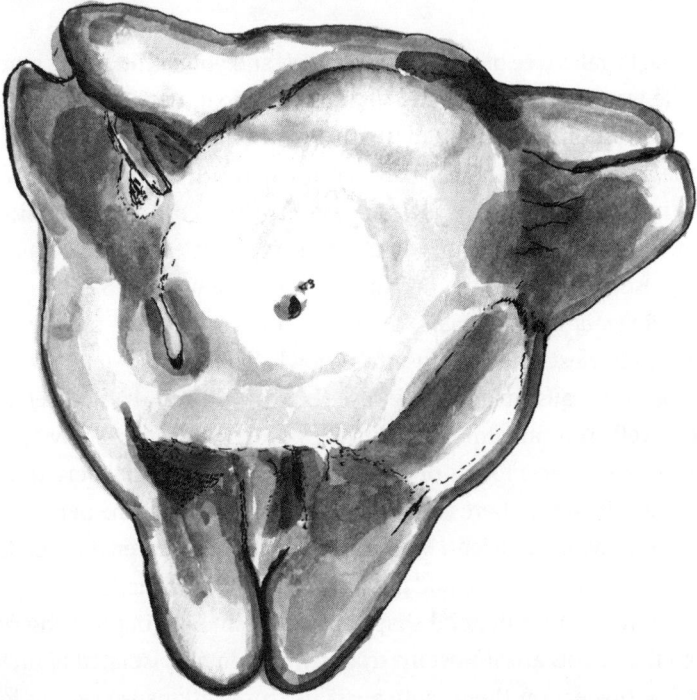

Nudopollis terminalis **pollen grain from the Middle Eocene.** *Nudopollis* is a member of the Normapolles group. This specimen comes from a quarry near Messel, Germany. Illustrated from a light microscope image provided by Johannes Bouchal.

Unfortunately, DNA breaks down more rapidly than fossils. The oldest sequenced DNA is from a preserved woolly mammoth that died about 1 million years ago. Cell nuclei have been found in dinosaur fossils, but there is no reason to suspect that the DNA they contained can be sequenced to yield an estimate of their evolutionary relationships. The most ancient flowering plant DNA sequenced to date is less than 10,000 years old. This is far removed from the mid-Cretaceous origins of the Fagales.

Fossils nonetheless shed light on the lives of lineages that have gone extinct and could never be known otherwise. In Fagales, fossilized pollen is an important source of our evidence about ancient history. Wind-borne pollen travels in clouds, millions of grains at a time. Its abundance and diversity in the fossil record provide clues to the distribution and abundance of clades or groups of species. Some of the oldest records of Fagales come from a collection of plants that we recognize under the name "Normapolles." The name Normapolles was coined for a group of genera and species whose pollen grains are found as fossils across eastern North America and Europe from the Late Cretaceous through the middle Eocene. The Normapolles were abundant from the eastern shores of the sea that cut through the center of North America, through Europe, all the way to western Siberia. Normapolles show up in the fossil record across this broad expanse beginning around 96 million years ago in such high numbers that the entire region is referred to in the Late Cretaceous as the "Normapolles Province." More than one hundred species are known in roughly eighty Normapolles genera.

Normapolles pollen grains on their own are difficult to classify. They have been assigned to different families at different times, some in the Fagales, others in the unrelated water milfoils (Haloragaceae) or elms and their relatives (Ulmaceae). But fossilized flowers and fruits that are associated with some of the Normapolles grains place them in a clade called the core Fagales. Core Fagales is the lineage that includes the Walnut, Birch, Cassowary, and Sweetbay Families (Juglandaceae, Betulaceae, Casuarinaceae, and Myricaceae), plus Ticodendraceae. Core Fagales is what is left of the Fagales if you prune off the Nothofagaceae and Fagaceae, the Southern Beech and Beech Families, respectively. Flowers associated with Normapolles pollen grains were tiny, like today's Fagales, but mostly bisexual. At least some of the species produced small nutlets. There are roughly twenty-five different species described from these flowers. Their variability suggests that Normapolles was probably not a single clade, but a collection of clades borne on various branches within core Fagales. Upland forests throughout eastern North America and Europe may have been dominated by Normapolles species throughout much of the Late Cretaceous, possibly extending to Central and East Asia. In some Central European fossil beds, Normapolles are the most diverse group of plants as early as 90 million years ago.

Normapolles began to decline in diversity and abundance after the Cretaceous. They were extinct by the end of the Eocene so far as we can tell, or nearly so. Today, one of the core Fagales has pollen that is very like Normapolles pollen: *Rhoiptelea*, which is endemic to the mountains of Southwest China and northern Vietnam. This solitary species is sister to the remainder of the Juglandaceae. It may be the last remnant of the Normapolles, as today's birds are the remnants—though much more successful—of the dinosaurs.

The Normapolles give us some ideas of what early Fagales looked like. Only 10–20 million years may separate the last common ancestor of the Fagales from the earliest Normapolles we know. But because Normapolles appear to be members of the core Fagales, using the Normapolles alone to understand the origins of the Fagales would be a bit like using fossils from a specialized dinosaur clade—perhaps the Theropods, which gave rise to *Tyrannosaurus rex* and, eventually, birds—or a few such clades to date the entire dinosaur clade. There are at least five known fossil flowers that might fall closer to the base of the Fagales tree of life. One of these is an unnamed collection of fossils from the eastern United States, dating to about 83 million years ago. They have small flowers and dry, indehiscent fruits. They were found near similar fossils that have cupules, though they do not have cupules themselves. They also have floral nectaries to attract insects, which is something today's Fagales do not. A different fossil genus, *Archaefagacea*, is known from an assemblage of flower and fruit fossils collected in northeastern Japan. It has pollen that my colleague Thomas Denk, a leading paleobotanist of oaks and their relatives, describes as indistinguishable from pollen of several of the modern Beech Family genera. But *Archaefagacea* has a three-seeded fruit, which would be highly unusual for the order, and was discovered without a cupule. This fossil has been argued by some to belong outside the Fagales. Others consider it a plausible early-diverging lineage from the stem of the order.

Of the remaining three fossils, two with the most evocative names— *Protofagacea*, suggesting that we're getting close to the root of the oaks and their kin; and *Antiquacupula*, named for its ancient cupule—were discovered in a pit quarry at the Atlanta Sand and Supply Company in Gaillard, Georgia. These two fossils were embedded in 83 million–year-old clay, which the researchers dissolved away with water, leaving the fossils naked in the sieve. The third fossil—*Soepadmoa cupulata*, named for both the cupules and for Engkik Soepadmoa, an Indonesian taxonomist who, among numerous other publications, wrote a monograph on Fagaceae and named numerous species of *Lithocarpus*, *Quercus*, and *Castanopsis*—was unearthed from the Old Crossman Clay Pit in Central New Jersey. It was embedded in amber that had dripped from a conifer, probably a relative of the junipers or cypresses, roughly 90 million years ago. The ancient inflorescence could be studied through the resin with a light microscope.

All three of these fossils have small cupules. Cupules surround the female flower and then the developing fruit in two of the first-diverging branches of the Fagales: the Beech Family (Fagaceae) and the Southern Beech Family (Nothofagaceae). Cupules are modified branches of the inflorescence, merged to form scales and valves, spines, and husks. They evolved into acorn caps, the spiny husk of a chinkapin or chestnut (*Castanopsis* or *Castanea*), one layer of armor around stone oak (*Lithocarpus*) fruits, and the thinner cupules of the beeches (*Fagus*) and the Southern Beech Family (Nothofagaceae). All species of the Fagaceae and Nothofagaceae have cupules. *Protofagacea*, *Antiquacupula*, and *Soepadmoa* have cupules reminiscent of Nothofagaceae and the earliest-diverging genus of Fagaceae, the beeches (*Fagus*). The fossil cupules are very small, however, about the size of a grain of sweetcorn, thinning to wafer-like scales at their margins.

All three fossil species formed nutlets—tiny, hard, single-seeded fruits, like one might find in a birch (*Betula*) or a southern beech (*Nothofagus*)—that developed from ovaries with two or three chambers. Fagales today essentially all have a single-seeded fruit: all ovules abort except for one, barring such oddities as the occasional two-seeded acorn. At least two of the fossil genera—*Soepadmoa* and *Antiquacupula*—were most likely insect-pollinated. In *Soepadmoa*, the floral parts formed a narrow funnel, with the styles slender and concealed within. This suggests that wind would have been an inefficient pollinator for the plant, whose stamens could not have bobbed around freely as in most of today's Fagales. A bee, fly, gnat, or butterfly could have reached into such a flower and gotten a little pollen on its face or the base of its proboscis as it searched for nectar. And *Antiquacupula* appears to have had nectaries between the swollen stamen bases, a reward to pollinators. Modern Fagales do not have floral nectaries, though several of the genera are still insect-pollinated, including the closest relatives of oaks.

While we do not know precisely where they fall on the Tree of Life, all three fossils appear to be ancient descendants of species that arose near the base of the Fagales. Each of the three fossil species has a mosaic of characteristics that allies it with Fagaceae, Nothofagaceae, or the stem leading to the Fagales. Based on these fossils, it is plausible that the species that gave rise to the Fagales was a tree with small nutlets and tiny cupules. It likely had female or bisexual flowers. There is a good chance that it was insect-pollinated, and perhaps opportunistically wind-pollinated on dry, warm days, at least in populations that had colonized open areas where wind could move freely. Its male flowers were laid out along ament-like spikes. It is unclear where in the Northern Hemisphere that species lived, however, or how widespread it was. Many of the early Fagales fossils were found in eastern North America, so perhaps the species originated there. Alternatively, if *Archaefagacea* is a stem Fagales, then East Asia might be the birthplace for the order. A hypothesis from the early 1990s suggests that the ancestor of the Fagales was a Southeast Asian population that gave rise to the Nothofagaceae toward the south, into Gondwana, and the rest of the order to the north. But there are no Fagales in Africa, and Australia was, at that time, across the globe from what is now Southeast Asia; this would be a very long way to jump (thousands of kilometers). A relatively small collection of fossils cannot stand in for all of biogeographic history.

It is also plausible that the ancestor of the Fagales was not an isolated species or population, but one of an interbreeding group of species, which may have been widespread across northern North America and Europe. Interbreeding among species is common in the Fagales today, and the Beech Family, Fagaceae, show a history of ancient hybridization among genera. The entire Walnut Family, Juglandaceae, may have arisen from a hybrid involving species from the Sweetbay Family, Myricaceae. What's more, essentially all families of the Fagales produce species with large populations and at least some species of continental range. The ancestral population of today's Fagalean families may have been part of a species complex that stretched broadly across a section of the Northern Hemisphere, with various species contributing genes to the ancestor of the entire Fagales.

At this point, we are in the realm of conjecture, so we might as well stay here for another paragraph. Imagine a population of trees established from seeds blown by a storm or carried by a flock of birds. Imagine it far enough outside the ancestral species' range that it can no longer exchange pollen with its relatives. It spreads and begins to adapt ecologically to new conditions. It forms new populations of its own. Imagine that over the course of hundreds of generations, the selective death of ovules evolves within one or several of these populations, setting up evolution of the single-seeded fruit. In other populations, the tepals on the minute male flowers evolve to be a bit smaller than they were in their ancestors. Each population would evolve solutions to ecological problems in semi-isolation from the other populations. Then, with long-distance dispersal or contact as species' ranges stretched and shifted, these evolutionary innovations could swap between populations. The adaptations that make the Fagales successful may have spread among populations by hybridization and backcrossing, with beneficial genes hustling around in pollen and seeds and introgressed alleles persisting in their new populations with help from natural selection.

This is, of course, speculation. A more traditional scenario would involve the evolution of the attributes of the Fagalean ancestor arising as a single, perhaps widespread, population in a single, isolated species. Given what we know of hybridization in the oaks and their relatives, as well as the evidence that their genera may have swapped genes when they were younger as freely as today's species do, the first scenario strikes me as more plausible. In either case, whether by evolution in a single population or by evolutionary brico-lage, the ancestor of the Fagales brought together all these innovations in one species.

Temperatures began dropping about 93 million years ago. Broad-leaved evergreen forests covered the midlatitudes of North America and Eurasia. Deciduous forests were mostly relegated to pockets in the Arctic, where they endured long, dark winters. A mix of tropical and temperate lineages grew across much of the northern reaches of the Northern Hemisphere. Plants and animals migrated between North America and Eurasia across land bridges spanning the northern Atlantic and Pacific Oceans. The early evolution of the Beech Order played out on this stage, as individual plant families peeled off the main trunk of the Fagales tree of life in turn.

The very first speciation event we know of in Fagales split the order into the ancestor of the Southern Beech Family, Nothofagaceae, and a northern lineage that gave rise to the majority of the order. The Southern Beeches colonized South America, Antarctica, Australia and New Guinea (which was part of Australia at that time), and New Zealand. These continents were shoulder-to-shoulder near the bottom of the globe, the last holdouts of the former supercontinent of Gondwana. As the Nothofagaceae were getting a start in the world, Gondwana was breaking up into separate continents, but plants could still migrate between them. Nothofagaceae cupules form clusters of woody or fleshy overlapping scales. Each cupule encloses three narrowly winged fruits that are not strongly adapted for dispersal: in some Nothofagaceae species living today, 90% of seeds germinate beneath the canopy of their own mother tree, with seedlings rarely straying beyond twelve meters from the tree. With only infrequent long-distance dispersals to shuffle populations across landmasses, the family largely tracked the shifting continents. South America and Antarctica separated from Australia and New Guinea about 50 million years ago, and the major Nothofagaceae clades each split into a South American and Australian clade, followed by some dispersals between Australia, New Zealand, and New Caledonia. To-day, *Nothofagus* forests can be found as far north as the mountains of equatorial New Guinea and as far south as Tierra del Fuego.

Castanopsis fissa **fruit, acorn-type.** Illustrated from a photograph provided by Charles Snyers.

While Nothofagaceae evolved in the Southern Hemisphere, the remainder of the order diversified in the northern supercontinent of Laurasia. The Beech Family, Fagaceae, forms one of the first branches of that northern clade. The beeches, *Fagus*, are the namesake of the Fagaceae. They closely resemble the Southern Beeches (Nothofagaceae) in their flattened, bristly cupules and male flowers clustered in the leaf axils. They split off at the same time as the remainder of the family and spread across the Arctic. The oldest *Fagus* fossils—beech pollen dispersed from flowers more than 60 million years ago—are widespread, distributed from Northeast Siberia to western Greenland. They are quickly followed in the fossil record by pollen, leaves, and fruits in northeastern Asia and western North America. Their early fossils range from Axel Heiberg Island, at the north edge of Canada, to Hainan in the South China Sea. This distribution of fossils suggests that the genus probably spread through arctic to subarctic deciduous forests soon after its birth. Then, as the climate cooled in the latter half of the Eocene, the ge-

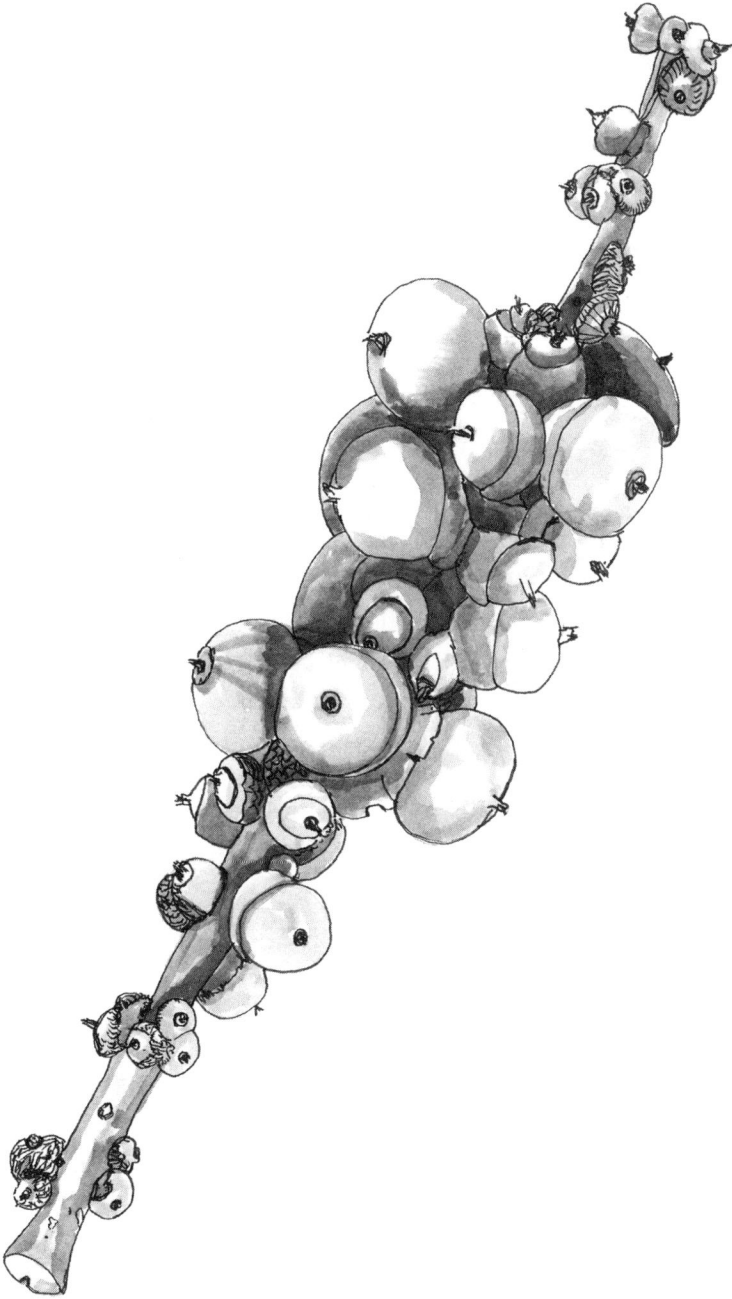

***Lithocarpus elegans* branch with acorn-type fruits**. Acorns are nestled in cupules that look like the acorn cap of an oak (though this is not an oak but a stone oak). Illustrated from a photograph provided by Charles Snyers.

nus radiated in East Asia and spread into Europe. One lineage descended from the Arctic into North America. This lineage led to the American beech (*Fagus grandifolia*), which remained genetically separated from the remainder of the genus as the continents moved farther apart and the Arctic became too cool and dry for beeches. The ghostly marcescent leaves of these trees hang white on the branch throughout much of eastern North America every winter. That species in turn gave rise to an eastern Mexican variety or subspecies. Beeches today are widespread in rich forests of eastern North America, Europe to northern Iran, and East Asia. They number only a dozen species, all of which bear genomic signals of a complex hybridization history.

The genus *Trigonobalanus* split off next and left us only three modern-day species, two in tropical East Asia and one in Colombia, South America. The fossil record is enigmatic, including extinct lineages that appear to have left fossils in Europe, western Russia, Iceland, and Greenland from the Eocene through the Miocene, and possibly Late Cretaceous Siberia. These fossils suggest that today's *Trigonobalanus* species may represent endpoints of a once-widespread genus that receded to two refugia. Grasslands spread to replace large tracts of forest near the beginning of the Miocene, as the climate became warmer and drier. All that was left, in this scenario, were the disjunct remnants of a once expansive distribution.

As the early Fagaceae were diversifying, so were the families of the core Fagales. The Walnut Family (Juglandaceae) originated in North America and spread across the Northern Hemisphere. Its wing-fruited lineages were mostly squeezed down into South and Central America and Southeast Asia by cooling temperatures. The Sweetbay Family (Myricaceae) probably arose in the Northern Hemisphere, but an early-diverging genus (*Canacomyrica*) headed to New Caledonia, and the genus *Morella* made its way from East Asia to South America and Africa. The Cassowary or "She-oak" Family (Casuarinaceae) left fossils in South America and South Africa and ultimately radiated in Australia and New Zealand. Its close relatives, the Birch Family (Betulaceae), spread across the Temperate Zone, becoming an important woody component of our northern forests alongside the oaks, but specializing in growing fast and invading open territory.

Then, in the early to middle Eocene, the Ticodendraceae showed up in the fossil record in the Clarno Nut Beds of Oregon and the London Clay flora of England. It arose as the sister group to the Birch Family. Ticodendraceae may once have been a widespread clade whose members migrated between the continents. Today, it is represented by a single species, *Ticodendron*

***Trigonobalanus excelsa* inflorescence in front of leaf fragment**. Illustrated from a photo provided by Paul Manos.

incognitum, which ranges from Chiapas to Panama. It is listed as near-threatened in the International Union for Conservation of Nature (IUCN) Red List of Threatened Species. The species was unknown to western science until 1989.

The next step in the evolution of the Fagaceae was a figurative explosion of species, which may have been coincident with a literal explosion. At the end of the Cretaceous, a meteor the size of Mount Everest struck the north end of Mexico's Yucatan Peninsula at nearly 30 kilometers per second, 90 times the speed of sound. The meteor left a crater 160 kilometers in diameter. It pulverized the Earth's crust and launched fragments high into the atmosphere. Molten shards of rock returned to Earth up to thousands of miles away. Almost everything living above ground within 950 kilometers of the impact was vaporized. A half-meter-thick layer of compressed and fossilized ejecta was deposited on what is now Haiti, 1,600 kilometers to the east. More than 3,200 kilometers away, in Hell Creek, Wyoming, ferocious winds and a rain of molten rock from the meteor killed dinosaurs and mammals in their tracks. Widespread volcanic eruptions in West-Central India were already ongoing when the meteor hit and likely exacerbated by the impact. They spewed out more than a million cubic kilometers of lava over the course of approximately 750,000 years and ejected smoke and sulfate particles, darkening the skies for months, perhaps even years.

Temperatures plummeted as a consequence. While the duration of this "impact winter" is uncertain, the effects on living organisms are evident: nonavian dinosaurs were driven extinct worldwide, along with 90% or more of the mammal species living at that time. In the aftermath, the remaining mammal lineages slid into the habits and habitats dinosaurs had occupied, diversifying to form many of the major lineages we recognize today. North American plant communities were decimated. While few large flowering plant lineages went extinct, many populations and species did. Those plants that survived suffered lower productivity under the darkened skies that blanketed the globe. Broad-leaved evergreen trees are believed to have suffered particularly, leaving room for deciduous tree species to gain a toehold in the high-middle latitudes. Earth would return to a tropical state for more than 20 million years following the end-of-the-Cretaceous cataclysm, but the meteor strike may well have helped set deciduous trees on their path to success in the Northern Hemisphere.

At about the same time, the ancestor that gave rise to the vast majority of the Fagaceae developed a novel innovation for the family: seeds that include thick, fleshy cotyledons as storage organs for food. You read about these cotyledons in gory detail in chapter 1. They form the mass of an acorn or chestnut. They are thick with starches packed in by the mother tree. They do not become the aerial, green, photosynthetic leaves that the cotyledons of *Fagus* and *Trigonobalanus* are and that, presumably, the cotyledons of the ancestral Fagaceae were. Instead, the seeds of oaks, chestnuts, and their relatives—the clade doesn't have a name, but we can call it the "big-seeded Fagaceae"—stay at or below the ground's surface as they germinate. The role of the nutrient-rich cotyledons is not to photosynthesize, but to carry food for the seedling. If the evolution of the big-seeded Fagaceae preceded the meteor strike, this trait would have proved exceptionally useful in the dark and difficult years that followed.

Fleshy cotyledons, as you already know, are also an essential food source for numerous rodents and birds, who move seeds around and cache them. By making themselves indispensable to animals who often go out of their way to save the seeds for later, these big Fagaceae seeds helped trees migrate. They decreased the risk of species going extinct and promoted the formation of new species through dispersal and the establishment of new populations. Biotic seed dispersal, which has evolved several times in the Fagales, is one of the most significant predictors of biological diversity within the order, correlating with roughly 98% of the Beech Family's species. It is part of the family's success.

The chestnuts (*Castanea*) and chinkapins (*Castanopsis*) together form one of the clades of the big-seeded Fagaceae. They radiated into the temperate deciduous forests and broad-leaved evergreen forests, respectively. DNA of chestnut species living today, combined with data on the distribution of extant and fossil species, suggests that the genus *Castanea* arose in East Asia and possibly western North America, then dispersed westward across Europe. A population of the genus likely migrated across the North Atlantic Land Bridge to reach eastern North America, where it gave rise to the shrubby dwarf chestnut (*Castanea pumila*) and the once-majestic American chestnut (*C. dentata*). American chestnut was widespread in Appalachian forests into the early twentieth century. Its spiny husks fell in abundance on the forest floor. Its nuts fed passenger pigeons, squirrels, humans, and others. Its wood was treasured for building; "wormy chestnut" is still salvaged from old barns for furniture and paneling. The species was decimated by chestnut blight, a pathogenic fungus that arrived on the continent around 1904, most likely via cultivated Japanese chestnuts. The disease has largely erased American chestnut from the forests of eastern North America. For-

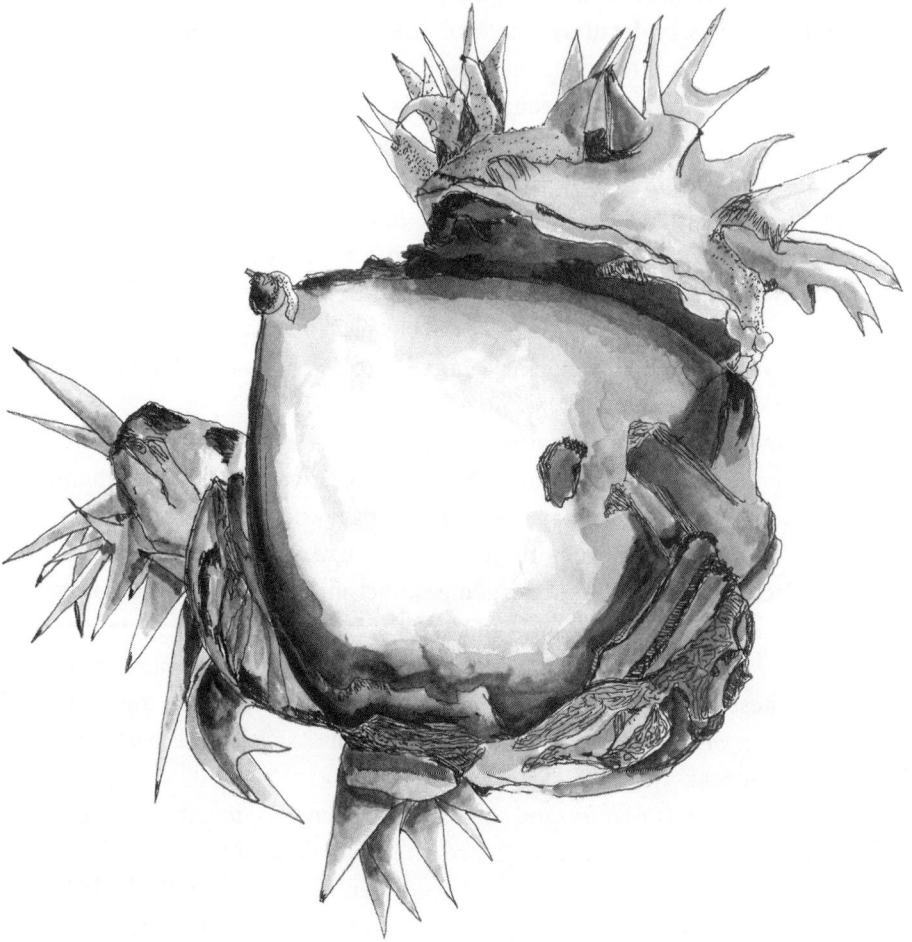

Castanopsis chinensis **fruit with spiny cupule**. Illustrated from a photograph provided by Béatrice Chassé.

merly huge trees persist today as rare reproductive individuals and a large number of young sprouts that regrow from old stumps. Resprouts become just large enough to be attacked by chestnut blight and knocked back to the ground. Most will never bear seeds. They are a reminder—along with the Normapolles lineages and the once-widespread Ticodendraceae—that ecologically dominant species can shrink in range, and even go extinct, in a geological moment.

The mostly spiny-fruited chinkapins (*Castanopsis*), meanwhile, spread rapidly across the Northern Hemisphere. The earliest known *Castanopsis* fossil is from the Clarno Nut Beds of central Oregon, which in the early Eocene was a paratropical forest: warm and wet, like a true tropical forest, but displaced northward. The forest that stood where the Clarno Nut Beds are today hosted crocodiles, cycads, palms, and bananas, along with chinkapins and tropical oaks. *Castanopsis* grew in semitropical forests across Europe as well through the early Eocene, moving into wet, cooler evergreen forests as temperatures dropped worldwide. While the genus is today limited to East Asia, where it is dominant in broad-leaved evergreen forests, the Eocene fossil evidence for the genus in East Asia is ambiguous. *Castanopsis* might in fact have arisen in North America or Europe and only subsequently dispersed to East Asia. From North America, the genus extended down into Patagonian South America, a southern lineage that was probably an evolutionary dead end. *Castanopsis* persisted in the United States and Europe possibly into the Miocene, then receded to Southeast Asia as the globe continued to cool. It remains a dominant forest species there. It is one of the most species-rich genera in the Beech Family, at roughly 140 species.

At around 60 million years ago, Fagaceae took another turn: evolution of the acorn. You might expect me to follow that sentence with, "and the acorn is what makes the oak genus *Quercus* unique," but the story is not that simple. To understand the acorn, it helps to start with the Fagaceae genera that did not evolve acorns. In those genera—*Fagus*, *Trigonobalanus*, *Castanea*, and most of *Castanopsis*—each cupule surrounds multiple flowers. The tissues that form the cupule are called "valves," and they typically split open to release the fruits. Think of a chestnut, usually with two or three nutlets peeking out from between the edges of the hedgehog-spiny valves. Consider bristly beech bracts spreading to expose the fruits paired inside. These are a little like the cupules of Nothofagaceae as well as the extinct *Protofagacea* and *Soepadmoa*, all with multiple nutlets nestled inside.

However, in one of these genera—*Castanopsis*—some species have valves that have fused to form a cap. The caps enclose three flowers, but two flowers abort. As the fruits ripen, they look a great deal like acorns. The remaining genera of the Fagaceae—*Lithocarpus*, *Notholithocarpus*, *Chrysolepis*, and *Quercus*—form a single clade in which there is only one flower per cupule or, in the case of *Chrysolepis*, a few (but all the *Chrysolepis* flowers get flaps of tissue between them, like little privacy curtains). Three of these genera produce acorns: many of the species of *Lithocarpus*, the solitary species of *Notholithocarpus*, and, of course, *Quercus*. In all four genera, the acorn cap

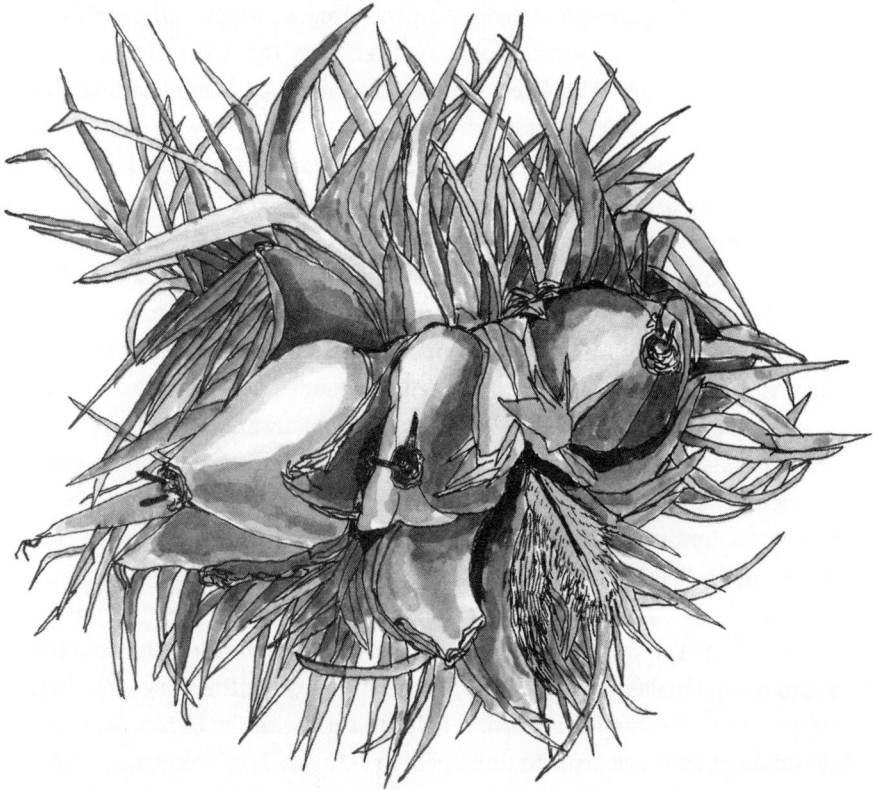

***Chrysolepis* fruit**. The fruit is three-seeded and surrounded by a spiny cupule. Between the three nutlets are internal cupule valves, made of leaf-like tissue that separates the individual seeds. Illustrated from a photograph provided by Sang-Hun Oh.

is a cupule, just like the spiny valves of a beech nut or the armed husk of a chestnut. This means that the acorn fruit type is present in four genera. "Acorns"—in quotes here, because it seems unlikely that the *Castanopsis* acorn has the same evolutionary origin as the *Quercus* acorn—probably evolved two or three times: once in *Castanopsis*, and either once at the base of the single-flower clade or once in *Lithocarpus* and once at the base of the *Quercus* plus *Notholithocarpus* clade.

The acorn-producing, four-genus clade that includes oaks and their close relatives got its start in the world at the same time that the chestnuts and chinkapins peeled off from the Fagaceae main trunk. The oaks themselves are the genus *Quercus*. They don't need much introduction at this point. They are the largest genus in the family at 425 species, and we'll follow their history in the next chapter. The stone oaks, *Lithocarpus*, are the second largest, with more than 320 species. They dominate Southeast Asian forests alongside *Castanopsis*. *Lithocarpus* fruits range from acorns to "enclosed receptacle" fruits, in which the cupule forms a bony covering around the fruit and the fruit husk thickens to protect the seed inside.

The acorn-producing clade also includes the two smallest genera in the family, both of which are North American. *Notholithocarpus* comprises a single species distributed along the West Coast of North America, roughly from Bandon, Oregon, to Los Angeles. While *Notholithocarpus* was named and described in 1840 as an odd *Quercus*, it was moved over to *Lithocarpus* in 1917 based on its overall similarity to that genus, and then separated into its own genus in 2008. From a phylogenetic standpoint, *Notholithocarpus* could just as easily have been left in the oaks, but the similarity of its leaves to those of *Lithocarpus* and the fact that it is insect-pollinated set it apart morphologically and ecologically.

The fourth genus, *Chrysolepis*, is endemic to the West Coast as well. It has, instead of acorns, spiny-husked fruits that resemble the fruits of *Castanopsis*. *Chrysolepis* comprises just two species on the West Coast of the United States.

It's hard to say for sure what accounts for the disparity in species diversity between *Lithocarpus* and *Quercus*, on one hand, *Chrysolepis* and *Notholithocarpus*, on the other. We could point to wind pollination in the oaks, which promotes gene flow between and within species, and thus migration and adaptive gene flow. Most of the rest of the Beech Family are insect-pollinated, including *Notholithocarpus*, the sister clade to the oaks. But so are the species of *Lithocarpus*, the second largest genus of the family. And the nine species of *Fagus* are wind-pollinated. Wind pollination can't explain all the diversification success of *Quercus*.

Unlike *Lithocarpus* and *Quercus*, *Chrysolepis* and *Notholithocarpus* have not moved beyond the western United States. Despite their low species diversity, these two small genera are ecologically successful. They both grow from less than 150 meters to more than 1,500 meters in elevation. They both range widely across the Sierra Nevada and Coast Ranges of California and, for *Chrysolepis*, through Oregon's Cascade Range. This clade of four genera is a fascinating evolutionary set piece. *Lithocarpus* specialized on the tropical habitats and broad-leaved evergreen forests where it dominates, and it

Notholithocarpus densiflorus acorns, one mature and one undeveloped.
Illustrated from a photograph provided by Béatrice Chassé.

has flourished in place as continents separated and temperatures dropped. *Quercus* diversified in the far north of the Northern Hemisphere and was poised to take advantage of the cooling world. *Notholithocarpus* and *Chrysolepis*, by contrast, ended up in a place where they had little room for new species to differentiate. They survived, but they did not radiate.

By the end of the Paleocene, 56 million years ago, all the families and nearly all the genera of the Fagales were in place. The climate had been relatively stable for about 10 million years following the cataclysm that wiped out the nonavian dinosaurs. But the world was no less dynamic. Clades shuffled across land bridges as the northern continents were pushed away from one another on shifting tectonic plates. *Rhoiptelea* grew in North America from the end of the Cretaceous to 38 million years ago, and possibly in Central Europe near the end of the Cretaceous. Today it is endemic to the mountains of Southwest China and North Vietnam. Wing-fruited walnut family members that we now find only in temperate East Asia and the tropics were widespread in North America as well. The world's forests were, without knowing it, gearing up for their transition out of tropical Earth.

The oaks, *Quercus*, were born into this game of musical chairs. Life on Earth had already passed through more than 98% of its history. The wait was worth it, and exactly the time oaks needed. From a northern ancestor, the oaks inherited a woody habit, wind pollination, delayed fertilization, and a nutrient-packed nut that developed in relative safety inside the acorn cap, partly protected from fire and drought and predators. Their evolutionary path through the Fagales primed oaks for success.

5

Radiation

Quercus

If we could head back in time 56 million years and spend a few weeks botanizing in the temperate forests of the Northern Hemisphere at the boundary between the Paleocene and the Eocene, we would be hard-pressed to find any oaks. We would find alligators and giant tortoises on Ellesmere Island, across from the northwest coast of Greenland. We could roam through flowering-plant-dominated forests whose diversity approached the plant diversity we might find in the modern forests of the southeastern United States. We would encounter a diversity of Fagales, lineages that would eventually give rise to walnuts, birches, sweet gales, beeches, chestnuts, chinkapins, and oaks. They were all spreading across the Northern Hemisphere by this time. The oaks themselves, however, were so few in number at that point that they left scant if any pollen in the mud and no acorns or leaves to be recovered by twenty-first-century botanists.

The world was about enter a heatwave, the Paleocene-Eocene Thermal Maximum (PETM). Over the course of 8,000 to 10,000 years, atmospheric temperatures would spike, increasing by an average of 8°C worldwide and reaching even higher levels in the Arctic. The PETM may have been triggered by a massive and protracted period of volcanic activity. Magma gurgling up through a fissure at the bottom of the North Atlantic drove a wedge between North America and Europe and poured a trillion kilograms of carbon into the atmosphere every year for several thousand years. Rising temperatures melted corpses out of the Antarctic permafrost, and the rotting sedges, sphagnum mosses, fungi and lichens, mollusks and marsupials

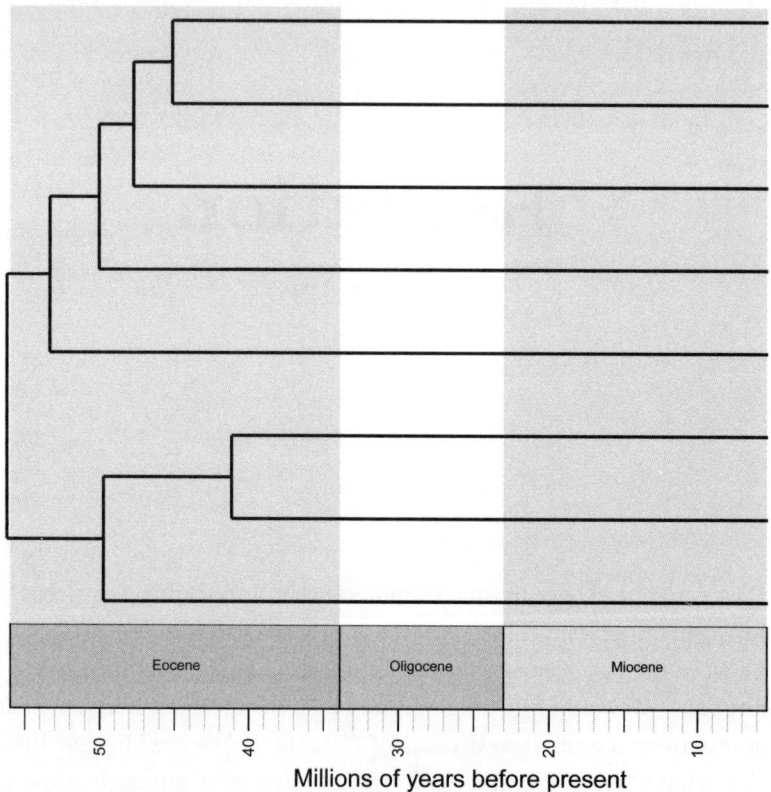

Oak tree of life. This phylogenetic tree includes one representative from each section in the genus *Quercus*. It was estimated using RAD-seq genomic data and time-calibrated using seven fossil calibrations (Hipp et al. 2020, 2023).

returned greenhouse gases—carbon dioxide and methane—to the atmosphere. Temperatures then crashed back to their original levels within about 120,000–220,000 years. That's barely enough for a double take in geological terms: when you look at a temperature plot for the past 100 million years, the PETM looks like a fencepost driven into the hillside at 56 million years ago. It goes straight up and almost straight back down.

The effects were dramatic. The PETM drove 30%–50% of deep-ocean-bottom foraminifera—single-celled organisms that populate the seas, eating plankton and detritus, feeding small fish and marine snails—extinct. Mammals, lizards, and turtles migrated widely across the continents in response to the changing climates, traveling between northern land bridges that would become too cold for regular travel by most of these species in the late Eocene. In northern South America, tropical forests were flooded with

sect. *Virentes*

sect. *Quercus*

Quercus subgenus *Quercus*

sect. *Ponticae*

sect. *Protobalanus*

sect. *Lobatae*

sect. *Cerris*

Quercus subgenus *Cerris*

sect. *Ilex*

sect. *Cyclobalanopsis*

Pliocene | Pleistocene

0

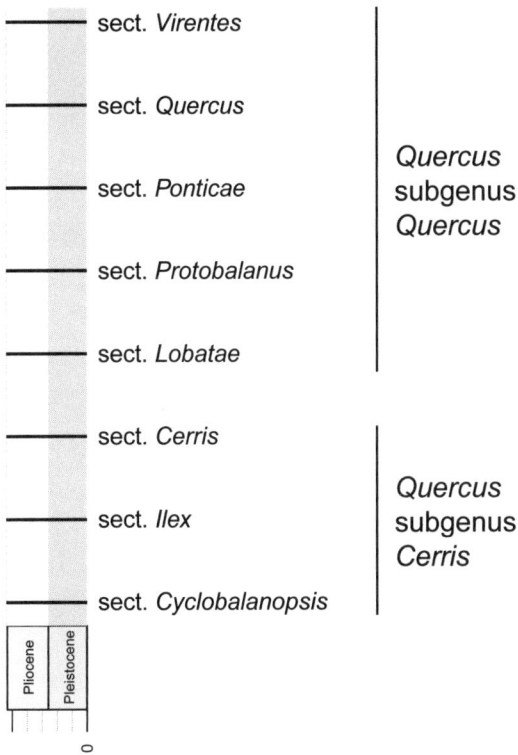

new flowering plants: palms, grasses, and the Bean Family (Fabaceae) all increased in diversity in the Eocene, and the Spurge Family—Euphorbiaceae, a global family that numbers about sixty-five hundred species today— showed up in northern South America for the first time during the PETM. Insect herbivores, particularly leaf miners and surface feeders, increased in abundance and became more specialized. Plants raced across the landscape: in Bighorn Basin, Wyoming, at least twenty-two species were extirpated at the onset of the PETM, only to return after the event was over. Some of these sojourners migrated an estimated 1,000 kilometers.

The first fossil oaks we know of appear in this uncertain world, along what is now a hiking trail running south of the Church of Saint Pankraz in Oberndorf, Austria. Fifty-six million years ago, this area of Europe was dissected into islands and peninsulas, which were warmed by the ocean. What is now

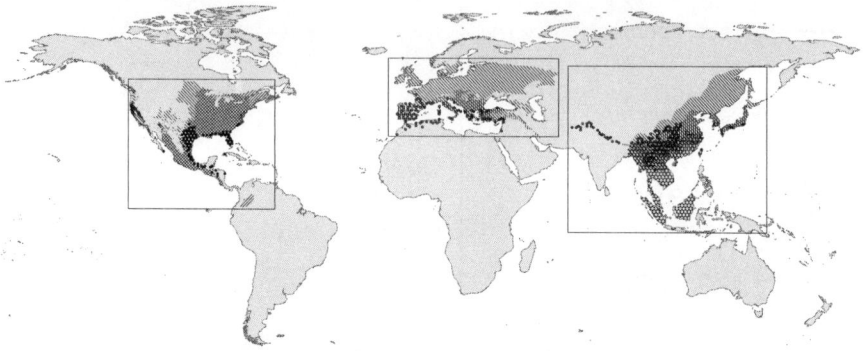

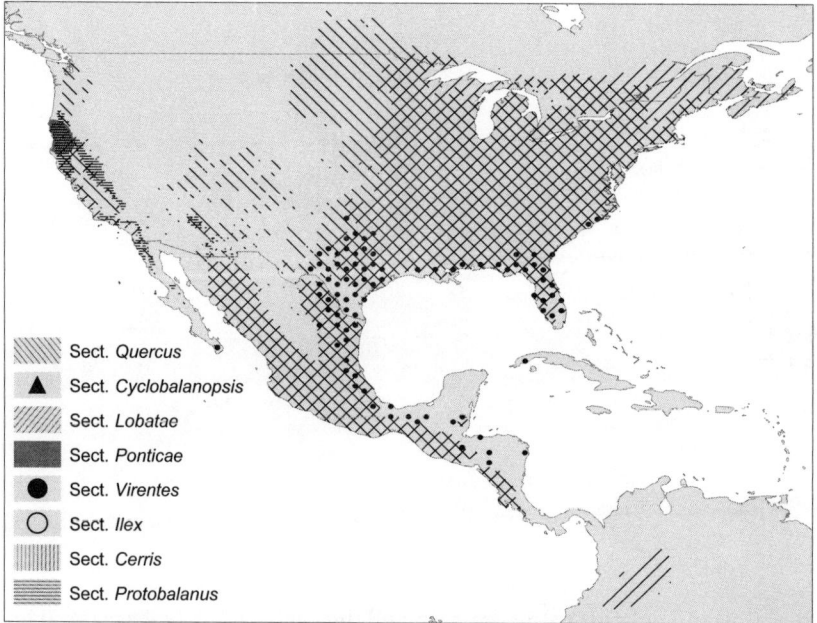

Oak distribution map. The range of each section is based on publicly available range maps for the component species of each section as well as range maps drawn by Kieran Althaus and Rubén Martín Sánchez for additional sections. The map is by Althaus and Martín Sánchez and used with their permission.

Saint Pankraz lay beneath shallow water at the edge of the sea. It became a repository for pollen from adjacent forests, deposited alongside oceanic plankton and dinoflagellates. The forest growing in the area was a mosaic of subtropical and temperate species, including members of the Restionaceae, a grass-like family that today is limited to the Southern Hemisphere tropics; *Eotrigonobalanus*, an extinct genus of the Beech Family that formerly ranged

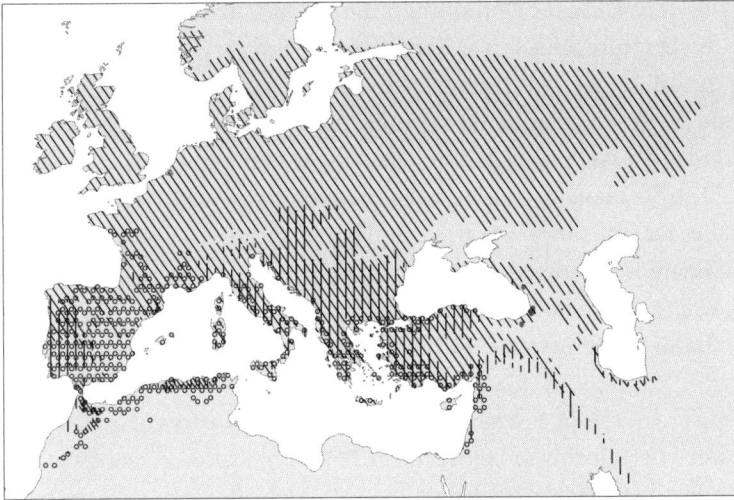

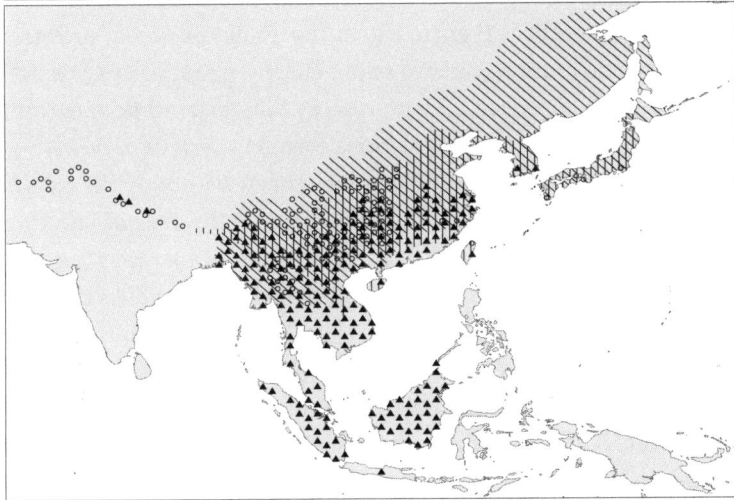

across eastern North America and Europe; and relatives of today's Cashew Family, Mallow Family, and the pantropical Sapotaceae.

The world was entering the last days of the nearly global tropics. For 4 million years after temperatures retreated from the PETM, the climate continued to warm. By 52 million years ago, the world hit the highest temperatures since the demise of the dinosaurs. This period of warmth is called the Early Eocene Climatic Optimum. If the PETM is like a fencepost driven into the temperature hillside, the Early Eocene Climatic Optimum is like the crest of the hill. Forests of tropical species growing alongside genera of the temperate forest—maples, elms, walnuts, birches, cherries, and eventually oaks—spread across the high Arctic. The long winter nights favored species that could go dormant for months at a time. Deciduous forests spread across upland sites that are now permafrost and boreal forest.

The climate was perched at the top of a long slide down to the Anthropocene, where we find ourselves today. Oaks were pioneers in what would become the largely temperate Northern Hemisphere.

The oaks were not born at a particular moment or in a particular place. Instead, somewhere during or before the PETM, a population of woody plants gradually became the oaks. Each seedling in this lineage looked like the trees that produced it. Had we been there to witness the evolution of that ancestral population, we could at no point have said, "There were no oaks yesterday, but today there are." We ended up with oaks by the steady work of natural selection acting on variable tree populations over long periods of time.

This lineage of individuals and populations slowly becoming the oaks is called the stem of the oak clade. It is represented on the Tree of Life by a single line. The population of trees that deposited the St. Pankraz pollen may represent a sprig sprouting from that stem or one that sprouted very near the crown of the oaks. In either case, the St. Pankraz pollen is, for now, our best bet about how the old the oaks are. Oaks probably go back at least a little longer than these fossils, older than the PETM: fossils are hard to find, so it's reasonable to suspect that we may have missed some older ones. But these fossils provide us a landmark by which to date the oak tree of life.

The first speciation event we know of in oaks likely occurred within 8 million years of the St. Pankraz oak fossil. It split the oaks into two lineages: one that is today limited to Eurasia and North Africa, and one that evolved in the Americas and only later returned to Eurasia. Sister clades—which are born as sister species—can arise in separated geographic regions when their ancestral population becomes physically subdivided. A mountain range, a river, a desert, an expanse of ocean, or any other barrier between the two portions of the population keeps seeds and pollen from moving between the two new populations. Speciation and the birth of new clades often result. The spreading Atlantic Ocean is a plausible explanation for this first oak speciation event. Magma spilling into the North Atlantic off the coast of Ireland at the beginning of the PETM added crust to the east edge of the North American (tectonic) Plate and the west edge of the Eurasian Plate. It continues to do so today, steering the continents apart at a rate of about an inch a year. As the Atlantic grew wider, the ancestral population of all of today's oaks may have been straddling the continents of the Northern Hemisphere. If so, the ancestor of the oaks we know today was a widespread population that was cleaved in half as North America inched westward.

We recognize the two continental clades that arose from this event as subgenera. Subgenus *Quercus* is the mostly American oak clade. It includes the Red Oaks (*Quercus* section *Lobatae*), the Golden-Cupped or Intermediate Oaks (*Q.* sect. *Protobalanus*), the Deer Oaks (*Q.* sect. *Ponticae*), the Southern Live Oaks (*Q.* sect. *Virentes*), and the White Oaks (*Q.* sect. *Quercus*). Subgenus *Cerris* is a mostly Eurasian clade today. It includes the Holly Oaks (*Q.* sect. *Ilex*), the Cork Oaks (*Q.* sect. *Cerris*), and the Ring-Cupped Oaks (*Q.* sect. *Cyclobalanopsis*). Each of these major clades arose from one of eight species that were in the right place at the right time. They won the lottery.

The taxonomy of oaks, like any taxonomy, is a set of nested dolls, clades tucked inside of clades: the oak subgenera are continental, and each one contains clades that we classify as sections. The sections each have their own histories that together form the outline of the oak classification. That classification looks like this, with the scientific names before the colon (e.g., "Section *Quercus*") and the common name after (e.g., "the White Oak Group"):

- Subgenus *Quercus*: the mostly American Oak Group.
  - Section *Lobatae*: the Red Oak Group; widespread in the Americas, though some species—perhaps nothing we would recognize as a modern species today—were present in Europe until the Pliocene.
  - Section *Protobalanus*: the Intermediate or Golden-Cup Oak Group; California and northern Mexico.
  - Section *Ponticae*: the Deer Oak Group; represented today by only two species, *Quercus sadleriana* in northern California and southwestern Oregon, and *Q. pontica* of the western Caucasus Mountains.
  - Section *Virentes*: the Southern Live Oak Group; southeastern United States to eastern Mexico, Central America, and Cuba; one species in Baja California, Mexico.
  - Section *Quercus*: the White Oak Group; widespread in the Americas, with one clade of about twenty-five species extending into and distributed across Eurasia to North Africa.

- Subgenus *Cerris*: the Eurasian and North African Oak Group.
  - Section *Cerris*: the Cork Oak Group; Europe and East Asia, most diverse in Europe.
  - Section *Ilex*: the Holly Oak Group; Mediterranean region via western Asia and the Himalayas, East Asia, and North Africa.
  - Section *Cyclobalanopsis*: the Ring-Cupped Oak Group; Southeast Asia, though perhaps also distributed in western North America in the early Eocene.

The oak classification recapitulates, in general terms, our current understanding of the evolution of oak diversity. It summarizes major events in the oak tree of life that shaped our forests: divergence of the genus on different continents at the base of oak phylogeny, making our subgenera, then separation into morphologically and genetically distinctive sections within each subgenus and continent. Thus, the classification of oaks is hierarchical, like a family tree: subgenera are nested inside the genus, and sections are nested inside subgenera. The oak classification goes even further. At lower levels, there is a subsectional classification for the American oak clade that groups, for example, eastern white oak and its relatives—from section *Quercus*—into subsection *Albae*, and the California Red Oaks—in section *Lobatae*—into subsection *Agrifoliae*. In Europe, the Cork Oak Group—section *Cerris*—has recently been classified into five subsections, each representing a clade. The oak classification of today encodes what we know about many ancient speciation events spanning the life of the genus.

But the oak classification is only a simplified sketch of the oak tree of life. It ignores the branching history that relates sections to one another within subgenera as well as most of the relationships between species within sections. The White Oak Group and the Southern Live Oak Group, for example, are more closely related to each other than either is to any other section, but we do not have a name for the clade that includes just these two lineages. Could a sub-subgenus be named to connect them? Yes, but at this point, no one I am aware of uses such a name. At some point, naming branches on the Tree of Life ceases to be useful for communication. The classification is itself an evolving entity: today's classification is closer to the actual evolutionary history of oaks than the classification of ten years ago was. Tomorrow's classification will probably be closer still, particularly in areas where rapid evolution makes it difficult to be clear about the order of ancient speciation events.

Our understanding of the Tree of Life and our taxonomies are interpretations of genomic evolution, fossil history, and morphological distinctions. The oak classification is imperfect. Nonetheless, it approaches what Charles Darwin anticipated when he wrote, in the last few pages of *The Origin of Species*, that in time "our classifications will come to be, as far as they can be so made, genealogies; and will then truly give what may be called the plan of creation."

Earth began to cool following the Early Eocene Climatic Optimum. The rising Alps, Rocky Mountains, and Himalayas exposed a massive amount of granite and other silica-based rock to erosion. Carbon from the atmosphere bound chemically to the weathering rock and washed downslope to the ocean. A massive colony of floating fern, *Azolla*, growing in the then-much-warmer Arctic Ocean, absorbed CO_2 through photosynthesis and then sank, sequestering the carbon it had photosynthesized. Atmospheric greenhouse gases dropped as a consequence. At the same time, the breakup of Gondwana allowed ocean currents to flow around Antarctica, cooling the continent and deep ocean waters worldwide. In combination, these effects drove temperatures down by roughly 8°C from the Early Eocene Climatic Optimum until the end of the Eocene, 34 million years ago.

Climates across much of the Northern Hemisphere became more seasonal as they cooled. Increased fluctuation in both temperature and precipitation from one season to the next favored deciduous species. Being evergreen is potentially a huge benefit: you can photosynthesize for up to twelve months of the year. But this opportunity comes with risks. Long-lived leaves have more time to photosynthesize, but they also have more time to become damaged. Freezing temperatures, heavy snows, deep droughts, or long periods of darkness make evergreen leaves relatively costly. The range of factors influencing the evolution of deciduousness is complex, and in fact, harsh climates favor a wide range of growth strategies. Deciduousness—ditching leaves when they're more trouble than they're worth—is nonetheless one solution to the problem of seasonal climates.

Deciduous forests colonized the relatively cool, temperate uplands and spread across the Northern Hemisphere. The broad-leaved evergreen forests that covered North America and Eurasia were squeezed into Southeast Asia, where they continue to flourish today; the southeastern United States, where the only escape route for tropical species into the newly evolving American tropics would have been along the Gulf Coast or across the waters

Quercus sect. *Ilex*: *Quercus alnifolia* leaves. Illustrated from a photograph provided by Béatrice Chassé.

of the Caribbean; and Mexico, which at this time was separated from South America. Chasing them southward were oaks and other species of the deciduous forests, including birches, beeches, basswoods, elms, hackberries, and a wide range of other genera familiar to us today. Accompanying the deciduous forest species we know from eastern North America today were

Quercus sect. *Ilex*: *Quercus ilex* acorns. Tree in cultivation, UC Davis Shields Oak Grove, October 2018.

now-extinct genera as well as some, like the wing-fruited walnuts (*Engel-hardia* and relatives), that eventually became limited to East Asia and the neotropics.

As temperate forest species of North America and Eurasia expanded southward, they continued to migrate between continents along northern routes crossing the oceans, at least once in a while. The Bering Land Bridge between Russia and Alaska had arisen during brief periods in the Late Creta-ceous to early Paleocene. It appeared and reappeared regularly throughout at least the early Eocene, and sporadically even later: humans would cross the Bering Land Bridge from Northeast Asia between 30,000 and 20,000 years ago, at or before the onset of the Last Glacial Maximum. Beringia is beyond the northernmost extent of oaks and many other temperate forest species today, but it was accessible for much of the evolutionary history of oaks. In fact, the connection between western North America and East Asia was stronger throughout much of the Eocene than the connection between

Europe and Asia, as Eurasia was divided into two major land masses by the Turgai Sea. On the other side of the continent, the distribution of the Normapolles across eastern North America and Europe reflects a complementary connection between eastern North America and Europe: the North Atlantic Land Bridge, which provided a northern route through Svalbard and Scandinavia and a southern route through southern Greenland and the British Isles.

Mammals could pass over the North Atlantic Land Bridge freely until about 49 million years ago, but temperate forest species continued to traverse it until as recently as about 5 million years ago. Occasional dispersals connected the major oak clades in their infancy, youth, and adolescence, as each explored the evolving northern climates. The evolutionary history of oaks is a complex of parallel stories with crosstalk between the continents and across regions within each continent.

The mostly Eurasian oaks—subgenus *Cerris*—appear to have arisen in East Asia, where they show up in the fossil record between 56 and 48 million years ago. The largest section within the subgenus is also the most distinctive group of oaks ecologically and morphologically: the Ring-Cupped Oaks of Southeast Asia, section *Cyclobalanopsis*, named for the concentric rings (*cyclo*) formed by the lamellae on the outsides of their acorn caps and the Latin word for "acorn" (*balanus*). The Ring-Cupped Oaks are so distinctive that they were long thought to be a separate genus. Their fossilized acorns, in fact, cannot consistently be distinguished from fossilized acorns of the stone oaks (*Lithocarpus*). Ring-Cupped Oaks are denizens of broad-leaved evergreen forests of East and Southeast Asia, which range from East China and Japan to Sumatra and Central Java. They are the only exclusively tropical to subtropical section in the genus. Their leaves are unlobed and have driptips, slender extensions of the leaf apex that help the leaves shed rainwater. Almost all known Ring-Cupped Oak fossils are within the range of the section as we know it now. Some range farther north, and one has been found as far west as Portugal. One fossil in the Clarno Nut Beds of Oregon may be a *Cyclobalanopsis*, but it may instead be a *Lithocarpus*. The Ring-Cupped Oaks were restricted to the paleotropics and the southern edge of the Tibetan Plateau as climates dried in the Oligocene. The uplift of the Himalayas provided greater diversity of habitats and more opportunities for genetic diversification. This, combined with the onset of the summer monsoon system in the Miocene, helped drive the early diversification of the section. Rising temperatures in the Middle Miocene Climatic Optimum may have allowed the section to expand its range in East Asia, where the broad distribution of warm, moist climates provided space for the evolution of today's 90–120 *Cyclobalanopsis* species.

As the Ring-Cupped Oaks were filling evergreen forests, their sisters, the Cork Oak and Holly Oak Groups (sections *Cerris* and *Ilex*), were forging a life in drier forests east of the Turgai Sea. Fossil data and chloroplast genomes suggest that both sections arose in northern Asia. When the Turgai Sea receded near the end of the Eocene, the Cork Oaks followed a northern route from East Asia to Europe. They dispersed westward along the north side of the Tibetan plateau. One small clade of three Cork Oak species persists today in East Asia. The populations that arrived in Europe radiated to produce the dozen species we recognize today.

Meanwhile, the Holly Oaks moved southward to evolve along the Tethys seaway at the southern margin of Eurasia. They radiated initially into tropical to subtropical forests of Eocene Southeast Asia. One branch of the section took a route westward into the newly forming mountains of western Asia and the Mediterranean. Very likely this group migrated through a forested corridor along the southern edge of the Himalayas or a valley running through Tibet, which would have been suitable for forests in the Oligocene. The high mountains flanking these migration routes may have subjected Holly Oak populations to the trials needed to adapt to cooling temperatures. One clade of Holly Oaks did, in fact, adapt to higher elevations, thriving in the Himalayas. Another branch of the section continued farther west into Europe, perhaps during the Miocene, to form the four or five Mediterranean Holly Oak species. Holm oak (*Q. ilex*) is widespread from western Spain and North Africa east to Turkey and probably the best-known species of the section. Orchards of Holm oak are planted to grow truffles. Its acorns sustain Moroccan jays in the Atlas Mountains and feed Iberian pigs, who can eat fifteen to twenty pounds or more per pig in a day. It is used for charcoal and firewood. Holm oak shaped Mediterranean economy and culture through the modern age only because its ancestors arrived in the Mediterranean from East Asian tropical origins.

Alleles migrated between the young clades as Eurasia filled with oaks. The Ring-Cupped Oaks, the Holly Oaks, and even the genus *Castanopsis* swapped genes in East Asia, where today many of them share a single chloroplast lineage. The Cork Oaks and Holly Oaks exchanged genes as well. So far as we know, neither the Cork Oak Group nor the Holly Oak Group ever expanded to North America. But they would meet up with their distant cousins from the Americas in the Miocene, sometime between 20 and 10 million years

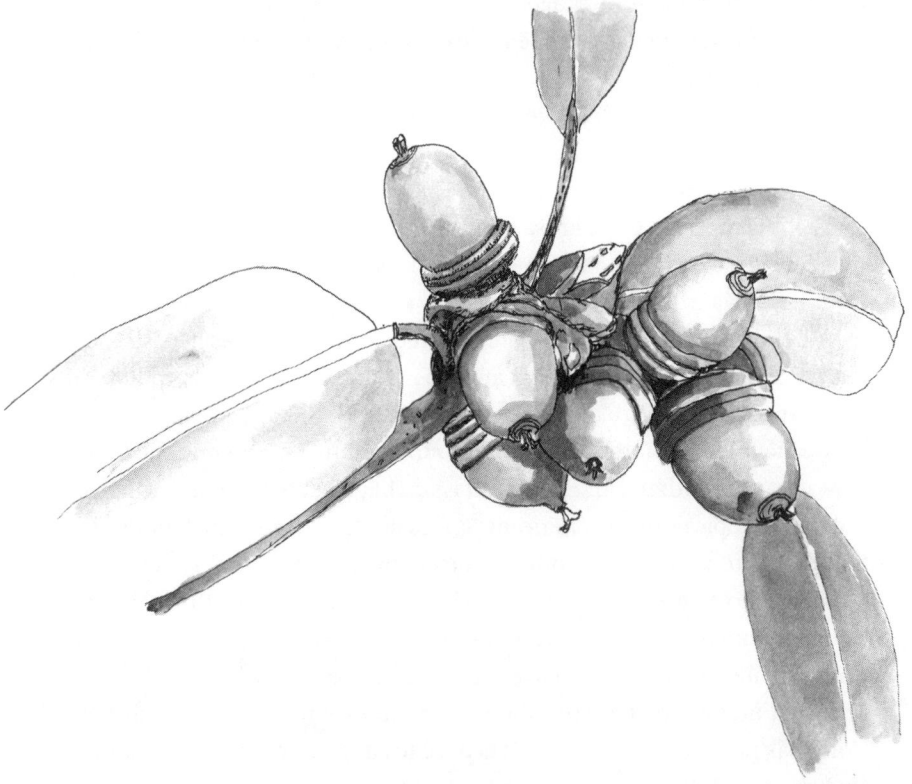

Quercus sect. *Cyclobalanopsis*: *Quercus glauca* acorns. Wuyi Shan, China, November 4, 2017.

ago. They would have to wait for the Americans to make the trip across the ocean.

North America crept westward as the Eurasian oaks diversified. By 45 million years ago, a disarmingly modern-looking White Oak (*Q.* sect. *Quercus*) species was growing on Axel Heiberg Island, situated in the Arctic at the north edge of Canada. The species is known from more than thirty slender, long-tapering fossil leaves. The margins have shallowly rounded lobes, which look a bit like a toy sawblade. They resemble the modern oak species of eastern North America and Eurasia that have chestnut-like leaves. By 42–40 million years ago, species from the Intermediate Oak Group (*Q.* sect. *Protobalanus*) were growing in Greenland. These fossils together tell us that early North American oaks were living and evolving in the Arctic. Even

Quercus sect. *Cyclobalanopsis*: *Quercus glauca* **leaves**. Tree in cultivation, preserved as a specimen in the Morton Arboretum Herbarium, Accession 98796.

earlier—48 million years ago—there was pollen of the Red Oak Group in the Princeton Chert of British Columbia. The timing of these Red Oak and White Oak fossils tells us that the American oaks split early into the major clades we know today.

Oaks would soon migrate south into a cooling, drying continent. The Western Interior Seaway drained from what would become the Great Plains. With no massive body of water running through it, North America became drier and more seasonal. The continent was further dried by the growing mountains themselves. At the beginning of the Eocene, the Rocky Mountains were dominated by tropical to subtropical plant communities. The

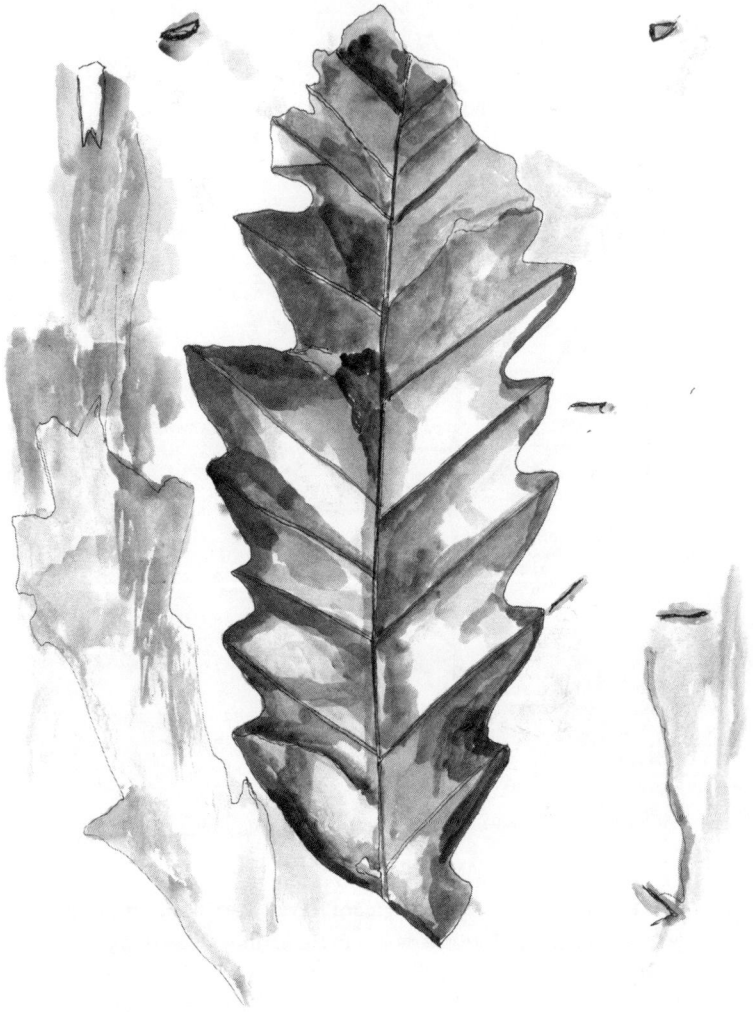

Quercus sect. *Quercus*: **Fossil of unnamed species from the middle Eocene, Axel Heiberg Island, circa 45 million years before the present.** Illustrated from a photo provided by Thomas Denk; specimen from the paleobotany collections at University of Saskatchewan.

Rockies grew throughout the Eocene, forcing oceanic winds blowing in off the Pacific Ocean upslope to cooler temperatures. The arriving winds were heavy with moisture, as they are today. Just as our breath condenses from fog to drops of water on a winter window, moisture in the air turns from fog to rain as it cools. Winds off the Pacific drop rain on their way up. They retain relatively little moisture as they glide back down the east slope of the Rockies. The resulting rain shadow that formed over the continent's interior would set the stage for the eventual spread of prairies over the course of the Miocene. Today, we see the drying effects of the Rocky Mountains on the North American Great Plains as a moisture gradient from the relatively short, dry prairies of western North America to the tallgrass prairies of the upper Midwest.

Oaks spread rapidly across the eastern half of the continent. Oak pollen was abundant on the Gulf Coastal Plain by the middle Eocene (48–38 million years ago) and increased in the southeastern United States by the early Oligocene (approximately 32 million years ago). By at least 30 million years ago, the White Oak and Red Oak Groups had reached the Gulf Coastal Plain of Texas.

As oaks were diversifying in the east, they were also spreading into California and climbing into the Rocky Mountains. High topographic diversity in the mountains—cool, moist slopes right around the corner from warm, dry slopes—may have allowed oak populations to separate from one another ecologically and escape to favorable sites as the climate fluctuated. Rains squeezed from the west winds made the Rockies a refuge for lineages that subsequently sorted westward to the coast: the Ring-Cupped Oaks or the stone oaks (*Lithocarpus*), both of which today are limited to East Asia; the genus *Notholithocarpus*; the Golden-Cup Oaks (*Q.* sect. *Protobalanus*); and ancestors of the California oaks, including California black oak (*Q. kelloggii*) and valley oak (*Q. lobata*), close relatives of which appear to have inhabited the mountains of Idaho until the Middle Miocene (16–11 million years ago).

By the end of the Eocene, 34 million years ago, the uplift of the Rocky Mountains was nearly complete. Their forests began to look more modern, dominated by pines, birches, and other conifers and deciduous genera that we find today. But they also held relicts of a moister, warmer past. Red Oaks, White Oaks, and Golden-Cup Oaks grew with ginkgoes, sequoias, elms, tree of heaven, and *Platycarya*, a walnut relative with small, winged nutlets that was once widespread across the Northern Hemisphere but is now restricted to East Asia. There is no modern plant community where you can find these genera all growing together.

The Rocky Mountains became a nearly insurmountable barrier to oak migration. Species that are today limited to East Asia went extinct, first on the progressively drier east side of the mountains, which came to resemble the pine-oak forests of northern Mexico. Because the North American oaks had already split into the major sections we know today, each existing American oak clade had to split around the mountains or choose a side. The Golden-Cup Oaks (*Q.* sect. *Protobalanus*) sorted westward to arid habitats, from there south into Baja California and east to Arizona. The Southern Live Oaks (*Q.* sect. *Virentes*) sorted to the east, though the clade today ranges from the southeastern United States to Baja California and Central America. A third clade—section *Ponticae*, the smallest section in the genus—has a curious distribution: *Quercus sadleriana* is restricted to the Siskiyou Region of northern California and southern Oregon, while its sister species, *Q. pontica*, is endemic to the Caucasus. There are no fossil records to tell us whether this lineage is the skeleton of a once-widespread species that spanned the continents or the outcome of a chance long-distance dispersal event.

The two largest oak clades, the White Oaks (*Q.* sect. *Quercus*) and the Red Oaks (*Q.* sect. *Lobatae*), each reflect a history of speciation on the two sides of the Rockies around 40 to 30 million years ago, as the mountains reached their maximum elevation. The rising Rocky Mountains were a barrier between clades of both the White Oaks and the Red Oaks. Today's Red Oaks are separated into an eastern North American clade and one that is endemic to the California Floristic Province. The same is true of the White Oaks. The ancestors on the west side of the mountains thus gave rise to today's nine or ten white oak species and four red oak species of the California Floristic Province. Simultaneously, the ancestors east of the Rockies gave rise to what would become today's seventeen eastern White Oak species and twenty-five eastern Red Oak species. Those eastern clades would, in turn, give rise to even more diversity.

Had we been able to traverse the globe at the end of the Eocene, we might have recognized the ancestors of many of the oaks we know today. They would have been intermixed with puzzling relatives—genera like *Paraquercus*, *Fagopsis*, *Berryophyllum*, and *Eotrigonobalanus*—which we would be hard-pressed to situate on the oak tree of life. Most likely, we would not have recognized modern species: the modern diversity we know today arose in the last 15–10 million years. But the major clades were all in place and poised for dropping temperatures.

Much of the world became dramatically colder 34 million years ago, at the beginning of the Oligocene. Atmospheric carbon dioxide was cut roughly in half. The deepening of the Drake Passage (between Antarctica and South

Quercus sect. *Protobalanus*: *Quercus palmeri* **acorn**. Illustrated from a photograph provided by Béatrice Chassé.

America) and the Tasmanian Passage (between Australia and Antarctica) allowed water to circulate freely around Antarctica. This lazy river encircling the southern continent blocked warm ocean currents that had previously staved off the cold. Ice expanded to cover Antartica, sopping up precipitation otherwise destined for the oceans and further forcing down temperatures and sea levels. In North America, air temperatures plummeted by an estimated 8°C in less than half a million years, driving many cold-blooded reptiles and amphibians extinct. Large land snails were replaced by smaller snails that could weather a drought. More than 90% of mollusk species were driven extinct in the U.S. Gulf Coastal Plain. Aquatic salamanders and turtles were progressively replaced by land tortoises. Air temperatures dropped in northern Europe by an estimated 4–6°C over the same period. Worldwide, the effects of decreasing temperatures and increasing seasonality—fluctuations of temperatures and precipitation from summer to winter—on plant communities ranged from relatively abrupt replacement of tropical or subtropical forest with temperate forest in some sites to little if any change in others.

Oaks were suited to the changing climate. Much of Eurasia was occupied by the Paratethys Sea, but oaks spread onto what would become the Eurasian Steppe. In the Americas, temperate forests extended as far north as Iceland and southern Greenland and spread onto what would become the North American Great Plains, where they would only later be outcompeted by grasslands. Oak species with lobed leaves, a hallmark of seasonal, temperate forests, became widespread in the Northern Hemisphere early in the Oligocene. Oaks' ability to adapt rapidly to climate change served them well at the onset of the Oligocene, as it does today. If the Eocene is a story of oak origins, the Oligocene is the history of their growing dominance.

Temperatures fluctuated until the Middle Miocene and then began a 15-million-year descent to the modern era. The Rocky Mountains became too cold and dry for oaks, which went extinct or nearly so in the northern Rockies between 15 and 11 million years ago. The American Oaks reached their peak abundance in California and eastern North America in the first half of the Miocene, spreading at least as far west as Texas. Now that the more tropical tree clades had been pushed out, temperate clades—including the oaks—evolved to fill the warm, moist, subtropical climates of southeastern North America: post oak (*Q. stellata*) and its relatives, blackjack oak (*Q. marilandica*) and its relatives, and the Southern Live Oak Group (sect. *Virentes*) are concentrated in this region. The foundations of North American oak diversity were in place. The temperate forests looked a lot like we know them today. We would have been the recipients of a rich set of oak forests if the oaks had stopped right there.

But oaks weren't done yet. All their prep work getting settled into North America primed them for three big moves that would lead to roughly 40% of the oak species we know today.

In October 2022, one of my graduate students, Kieran Althaus, and I flew to northeastern Mexico to collect oaks with twelve of our colleagues. Our flight took us southward over glaciated territory for about 240 kilometers. We passed the succession of moraines that mark the end of the last northern continental glaciation. Prairie gave way to a reticulum of gallery forests as we reached the Illinois River, then the Mississippi River. The forest closed in below us as we passed over the Ozarks. We left the eastern North American forests behind once we were northeast of Dallas. The landscape below us opened into a mosaic of desert, grassland, and shrubland—now almost all rangeland except where cities have taken over. Then forested mountains slid into view beneath us, showing the northern extent of the Sierra Madre Oriental, the easternmost mountains of Mexico. We landed in Monterrey.

Our flight between the temperate forests of eastern Texas and the Sierra Madre Oriental passed over the sites of ancient forests, now long gone, through which oaks and a host of other eastern North American species migrated in the Miocene and probably intermittently during glacial periods of the Pleistocene. More than one hundred species or clades of seed plants live in both the forests of eastern North America and the montane forests of eastern Mexico. Between those two forested regions is a gap of grassland and desert in Texas and northeastern Mexico that many temperate forest species cannot tolerate. This gap separates two halves of a widespread forest that was divided around 15 million years ago, as the first grasslands were getting a foothold in Central North America and Eurasia. Drying climates allowed the Chihuahuan Desert to spread in northeastern Mexico, Texas, and adjacent areas of New Mexico and Arizona. In some cases, single species were cut off from their eastern North American sources: American sweetgum (*Liquidambar styraciflua*), for example, is widespread in rich deciduous forests of eastern North America, then drops out in southern Texas and northern Mexico. It picks up again in the cloud forests of eastern Mexico through Central America. The spreading deserts and grasslands severed the once-broad distribution of this species. In other cases, entire clades were divvied up: groups of Mexican oaks, viburnums, and pines arose from eastern North American ancestors in this way.

Both the Red Oak and White Oak Groups left a calling card as they sent emissaries through Texas into eastern Mexico. *Quercus polymorpha*, a member of the White Oak Group, grows in gallery forests, pine-oak forests, and cloud forests of northern through Central Mexico and just squeaks into southwestern Texas. Its close relative, *Q. mohriana*, grows on limestone in western Texas and adjoining counties of Oklahoma, New Mexico, and Mexico, while *Q. pungens* grows on dry slopes of West Texas, southern New Mexico and Arizona, and the Sierra Madre Oriental. In the Red Oaks, *Q. gravesii* is centered on the mountains of western Texas, where it grows with the elusive but recently rediscovered *Q. tardifolia*. It has scattered populations in Coahuila and Nuevo León. Its relatives *Q. cupreata* and *Q. hypoxantha* live in the northern Sierra Madre Oriental. Two other relatives, *Q. rysophylla* and *Q. mexicana*, extend much of the length of the Sierra Madre Oriental, and the latter makes its way all the way down to Oaxaca. These White Oak and Red Oak species evolved from clades that budded off around the time eastern North American oak populations were cut off from Mexico.

The earliest reported Mexican oak fossils are pollen grains from northern Chiapas that date to the Early or Middle Miocene. These are followed by oak fossils at the boundary between the Miocene and Pliocene in Guatemala and Panama. DNA data also suggest that both the Red Oak and the White Oak Groups in Mexico split from Texas in the Early to Middle Miocene. Oaks dispersing into Mexico helped form temperate forests in cloud forests primarily in the eastern mountains, with species such as basswood (*Tilia americana*), redbud (*Cercis canadensis*), shagbark hickory (*Carya ovata*), sweetgum (*Liquidambar styraciflua*), and *Acer skutchii*, a close relative of the sugar maple. These eastern North American forest species or close relatives grow in temperate forests of eastern Mexico with *Podocarpus*, the plum-pine, a predominantly Southern Hemisphere genus of evergreen conifer that reaches its northern limit in the northern Sierra Madre Occidental. Bromeliads and epiphytic orchids cling to the trunks.

In mountain ranges throughout the country, oaks remained at relatively high elevation, often with pines. Climbing their way into Mexico enabled oaks to reach relatively moist, cool sites for which they were preadapted by their birth in the north. They came to dominate a diverse range of montane habitats and climates they encountered, remaining a mostly temperate genus at heart as they moved south. Oaks churned out species rapidly in the Mexican mountains, where speciation rates increased to approximately double the rate of northern oak lineages. Populations became separated from each other by differences in slope, soil, moisture availability, and temperature. This environmental variation stimulated the isolation of populations and the production of new species: the rate of ecological evolution increased in the oak lineages that moved into Mexico, and elevation is one of most important environmental factors differentiating Mexican oak communities and species today.

From Mexico, White Oaks and Red Oaks each moved into Central America. One White Oak, *Q. insignis*, became a dominant in the cloud forests of southern Mexico and Central America. It produces some of the largest acorns in the world, to more than three inches in diameter. It is critically endangered in Mexico and threatened elsewhere in its range. The near-threatened *Q. skinneri* of the Red Oak Group shares many of the same forests but does not extend as far south. It has acorns nearly as large as those of *Q. insignis*. One population from the Red Oak Group leapfrogged from Costa Rica to eastern Panama and South America roughly 430,000 years ago to give rise to *Q. humboldtii*, the only oak species in South America. The species spread through the Colombian Andes, where it has persisted and thrived through multiple glacial cycles that repeatedly squeezed its distribution and then allowed it to expand. The populations on different cordilleras have diverged from each other genomically as the species' range has pulsed, accordion-like. But *Q. humboldtii* has never produced another species as far as we know. Give it another million years or so. This "lonely oak" may one day become the progenitor of a whole clade of South American oaks.

As the Red and White Oak Groups radiated southward, they were joined by *Q. fusiformis* and *Q. oleoides* of the Southern Live Oak Group (sect. *Virentes*). The latter species ranges from Tamaulipas to Costa Rica and grows in some of the most tropical habitats in which any American oak may be found. Today, an estimated 160 oak species live in Mexico. Our estimate may increase as we learn more. However, we may instead find that Mexican oaks are more diverse in ecology and morphology within species than expected, rather than more diverse in number of species. In either case, the Mexican oaks represent one of the most rapid species diversifications in the genus. And oaks did not stop in southern Mexico and Central America. The Mexican Red and White Oaks both gave rise to a few lineages that migrated back northward into the southwestern United States, leading to little clusters of White and Red Oaks in Arizona and New Mexico that are not closely related to most of the Texas oaks. These species flourish in the Gila Mountains, the Chiricahua Mountains, the Pinaleño Mountains, and other mountains of the Southwest. One, *Q. oblongifolia*, likely expanded westward into California by at least the Late Miocene and left *Q. engelmannii* as an isolated endemic there. The two are sister species now. Some relative youngsters migrated east into Texas and now live with the earliest-diverging Mexican oak groups. If you get the chance to visit them, it's worth remembering that the oaks of Mexico and Central America arose from eastern North American ancestors but many of the oaks of Arizona arrived by way of Mexico.

Around the time oaks were entering Mexico, an ancestor of three eastern North American oak species—the widespread eastern white oak (*Q. alba*), the bottomland chestnut oak (*Q. michauxii*), and the mountain chestnut oak (*Q. montana*)—was heading toward Eurasia. We don't know whether it was growing as a population that spanned the North Atlantic Land Bridge or the Bering Land Bridge. Beringian fossils seem to favor the western route: White Oaks closely resembling a few of the chestnut-leaved East Asian White Oak species were deposited in Alaska in the Middle Miocene, approximately 16–12 million years ago, whereas oaks appear to have been absent from the North Atlantic Land Bridge—at least from Iceland, which is part of the southern route they would have been more likely to take—from 15–10 million years ago, only appearing at about 9 million years ago. Molecular dating suggests that the East Asian and European white oak clades separated around the Middle Miocene. These data together suggest that White Oaks may have dispersed westward across far northern North America to East Asia via the Bering Land Bridge between 23 and 12 million years ago, then spread rapidly across Europe and Asia, where they diversified rapidly and roughly simultaneously in western Europe and East Asia. It will take additional work to figure out this migration history.

The Eurasian White Oaks diversified to form roughly twenty-five species. This is striking when you consider that oaks arriving in Eurasia from eastern North American had to duke it out with all three of the Eurasian oak groups. Eurasia became a family reunion between five lineages who had parted ways millions of years earlier: sections *Quercus* and *Ponticae* of the Americas; and sections *Ilex*, *Cerris*, and *Cyclobalanopsis* of Eurasia. The White Oaks were seasoned upstarts, coming from the oak-packed forests of eastern North America. One way of evaluating how rapidly a clade has speciated is to compare it with its sister clade. Sister clades share a single ancestor and are thus the same age. By this comparison, the Eurasian diversification stands out: three species of eastern North American White Oaks—eastern white oak, chestnut oak, and swamp chestnut oak—form the sister group to the twenty-five species of Eurasian White Oaks we observe today. This comparison is even more impressive when you consider that the European White Oak species count may be lower than its initial diversity, coming as it does at the tail end of roughly 2 million years of glaciation, which repeatedly squeezed the European flora against the Alps.

Today's White Oak Group overlaps with nearly all of the modern-day Eurasian oak range, from subtropical forests of Central China to low mountains in Mongolia, and from cool northern European forests to the Mediterranean. How did the White Oaks pull this off? Evolution in the forests of eastern North America may have laid the foundation for White Oaks to thrive across Eurasia: freezing-tolerance and drought-tolerance involve some of the same adaptations. White Oaks were already a successful clade, and they found a continent with no close relatives to compete against. But it is also possible that their radiation across the continent was propelled in part by previously adapted genes from the continent. After they arrived in Eurasia, the white oaks hybridized with *Q. pontica*. The record of that ancient gene flow is imprinted in relatively long blocks of the genome shared between *Q. pontica* and the Eurasian White Oaks. Moreover, all the Eurasian White Oaks surveyed show this shared history with *Q. pontica*, suggesting that gene flow happened early in the origin of the Eurasian White Oaks, very likely soon after they arrived in Eurasia. It may well be that Eurasian-evolved genes passed from *Q. pontica* to the newcomers helped set the White Oak Group up for success in this brave Old World.

Despite the fact that they are a young clade, the geographic range of the Eurasian White Oaks roughly equals the size of the Cork Oak, Holly Oak, and Ring-Cupped Oak ranges combined. In fact, the range of the White Oak Group almost exactly traces the boundaries of the Northern Hemisphere temperate forest biome. If *Q. alba* is the "king of kings" in eastern North American forests, this final move across the ocean made the White Oak Group king of the mountain.

Today's oaks look quite organized. The two major clades—subgenus *Quercus* and subgenus *Cerris*—are mostly sorted onto separate continents. Western North American oaks separate cleanly from the oaks of eastern North America, which in turn separate from the oaks of Mexico, Central America, and the southwestern United States. But this seemingly orderly diversification belies a more complicated history. Species of the Red Oak Group, which is today limited to the Americas, were growing in riparian forests along the Baltic Sea 38–34 million years ago along with relatives of bald cypress (*Taxodium*) and the now-extinct Beech Family genus *Eotrigonobalanus*. In mixed mesic forests in the same region, species of the Red Oak and Intermediate Oak Groups were growing with firs, the pine genus *Cathaya*, and golden larches (*Pseudolarix*). Those oaks, also, are long gone, members of clades that are relegated to the Americas now; *Cathaya* and *Pseudolarix* are each represented by a single living species of eastern to southern China. Relatives of today's Red Oaks, Ring-Cupped Oaks, and the North American White Oak Group (the ones who stayed, not the ones who crossed over and made such a splash in Eurasia) grew together in Portugal until the Late Pliocene. The Ring-Cupped Oaks may also have lingered in western North America and

Quercus sect. *Cerris*: *Quercus ithaburensis* acorn. Tree in cultivation, UC Davis Shields Oak Grove, October 2018.

Europe until the Pliocene. None of those lineages persist in Europe. Oaks and their relatives were, in fact, widespread across the northern continents through much of the Eocene, and cold-hardy oaks migrated between the continents until roughly 5 million years ago.

The most recent common ancestor of the oaks turned out to be well suited to the modern world. Some oaks were not so lucky. Some are evident in the fossil record. Others went extinct before they could make an imprint. But even lineages that we will never know have likely left us something. Genomic data show us that there was widespread movement of genes between species in the early days of the oak genus and even the Beech Family, just as there is today. Oak genera in the Americas interbred in the deep past, the ancestors of *Notholithocarpus* and *Chrysolepis* with the ancestors of *Quercus*.

In Eurasia, some species of the Holly Oak Group carry an ancient chloroplast that predates the split between the chestnuts, the stone oaks, and the oaks. The species of today's forests bear this history of ancient gene flow.

Every branch on the Tree of Life was, at one time, an experiment. Oaks are no exception. They just happen to be, like humans, a successful one.

6

"Pharaoh's Dance"

The Oak Genome

Miles Davis walked into Columbia Studio B on August 19, 1969, with an engineer, twelve musicians, and his producer, Teo Macero. There was very little music written out ahead of time. The musicians formed a semicircle around Davis. Davis started ideas, stirring the music together as the recording tape rolled, adding a "dash of Jack DeJohnette [on guitar] . . . a pinch of Lenny White [on drums] . . . a teaspoonful of Bennie Maupin playing the bass clarinet." The musicians would play, lay out, exchange ideas, play again. The tape rolled from 10 a.m. to 1 p.m. Then everyone headed back to Davis's house in the afternoon to listen through the unedited tape. They returned to the studio the next morning.

The musicians went home after three days. The recording sessions for the album *Bitches Brew* were complete. Joe Zawinul, the oldest and perhaps most experienced musician recording in these sessions aside from Miles himself, said of the sessions, "I didn't think they were exciting enough." Dave Holland, one of the bassists, said he wasn't sure at times whether they were rehearsing or recording. Bob Belden, who produced the 1999 reissue, listened to all the tapes made over those three days, and wrote of them, simply, "There were some low moments, some starts and stops."

The album we know today is, fortunately, not just a batch of these raw recordings. Instead, it came together in the editing room over a few months of 1969 and 1970 as Teo Macero and Davis worked through hours of recording time, distilling and running them through effects, looping them, cutting and splicing the sessions into minutes and seconds to make the final piece.

A few seconds of recording captured halfway through the morning could show up at the beginning of the piece. The collection of moments captured in the recording studio became the final music only as it was remixed and recombined in the editing room.

The opening piece, "Pharaoh's Dance," starts like a creek running through the forest. Everything is in motion, the drum ticking along, carrying leaves and debris with it. The electric piano jumps in and steps around the beat. Bass clarinet bubbles up from beneath. At 15 seconds the rhythm submerges as the bass and electric piano rise to the surface and tumble around each other. The drum kicks back in. Everything is fluid and brisk.

Then, at 1 minute, 38 seconds, the track abruptly starts over from the beginning, as though the musicians had nodded at each other and started again. In a sense, they had, but only in the editing room: this is a splice. At 2 minutes, 32 seconds, Davis's trumpet glides in low like a harrier: it's been spliced in. Twenty-two seconds later, the opening theme starts again—another splice. The music dissolves into the understory, then electric piano rains down at 7:55, spliced in from who knows where. There are long pauses. Themes replay in different forms. An interlude cuts in with delay. A keyboard vamp repeats several times at 8:50, then breaks out. This sounds intuitive and natural, but it's another splice.

"Pharaoh's Dance" was assembled from eighteen different edits. Once you know about the splices and the individual fragments that make up "Pharaoh's Dance," you find yourself waiting for them. But until you realize they are there, and sometimes even after you know it, when you are just sitting back and listening, what you experience is the whole piece. It expands to fill the space around you. The musicians themselves did not all recognize the final recording when they first heard it. Joe Zawinul, returning to the studio offices a few months after the recording session, sat listening to a recording the receptionist was playing, perhaps over the waiting room speakers. He thought it was "incredible . . . smoking." He asked, "Who the hell is this?" The receptionist replied that it was the album Zawinul himself had composed pieces for and recorded in August of 1969.

"Pharaoh's Dance" is cobbled together from potentially discordant parts, but you forget this when you listen to it. It is, for 20 minutes and 5 seconds, a coherent whole.

When I watch a robin flipping oak leaves over on the forest floor, I experience the cumulative effect of about 35,000 protein-coding genes in the oak genome (not to mention the genes of the robins and every other organism in the forest). Each protein-coding gene strung along the twelve paired chromosomes of an oak tree encodes proteins needed to build cell walls, photosynthesize, metabolize nutrients from the soil, survive a drought or a flood, fight off diseases, or collaborate with fungi. At its most basic level, life is a vast improvisation composed of genes building proteins, each kicking in when it is needed, laying out when it is not.

But the genome is much more than a collection of genes. Between genes and even between individual sections of genes stretch expanses of DNA that are never translated into proteins. Regulatory regions turn genes on when they are needed. Transposable elements replicate themselves and hop around the genome, sometimes breaking up genes or moving them around. Repetitive DNA regions form centromeres, regions of the genome where the chromosomes attach to each other and to the fibers that drag them into one daughter cell or another when cells divide; telomeres, the caps at the ends of the chromosomes; and wide stretches in between, amounting to more than 50% of the oak genome. All this DNA together constitutes the genome.

As though the riches of a single genome weren't enough, every cell in the offspring of two organisms carries one genome copy inherited from its mother—by way of the egg cell—and one from its father—by way of sperm borne in a pollen grain. Each pollen grain or egg cell recombines chromosome sections from its own mother tree with chromosome sections from its father. This happens in five basic steps every time an egg cell or the generative cell of a pollen grain is created. First, the chromosome set inherited from each parent is duplicated. Then the chromosomes from the parents line up, gene by gene and nucleotide by nucleotide, to form a near-perfect match. Third, the chromosomes cross arms with one another, and where the arms are crossed over, they break and reconnect. This process of crossing over shuffles alleles. Crossing over dices and splices both the maternal and paternal copies of the genome, making two copies that both—ideally— have the same number of genes. The maternal and paternal copies are remixed to make a novel combination of alleles in every pollen grain or egg cell. Finally, the chromosomes separate from one another and segregate into cells once—step four—and then a second time—step five—to produce the generative cell or egg cell. This process, called meiosis, is the means by which DNA is readied to move from one generation to the next. It is the same process that goes on in every eukaryote, humans included. The difference in flowering plants is that the generative cell nestles into a pollen grain and splits (by mitosis) to form two identical sperm cells that actually fertilize the egg.

Genetic recombination is the original "Pharaoh's Dance." If crossing over were only possible in 3,000 positions of the oak genome, the maternal and paternal copies of the genome could be permuted 2.46×10^{903} different ways. That's 246 with 901 zeros after it. It is an impossibly large number, about 6.16×10^{822} times as large as the number of particles in the visible universe. The possibilities are practically infinite. And in fact, oak chromosomes can, as far as we know, cross over in far more than 3,000 different locations. There are many more possible combinations than could ever be achieved in the real world. But the plausible ones are more than enough. Some of the new gene combinations are ready-made to solve problems. They may result in trees that grow faster or have greater capacity to withstand the long, wet springs and droughty summers of the past few years. Others are unfit, producing trees that may be stunted or photosynthesize inefficiently—trees that for any of a variety of reasons won't get many if any offspring into the next generation. Most gene combinations realized in the trees of a forest are not much fitter nor less fit than the others, but just variations on a theme: they'll do, despite their differences. Every combination of alleles introduces genetic variation into the population.

Because of crossing over and recombination, each region of the genome represents just a few seconds of recording time. When these regions anneal together within an embryo, they form a cohesive composition: the individual tree, adapted to its place in the world. Each oak genome might have been reassembled in any of trillions upon trillions of different ways. Yet each ended up just the way we find it today, conspiring with the wind, the seasons, the jays, and the squirrels to get its genes into the next generation.

The spool of tape in the Columbia Recording studio, with all the cuts and splices, is not the music itself. "Pharaoh's Dance" becomes music when the recording is played and the sounds interact with our ears and neurons. Likewise, an oak genome is not an oak. Oak trees arise from oak genomes when their cells encounter an environment that triggers the translation of the protein-coding genes needed at that particular moment. DNA is translated into proteins, and the proteins work together to form cells, organs, and entire organisms. Proteins run organisms through the moments that make a life. Understanding the genome enables us to connect the phenotypes of trees—their appearances, physiology, behaviors, even ecological interactions—to evolutionary history.

The first map of the oak genome was published in 1996. It showed the positions of only a few genes on two pairs of chromosomes. Within a couple of years, we had maps to all twelve chromosome pairs. For about twenty years, the maps filled in like short snippets of music, with long gaps between the mapped regions. These first maps were linkage maps, created by taking advantage of crossing over and recombination. Oaks cross over at only one to three locations on each chromosome during each meiosis. Crossing-over points—the chiasmata, X-shaped linkages that form between paired chromosomes as they break and reconnect—fall in nearly random locations along each chromosome. If you study enough offspring of a single pollination, you'll find that two genes lying right next to each other on the same chromosome stand a good chance of being carried along together into the next generation. Two genes that are distant from each other have a higher chance of being separated by crossing over. If genes lie near opposite ends of a chromosome or on different chromosomes, the chance of alleles for those two genes from a single individual moving together into the next generation is about as a random as a coin toss.

A linkage map is created by first generating many offspring from matings between two individuals. Some researchers use natural offspring: the first northern red oak (Q. rubra) linkage map was based on offspring from a tree on the campus of Purdue University. Each acorn was fertilized by wind-born pollen from a neighbor tree just a few feet away and tested using genetic markers to make sure all the genotyped offspring were full siblings. More often, researchers manually dust individual flowers of one tree with pollen from another to control what offspring they get. The seedlings that result from these crosses form a mapping population, a group of siblings whose parents are predetermined by the researchers, making it possible to trace the inheritance of each section of their chromosomes. A mapping population forms a record of the crossing-over events in each of the parents that produced the many pollen and egg cells leading to the mapping population. We read this record by comparing gene copies from scores or hundreds of offspring in a mapping population with gene copies in their parents.

A linkage map summarizes the genomic exchanges that happened during many meioses within the mapping population. Estimating the chromosome distance between any two genes or genome regions entails calculating the percentage of individuals that carry two maternal or paternal alleles together from those different genes. The more often two maternal or paternal alleles travel together into the next generation, the closer the genomic regions must lie on a single chromosome. Inferring gene order in this way sounds simple. The first linkage map was, in fact, calculated by hand in 1913 by Alfred Sturtevant, an undergraduate in Thomas Morgan's famous fruit-fly lab. Sturtevant's genetic markers were six morphological characters of mutant flies, related to body color, eye color, and wing size. Each of these attributes of the flies' phenotype was coded by one gene, or at least one position on the genome. Sturtevant's analytical tools were arithmetic, pencil and paper, and concentration: he is said to have worked the map out in his dorm room over the course of one night. But to reconstruct a linkage map that densely covers all chromosomes requires a set of molecular markers that probe the genome directly—using DNA sequences, not the phenotypes that arise from them—and a lot of work and computing power to assemble the map from individual markers.

Creating such a linkage map is a little like listening to each of hundreds of randomly assembled versions of "Pharaoh's Dance" and then identifying, from all the differences and similarities, which parts of the piece were actually recorded together and which others were spliced together in the editing room. The result would be a chart of break points for each edited version. You could then relate these break points to each other and to the recording sessions from which all these individual versions were remixed. This task would be coarse at best with "Pharaoh's Dance" because only two versions were released, one in 1970 and a second in 1999, when the original tapes were remixed and remastered for compact disc. There aren't enough crossing-over points to reconstruct the original recording sessions. In oaks, by contrast, every ovary or anther is an editing room. The process of remixing happens millions of times in the lifetime of an organism, every time a generative cell or egg cell is made. A linkage map pulls together many individual moments of recombination into a composite whole. It is like what Kurt Vonnegut's Tralfamadorians, who could see all of time at a glance, experienced when they read a novel: many marvelous moments seen all at one time.

Catherine Bodénès and her colleagues at INRAE-Pierroton constructed an intricate linkage map in 2016 from more than 4,200 mapped nucleotide variants. Their mapping populations were the offspring of pedunculate oak (*Q. robur*), sessile oak (*Q. petraea*), and hybrids between the two. You probably have a sense by this point in the book that *Q. robur* and *Q. petraea* are two of the best-studied oak species in the world. They hybridize, yet they are genomically, ecologically, and morphologically distinct species. Bodénès and colleagues used the linkage map they created to ask whether the alleles inherited by the offspring of different mother trees follow a basic rule of genetic inheritance: alleles should shuffle off to different egg cells or pollen cells with 50% probability in both directions. When they don't, we say that alleles exhibit *segregation distortion*. Encountering segregation distortion is like finding a loaded coin that turns up heads 70% of the time instead of 50%.

Bodénès showed that segregation distortion was common in her map when the experimental offspring were hybrids of the two different species, *Q. robur* and *Q. petraea*. Segregation was also particularly common in paternal maps. Something, in other words, was weeding out pollen grains unevenly, based on what alleles they were carrying, and this effect was strongest when the father was not the same species as the mother. This finding is what we expect if pollen competition in the female flower makes a pollen tube's trip through the stigma, style, and micropyle, and the success of fertilizing the egg cell, particularly difficult for pollen grains that come from a different species. Bodénès's linkage map thus provides evidence that pollen competition in the female flower is at least part of how *Q. robur* and *Q. petraea* remain distinct as species. Her linkage map gives us a gene's-eye view of the decisions made in hundreds of oak female flowers, which shape what trees compete on the forest floor every year.

Linkage maps have also been used to tie traits to the genes that control them. The powdery mildew study discussed in chapter 2 was only the tip of the iceberg. Oak linkage maps have been used to show that the date at which leaf buds begin to grow or expand is controlled by at least a dozen genome regions, each with small effect. By contrast, genes (or maybe one gene) in a narrow region of a single chromosome predict 20% of the variation in the composition of a leaf's carbon isotopes. Carbon isotope ratios allow us to estimate how efficiently plants use water. These studies thus give us insights into how particular genes may help fit organisms to their environments. They also give us insights into how genes shape species. The genes that define oak species, for instance, appear to be clustered in numerous but relatively small islands widely scattered across the genome. And the very regions of the genome that differentiate species most strongly encode, at least in some species, the blueprints for the morphological characters by which we recognize those species: leaf and petiole traits that we use to distinguish *Q. robur* from *Q. petraea* are shaped by variation in regions of the genome that are more genetically distinct than most of the rest of the genome. When we use plant identification guides, we are reading the long story of evolutionary history inscribed in species' genomes.

Linkage-mapping studies show us how oaks work by pinpointing genes and documenting the outcomes of pollen selection that we could never hope to observe directly. They tether chromosomes and genes to the world we can observe.

If a linkage map is a notation of the points where the raw recording has been cut and spliced back together, then an assembled reference genome is, at its best, the completed album. A reference genome is assembled by first sequencing millions of individual pieces of the genome. Researchers will commonly use a mix of short sequence reads, often 150 to 300 nucleotides in length, with long sequence reads of 20,000 to 200,000 nucleotides or longer. The short reads are less expensive, more numerous, and more accurate. The long reads help knit the short reads into chromosome-length assemblies. These individual sequences are then collaged together along areas of overlap, using the overlap to tile the sequences together into longer sequences.

Each genome assembly is a complex puzzle. The Human Genome Project illustrates how complex the problem can be. The project was launched officially in October 1990 and declared complete in 2003, but at that point about 8% of the genome was still unknown. It was only in March 2022 that the human genome was published in its entirety, telomere to telomere—almost. Even then, the assembled genome of the Y-chromosome—the chromosome that, with the X-chromosome, distinguishes the sexes in humans and other mammals—was incomplete. An additional 10 million nucleotides of the genome had yet to be deciphered.

Genome sequencing has become a lot easier and less expensive in the past few years, but it is still work, particularly with plants. Plant genomes tend to be more complex than animal genomes. Nongenic, repetitive DNA proliferates, particularly in plants; these repeated regions, like the background pieces of a jigsaw puzzle, are hard to assemble correctly. Genes become duplicated when chromosomes mismatch during recombination or when transposons move around the genome. The whole genome has often been duplicated as an outcome of a glitch in the meiotic machinery, often in association with hybridization; this is more common in plants than in animals. Gene duplications are an important source of fuel for evolution. Duplicate genes become the ancestors of new genes, allowing organisms to evolve responses to new challenges. They also make assembling a genome challenging. There have been no whole-genome duplications within the oaks so far as we know, but individual genes have proliferated.

There is a 2,000-fold difference in size between the smallest and largest known angiosperm genomes. There is only a 5-fold difference between the smallest and largest known mammal genomes. Genome complexity is compounded in oaks and many other trees by high heterozygosity. Heterozygosity is the condition of having different alleles of a particular gene from one's parents. Variation among these alleles can be hard to distinguish from variation among gene copies on the same genome.

If every sequencing read were 10,000 bases long, the oak genome would span about 80,000 sequence reads lined up end-to-end. If you had 100 sequence reads, on average, for each region of the genome, you'd have a total of about 8 million sequencing reads. If you applied these same ratios to reconstructing the 20 minutes of "Pharaoh's Dance," each scrap of tape you sequenced would be less than 1/50th of a second in length. It would be a great achievement to assemble even a few seconds of the piece.

Assembling the first oak reference genome was a complicated project led by Christophe Plomion of INRAE-Pierroton, with numerous collaborating researchers and institutions. The subject of the reference genome was a tall but unassuming hundred-year-old pedunculate oak that leans over a service road through the Pierroton research station. That tree was instrumental in many earlier studies, including the linkage map created by Catherine Bodénès. Its code name (*Q. robur* "3P") shows up in several papers. The first partial assembly of the *Q. robur* "3P" genome was published in 2016, and the first full assembly in 2018.

This first assembled oak reference genome yielded a view into the order and identity of nearly 26,000 protein-coding genes on all twelve oak chromosomes. (More recent work in a variety of oak species estimates the number of protein-coding genes at 31,000 to 40,000.) The assembled *Q. robur* genome suggested that genes that help trees resist diseases and insects have duplicated and diversified in oaks and other trees: longevity in oaks—in trees generally—may select for gene diversity that helps them survive stresses of various kinds. It also illustrated that the genome is variable even within individuals. Whole-genome sequencing of bud tips from branches that originated 15, 47, and 85 years after the germination of pedunculate oak "3P" carried forty-six mutations that the researchers could confirm. These mutations made the branch tips from different parts of a single tree genetically distinct from one another. And these mutations had a potential impact on the next generation: at least some were carried along in the acorns produced by those branches, meaning that the acorns on one branch of a tree could in principle have genes that give it a competitive advantage over the acorns produced on a different branch. Most likely, the mutation rate within individuals is even higher than this first study suggested, as these kinds of mutations are hard to find and potentially tricky to distinguish from sequencing errors, and only a handful of buds were studied. An oak tree contains whole populations of genomes, whose origins span decades or centuries.

Oak genomes have come out more quickly since 2018, as sequencing technology has improved. Chromosome-level reference genomes were published for California valley oak (*Q. lobata*), Mongolian oak (*Q. mongolica*), sawtooth oak (*Q. acutissima*), red-bark oak (*Q. gilva*), Chinese cork oak (*Q. variabilis*), northern red oak (*Q. rubra*), daimyo oak (*Q. dentata*), Holm oak (*Q. ilex*), and several related genera of the Beech Family—beech, chestnut, Asian chinkapin—just during the period while I was writing this book. By the time you read this, there will surely be more. These genomes show, among other things, that California valley oak, like pedunculate oak, has undergone a proliferation of genes that play a role in disease resistance. The California valley oak genome also illustrates two unexpected similarities between grasses and oaks: both genomes have particularly long stretches of repetitive DNA, some of which may serve as controls for when genes are turned on and turned off and what proteins are produced by a given gene, and evenly distributed regions of dense chromosome packing called heterochromatin, which also plays a role in gene regulation but is focused near the centromere in most plants. These genomic particularities, which appear to give grasses and oaks an additional dimension of genomic complexity, may help explain why both grasses and oaks were so successful during the rapid climate changes of the Miocene and Pleistocene and continue to thrive in diverse environments.

Genomes are our windows into variation at every level of an oak: within individuals, within and among populations and species, and among distantly related lineages. But even if we had sequenced genomes for every known species on Earth, we would still be far from being able to build an oak tree from genome sequences alone. Scientists will be studying the connections between genomes and their expression in individuals for lifetimes to come.

You and I are modern humans, members of the species *Homo sapiens*. But some of our alleles come from Neanderthals (*H. neanderthalensis*) or their close relatives, the Denisovans, rather than our modern human ancestors. Neanderthal alleles initially found their way into modern human genomes when populations of *H. sapiens* expanded across Europe and Asia and back into Africa. Neanderthals are extinct, but they live on in us today as genes that influence the shapes of our faces, skin color and freckling, lipid metabolism, blood coagulation, brain function and development, and adaptations to high elevations.

Alleles moving between species become part of "Pharaoh's Dance." Recombination at meiosis in generation after generation separated blocks of the genome containing Neanderthal genes from the background of the modern human genome. Natural selection could then give a slight edge to modern human offspring who carried Neanderthal alleles if those alleles were beneficial. Many Neanderthal alleles were harmful or neutral and were weeded out. In small populations or in genes that are linked to useful genes, some maladaptive alleles can hang on, fixed in the population by chance; some Neanderthal alleles, for example, increase disease risk while others provide disease resistance. But modern humans formed a sufficiently large population that natural selection was mostly an effective filter. Natural selection thus retained many useful stretches of the Neanderthal and Denisovan genomes while much of the rest fell away like unused scraps of recording tape.

An anatomically modern human who lived near Ust'-Ishim in western Siberia 45,000 years ago recorded a snapshot of ancient hybridization. The genome recovered from a fossilized femur of this individual carries longer blocks of Neanderthal genome than today's European and Asian populations do, because the Neanderthal regions of his or her genome had only 12,500–6,700 years since hybridization to get whittled down by recombination. Ours, by contrast, have been shortened by crossing over and selection over the course of 60,000–40,000 years. Another fossil from Romania is even closer to the moment of hybridization: its genome shows that this individual had a Neanderthal ancestor as recently as four to six generations earlier. This individual might have said, with no hyperbole, that one of their great-great grandparents was a Neanderthal.

These humans of whom we have only fragmentary fossils were the descendants of matings between *H. sapiens* and *H. neanderthalensis*, followed by generations of backcrossing to *H. sapiens*. The plant geneticist Edgar Anderson, whose work helped show that gene flow had the potential to connect species into adaptive networks, asked in 1949, "How important is introgressive hybridization? I do not know. One point seems fairly certain: its importance is paradoxical. The more imperceptible introgression becomes, the greater is its biological significance." The initial hybrids between modern humans and Neanderthals did not, on their own, change the course of human evolution: they only created a conduit for gene flow. Introgression in the following generations, accompanied by natural selection for beneficial alleles, made Neanderthals a force in the evolution of modern humans.

Evolutionary biologists argued about the evolutionary importance of hybridization and gene flow between oaks for a century before we could sequence DNA. It became easier to investigate adaptive introgression—gene flow that moves beneficial alleles from one species into another—when we could directly study evolving genotypes. The first DNA studies of adaptive introgression in oaks used a relatively small number of molecular markers. A study based on an early DNA genotyping method (RAPDs) and eco-physiological data showed that hybrids between Gambel's oak (*Q. gambelii*) and gray oak (*Q. grisea*) in the southern Rocky Mountains exhibit greater drought-tolerance than either pure species. Even seemingly pure *Q. grisea* populations appear to harbor some alleles from *Q. gambelii*. In the upper Midwestern United States, a study based on microsatellite data showed that alleles specific to Hill's oak (*Q. ellipsoidalis*), which is tolerant of dry soils, have introgressed into the more moisture-loving northern red oak (*Q. rubra*) in dry outwash plains. In this case, natural selection may maintain alleles from Hill's oak within northern red oak growing in drier habitats because the Hill's oak alleles are adaptive in the drier habitat.

One of the strongest molecular studies of adaptive introgression in oaks prior to the advent of affordable genomic sequencing data was conducted in the California species of the Red Oak Group. The study showed that interior live oak (*Q. wislizeni*) individuals vary in the amount of introgression they have from two other species, Shreve oak and coast live oak (*Q. parvula* var. *shrevei* and *Q. agrifolia* respectively). It also showed that interior live oaks with introgressed alleles can be found 300 kilometers or more from the nearest known population of the species with which they have exchanged genes. There are a few possible explanations. Long-range gene flow—the chance cloud of pollen carried by a storm—could move introgressing alleles around the landscape. Alternatively, introgressed alleles located far from home today might just be hangers-on from ancient introgression, persisting after species migrated away from their old ranges. A third possibility is that introgressed alleles from other species move around by gene flow between interior live oak populations for many generations after hybridization and end up in novel environments because they are favored by natural selection. Consider these three possibilities alongside another finding from the study: ecology does a particularly good job of explaining introgression in the California Red Oaks. Populations in cooler areas with more humid summers—the kinds of places where Shreve oak and coast live oak grow—show more evidence of gene flow than populations in warmer, drier areas. In other words, it is not being close to another species that predicts genetic similarity in these species: it is being in the right climate. That's exactly what we would expect if gene flow is adaptive. Option one seems unlikely; but either option two or option three could be due to adaptive gene flow. Moreover, the chemical signature of waxes in the leaf surface—cuticular hydrocarbons—told the same basic story of gene flow associated with ecology. These waxes help trees adapt to different levels of moisture and drought, and thus the sharing of traits between the taxa suggests that natural selection plays a role in maintaining introgressed genes within interior live oak.

These first studies gave us reason to believe that useful alleles might be shared between oak species. But it took genomes to demonstrate that the genes moving between species were actually tied to traits or ecology. One of the first papers to do so was a study by Thibault Leroy and colleagues at INRAE-Pierroton. Leroy and his colleagues investigated four European white oaks: the widespread pedunculate and sessile oaks (*Q. robur* and *Q. petraea*), pubescent oak (*Q. pubescens*), and Pyrenean oak (*Q. pyrenaica*). Their work demonstrated that while the four species diverged from each other millions of years ago, they only reconnected and started exchanging genes about 40,000 years ago. Genes seem to be moving rather freely among these species across much of the genome: the ratio of genetic variation among the four species to genetic variation within each species is less than 0.08 when averaged across the genome.

But between the genes that are largely shared among species, Leroy and colleagues discovered genomic islands that differentiate species about ten times as strongly as the genome average. These islands of differentiation are carved down to size by crossing over and recombination, which makes them distinct from the adjoining, more freely exchanged, regions of the genome. Within or close to the genomically distinct islands were an estimated 227 genes that control species' responses to habitat and climate. These genes have the potential to shape species' ecology. Other genes they detected were involved in plant responses to stress and interaction with symbiotic organisms. Still others influenced flowering time, pollen and embryo growth, and other factors that may explain reproductive compatibility or incompatibility between species. These all serve as genes that could explain how one species differs from another. Natural selection hones gene flow and makes these four oaks' genomes a mosaic. At the same time, species-distinguishing regions of the genome were not the same between all species pairs. Differentiation and hybridization operate differently on different regions of the genome for every pair of species.

Leroy and his colleagues then dove into the genomes of sessile oak along a latitudinal gradient and two elevational gradients to discover what genes are shared between species in different environments. The northern populations and the high-elevation populations are cooler on average than the southern and lower-elevation populations. Genome sequencing showed that *Q. robur*, which favors cooler and wetter sites, shares more alleles with *Q. petraea* populations at high elevation or northern sites than with those in the warmer sites. Some of the alleles that make their way into *Q. petraea* by hybridization and backcrossing in the high-elevation and northern sites help control stomata, which regulate the movement of carbon dioxide, oxygen, and water vapor into and out of the tree. These genes may influence how efficiently a plant can use moisture. Other genes retained in *Q. petraea* after hybridization are associated with the timing of leaf emergence or embryo development. These genes may help plants adapt to their habitat and climate, though more experimental work would be needed to be sure. Leroy's work shows, in other words, that oaks are able to share ecological solutions through introgressive gene flow: the regions of the genome that fit species to their environments aren't good just for one species at a time. The alleles that *Q. petraea* gained by introgression may be part of what allowed the species to migrate nearly to the Arctic Circle.

Asymmetric gene flow from California white oaks may have had an analogous effect on migration of the southern California endemic Engelmann oak (*Q. engelmannii*), which arose from the Mexican White Oaks. Its sister species is *Q. oblongifolia*, which extends north to southeastern Arizona.

Quercus engelmannii expanded into California when climates were moister and was stranded as climates dried; it showed up in Southern California and the vicinity of Reno, Nevada, by the Late Miocene. Leaving behind its closest relatives and the habitat in which its ancestors had evolved, Engelmann oak found itself surrounded by California shrub oaks. There, its offspring were pollinated by *Q. berberidifolia*, a distant relative from which it had evolved little if any physiological reproductive isolation. Scott O'Donnell, Victoria Sork, and colleagues used genome sequencing to demonstrate that windows of the genome that had introgressed from California *Q. berberidifolia* into *Q. engelmannii* contained thirty-two genes of known function. Several of these genes are associated with plant stress responses or acclimation. This asymmetrical gene flow—from the local species into *Q. engelmannii*—may be part of how a species of Mexican origin remained and thrived in southern California.

A genomic study of the widespread East Asian Chinese cork oak (*Q. variabilis*) and sawtooth oak (*Q. acutissima*) by RuiRui Fu and colleagues has shown that it is not always movement of the genes themselves across species boundaries that shapes species, but sometimes the movement of gene regulators. Repetitive regions between genes often serve in part as switches that flip genes on and off. Some of these gene switches are shared between *Q. variabilis* and *Q. acutissima* by introgression. This gives evolution another tool with which to modify oak populations: controls on the combinations of genes that are expressed in different environments and the total levels of gene expression may represent novel solutions to environmental problems, solutions that oak species can trade through hybridization and introgressive gene flow. The study shows as well that recombination maintains large blocks of introgressed genome around regions that are adaptive, presumably because natural selection drags whole regions of the genome across species boundaries. And those areas of the genome that cross species boundaries are replicated in different pairs of populations for *Q. variabilis* and *Q. acutissima*, connecting population pairs that have similar environmental conditions. The result is that while geographic distance drives genetic divergence between populations—geographically distant populations of a given species are genetically differentiated from each other—ecology has an even stronger effect.

We are only beginning to understand the role adaptive introgression may have played as oaks colonized the Americas and East Asia or chased receding glaciers northward at the end of the Last Glacial Maximum. It may be that the genomes of extant species are riddled with genome fragments of extinct species. And it may turn out that introgression is one of oaks' most important tools for responding to rapid climate change. Tying the evidence for adaptive introgression back to the evolution of the oak genome will be essential to understanding how best to conserve oaks as we rapidly alter the world in which they evolved.

Some genes track branches of the Tree of Life. Others track allele histories that evolve more or less independently of speciation, and that do not follow species divergence history. Still others track pollen traveling between distant lineages. These three kinds of histories—speciation, gene diversification, and hybridization—abut one another in the mosaic oak genome. No single gene can tell all the stories.

Discovering the shape and timing of the oak tree of life has been a process of teasing these histories apart. The first molecular phylogeny of oaks, published in 1999, was based on essentially two genes: the chloroplast and one section of nuclear ribosomal DNA. It showed that the Oaks of the Americas (*Quercus* subgenus *Quercus*) were a clade distinct from the Eurasian Oak Group (*Q.* subgenus *Cerris*). It revealed relationships among major clades. But it could tell few of the fine details about the relationships among closely related species. Over the next fifteen years, researchers made inroads into the oak tree of life using a handful of nuclear genes, a tangle of nuclear ribosomal repeat regions, and DNA fingerprinting techniques. These techniques in combination identified the major lineages, but many branches were obscure at best. Moreover, some well-known groups—the Southern Live Oak Group (*Q.* sect. *Virentes*), for example—did not consistently form clades in these analyses.

The oak tree of life—in fact, the whole Tree of Life—became more accessible as genome-sequencing technology evolved. In 2009, Paul Manos emailed me an article about a method called restriction site–associated DNA sequencing, abbreviated RAD-seq. RAD-seq uses next-generation DNA sequencing, which first became commercially available in 2005. The costs of DNA sequencing dropped from $500 per million nucleotides in 2007 to 50¢ per million in 2010. Today, you can sequence about 2 million nucleotides of DNA for a U.S. penny, and most people expect that prices will continue to go down. The implications have been huge for scientists working on any aspect of plant genetics. During my PhD work, students typically generated data from two or three genes for their projects. Today, students commonly generate data for hundreds of genes or even whole genomes in the same amount of time. Paul and our colleagues and I were hopeful that genomic data, when lined up for hundreds of oak individuals, would give us a new view into the histories that make up the oak tree of life.

The histories revealed by our work were complicated in ways that we hadn't expected. The Eurasian White Oaks are a great case study. You might not expect this group to be tricky: there are only about twenty-five White Oaks living in Eurasia, and they include some of the best-studied oak species anywhere in the world. But they kept flitting across the oak phylogeny. In the 1999 two-gene phylogeny, the European pedunculate oak (*Q. robur*) fell sister to all the North American White Oaks. In 2008, a study that added in an East Asian White Oak species (*Q. griffithii*) and an additional nuclear gene found the Eurasian White Oaks swinging to a position sister to the eastern white oak (*Q. alba*). In 2009, the Eurasian white oaks all fell together in an amplified fragment length polymorphism (AFLP) study that we published, but they were back where they had been in 1999, sister to all the American White Oaks. In other studies published over the next five years, the Eurasian White Oaks appeared to be entangled with the North American White Oaks, but they could not be pinned down to one position.

Our colleague Jeannine Cavender-Bares had numerous oak species growing in the greenhouse, and we selected twenty to sequence for a first trial using the RAD-seq method. Cavender-Bares's group extracted the DNA and sent it off to a commercial lab, and we had data within a few months. The first results seemed conclusive: they showed a single Eurasian white oak to be sister to eastern white oak (*Q. alba*) and the closely related swamp chestnut oak (*Q. michauxii*) with very high statistical support. But then John McVay, the postdoctoral researcher on the project, began adding more species and discovered that in many analyses, the Eurasian White Oaks moved once again to a position sister to all the remaining White Oak Group. As he analyzed the data in different ways, the Eurasian White Oaks toggled back and forth between these two positions, with some variation. It seemed they wanted to fall either outside of all the other White Oaks or embedded within the eastern North American White Oaks, generally as sister to *Q. alba* and its cousins. Which was the correct answer?

It turns out that both answers were correct, but only for portions of the genome. McVay interrogated each region of the genome to identify what different stories might be embedded and discovered an unexpected history. You may remember the Deer Oak Group (*Q.* sect. *Ponticae*) from chapter 5.

Quercus sect. *Ponticae*: *Quercus pontica* **leaves and acorns**. Illustrated from a photograph provided by Béatrice Chassé.

The Deer Oaks are sister to a clade made up of the White Oak and Southern Live Oak Groups (*Q.* sect. *Quercus* and *Q.* sect. *Virentes*, respectively). One member of the Deer Oaks is *Q. pontica* of the Caucasus. The other is *Q. sadleriana* of the Siskiyou Mountains in Northern California and Southern Oregon. That weird biogeographic history became a key to understanding the evolutionary history of the Eurasian White Oaks.

Because *Q. pontica* and *Q. sadleriana* are sister species, they are equally closely related to all other species. *Quercus pontica* should be no more genetically similar to the Eurasian White Oaks than to any other white oaks, and the same is true of *Q. sadleriana*. But McVay found that some regions of the genome placed *Q. pontica* sister to the Eurasian White Oaks. These regions of the genome were saying that *Q. pontica* was more closely related to the Eurasian White Oaks than it was to any other oak clade. *Quercus sadleriana*, however, showed no such discrepancy. Because *Q. pontica* and *Q. sadleriana* are sister to each other, there was only one plausible interpretation of the result: gene flow between *Q. pontica* and the Eurasian White Oaks was recorded in the genome alongside the speciation history. Pulling out the regions of the genome that recorded gene flow history made the Eurasian White Oaks fall sister to eastern North American *Q. alba* and its relatives (which is the speciation history recounted in chapter 5 of this book). Subsequent analyses using whole genomic data further supported this complicated, compound history of hybridization and gene flow layered on top of speciation and biogeography, and showed, moreover, that gene flow was predominantly among the European species and *Q. pontica* and bore the signature of adaptation. The genes moving between lineages appear to be maintained in their new genomic context by natural selection.

McVay went on to demonstrate that in Gambel's oak (*Q. gambelii*) of the southern Rocky Mountains, the genome is split between a portion that allies with eastern North American bur oak (*Q. macrocarpa*) and a portion that allies with the California valley oak (*Q. lobata*). Moreover, *Q. gambelii* may have served as a conduit between the two: in some datasets we find *Q. lobata* falling right next to *Q. macrocarpa*, a California oak sister to an eastern North American oak. McVay's work supported gene flow hypotheses that Jack Maze and John Tucker had advanced regarding *Q. gambelii* and *Q. macrocarpa* in the late 1960s, but *Q. lobata* had not previously been brought into the conversation. A subsequent study using more detailed genomic data showed that the position of *Q. lobata* in the phylogeny can be shifted from California to eastern North America by analyzing first one portion of the genome, then the other: two stories are interdigitated in the genome, creating a history carried across the continent through introgressive gene flow. Genome histories, these studies showed, are shaped by both the Tree of Life and by local and regional gene flow.

Treating the genome as a unitary whole is like taking two musical lines and trying to wring a single melody out of them. What do you get from that kind of mash-up? Nothing that makes sense. The oak tree of life is colored by disparate themes, with each allele punctuated by mutations that wind back through time and connect to others along varying paths. Some follow histories of species coming into existence, evolving solutions to ecological problems. Others are histories of alleles moving between unrelated species. The oak genome is dense with counterpoint.

Intrinsic reproductive isolation has evolved between at least some pairs of related oaks. Cornelius Muller observed in 1952 that many potentially interbreeding oaks rarely if ever hybridize in the wild. The European *Q. robur* and *Q. petraea* exhibit a cascade of isolating mechanisms from the moment the pollen lands until the growth of the seedling. And *Q. lobata*, present in California for tens of millions of years, hybridizes with co-occurring shrub white oaks much less than the much younger *Q. engelmannii*. In both these cases and probably many others not yet studied, reproductive isolation has evolved between oaks that have grown together for millions of years.

Yet many closely related oaks have been hybridizing for generations. One question is how they manage to do so without merging. Another question is why reproductive isolation isn't complete in these species, some of which are separated by 30 million years of evolutionary history or more and have probably grown in forests together for much of the life of the species. The answer may be related in part to their breeding system: oaks produce billions to tens of billions of pollen grains per tree, their pollen can travel for kilometers, and they are loath to self-pollinate. They outbreed readily with even distant populations. Delayed fertilization may also facilitate hybridization in some species by giving pollen tubes a chance to catch up to each other at the base of the style.

Beyond the traits that we can study by eye or with a microscope, the oak genome itself is part of why oak species can hybridize. Nothing forces closely related plant species to have conserved genomes. To the contrary, the genomes of many flowering plant species vary wildly. In sunflowers, chromosomal inversions and translocations have been key to the formation of species. These rearrangements can protect species-specific gene sequences even if species hybridize. In grasses, cotton, goatsbeard, and numerous other taxa, hybridization associated with duplications of the whole genome—polyploidy—may lead to new species in a single generation. In sedges, chromosome rearrangements accrue slowly and at first seem to have little effect on reproduction, but they work together with ecology to drive population divergence and species formation: mismatches in gene order can cause a meiotic train wreck when chromosomes try to pair up. If Mom's

Quercus sect. ***Lobatae***: ***Quercus ellipsoidalis*** acorn. Natural population, Taltree Arboretum, Indiana, October 5, 2005. Photo of live specimen, also vouchered at the Morton Arboretum, Accession 173099.

chromosomes don't line up with Dad's, their offspring are often sterile or inviable.

By contrast, all species in the Beech Family have twelve pairs of chromosomes, except for some uncommon triploid individuals. The order of genes on the chromosomes of an oak and a chestnut align very closely. Within the oaks themselves, you can draw a line between each gene in *Q. gilva*, a member of the Ring-Cupped Oak Group, *Q. ilex*, a member of the Holly Oak Group, and its match in California valley oak (*Q. lobata*, a member of the White Oak Group), and at a coarse level you'll find only a handful of rearrangements and inversions. This is remarkable, given the fact that the most recent common ancestor of these oaks dates to roughly 50 million years ago. There are, in fact, small-scale rearrangements in oaks, and their importance to hybridization dynamics and gene flow within as well as between species remains to be studied with more detailed genome sequencing. Large-scale shuffling and flipping of chromosome sections appear to be rare, however. The structural similarity of oak genomes across wide expanses of space and evolutionary time is part of what makes it possible for chromosomes to line up in hybrids that straddle disparate lineages of the oak tree of life.

We are just at the point of using the genome to learn how important adaptive gene flow between species may be to the ongoing evolution of oaks, poplars, irises, beeches, ground beetles, butterflies, mosquitoes, corals, tropical trees, deer, bears, humans, or any other group of interbreeding, hybridizing species. Each species occupies a portion of the world that is distinguishable from the portions of the world occupied by its close relatives: this is its niche. Because of these niche differences, each species becomes a testing ground for adaptations. Those alleles that serve the species in its own niche will flourish and spread within the species; those that don't will remain at low levels indefinitely or go extinct. Eventually, if hybridization is common enough and if hybrids are fit and fertile, these adaptive alleles will spill over into other species. As Hardin hypothesized in 1975, "Some of the breadth in ecological amplitude in many species could possibly be due to introgression rather than to something entirely inherent within the species." Genetically diverse and able to swap genes, long-lived oaks can bridge ephemeral shifts in climate while providing a bank of ancient alleles that may help each generation adapt to annual changes.

"Pharaoh's Dance" could have been assembled in twenty different ways, but oaks have "Pharaoh's Dance" beat: unique oak gene combinations number in the billions or trillions, one for every oak that has ever lived. Each of those individuals was assembled differently. Yet the species that a group of novel individuals forms in combination is an ecologically coherent whole. Each oak species is itself a novel subset of the possible combinations of oak genes, assembled through divergence between populations and gene flow between species. Species are theaters of adaptation and competition among mutations and recombined alleles. A complex of interbreeding oak species may be, in the words of evolutionary biologists Chuck Cannon and Rémy Petit, more than the sum of its parts. The mosaic of oak evolutionary history appears to be tailored to our rapidly changing world.

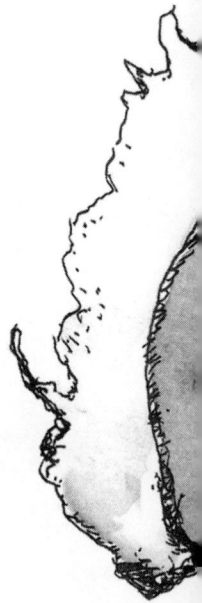

7

Oak Communities

Woodcocks migrate through the upper Midwest in the latter half of October. They stay low, banking over carcasses of fallen white ash and around shrubs in the neighborhoods. Juncos rattle in the bushes. Golden-crowned kinglets lisp in the subcanopy. Crown leaves of black oaks (*Quercus velutina*) fall, exposing the odd squirrel nest. Sugar maple leaves yellow and tumble. Solomon's plume becomes variegated and tattered. The skeletons of pale jewelweed sprawl, interspersed with wild ginger leaves yellowing along the margins. Bur oak (*Q. macrocarpa*) acorns fall to the ground quiescent and nestle in for the winter. Eastern white oak (*Q. alba*) acorns bury their taproots in the mulched trails. Northern red oak (*Q. rubra*) seedlings that germinated in the spring brace themselves for their first winter. Black walnut husks rot in hollows and ditches.

The oaks of eastern North America draw on more than 40 million years of inheritance to form this juncture between forest years. California's oak forests, savannas, and chaparrals reach back as deeply into oak evolutionary history, bringing together Red, White, and Intermediate Oak Groups (*Quercus* sections *Lobatae*, *Quercus*, and *Protobalanus*, respectively) that join in the middle of the Eocene. Oak communities of Europe and East Asia reach back further, more than 50 million years, to when the ancestral population that gave rise to all the world's oaks split in two across the separating continents.

The effects of the oak tree of life on other organisms reveal themselves in every walk through the woods. The diverse, oak-dominated communities we know today reached kindergarten age as grasslands expanded in the

Miocene and Pliocene. They went through adolescence as glaciers shuffled the ecological deck repeatedly over the past 2.6 million years. They matured in fluctuating fire regimes over the past 20,000 years following the recession of the last glaciers. Bacteria, fungi, nematodes, and slime molds feeding on fallen leaves, and deer mice and gray squirrels pilfering acorns as I bike through the woods on my way into work, are a snapshot in evolutionary time. Today's forests—all biotic communities—stand at the intersection between the environment and large swaths of the Tree of Life.

Give natural selection a few populations of bacteria and 4 billion years, and it may hand you back an orangutan and an oak tree. Evolution is surpassingly creative. But it can only work with the variation at hand, and it is often slow to discard what is no longer needed. Whales retain pelvises and hind limbs inherited from their land-bound ancestors. Cavefish numbering more than 150 species worldwide are blind, but they maintain vestigial eyes that degenerate in each individual. Our ancestors ground up tough vegetation using their backmost molars, but those wisdom teeth are mostly a bother for modern humans. All species have such holdovers from their ancestors. Each generation is assembled from the raw material of the previous one, with blind luck—recombination and mutation—to provide it with new variation. Combine this variation with natural selection, and the result is the collection of genes, traits, individuals, populations, species, and lineages exactly as they are today, not as we might design them if we were starting from scratch. Lineages, species, and organisms are all beautifully imperfect.

Of course, many if not most of the traits that evolve on the Tree of Life are exceptionally useful. As I write this, I'm thankful for my eyes and my opposable thumbs, as well as the intricate machinery of my inner ears (which are busy relaying the songs of late-afternoon cicadas). Oaks are packed with useful traits that they inherited from a shared ancestral population. Every oak has wind-dispersed pollen and acorns. Every oak is a woody plant, either a tree or a shrub. Every oak has xylem rays that follow the radii of its trunk and branches, ferrying nutrients, water, and secondary compounds around in the wood and helping to make the tree strong. All oaks share these traits not because the traits are perfect adaptations, but because the ancestors of the genus happened on this suite of traits that worked well.

Woody plant communities the length of the Americas, from the U.S.–Canada border to Tierra del Fuego, are structured by traits that evolved in the Cretaceous and the Eocene. Some tree families and genera grow mostly in the temperate North, others mostly in the temperate South. Still others

Oak root tips colonized by ectomycorrhizal fungi. Illustrated from a photo provided by Eduardo Pérez-Pazos.

thrive in the tropics. This latitudinal turnover—the change in which tree genera or families dominate as you travel from Calgary to Costa Rica—is determined in part by how trees respond to frost. When temperatures become low enough, the water coursing through a tree freezes. Freezing water expands. Bubbles form in the tree's water lines as dissolved air is forced out of solution, just as air bubbles form in the center of an ice cube. When the water thaws again, the bubbles may dissolve back into the water. That is a good outcome for the tree, because water can only flow from the roots to the leaves as a continuous column. But the same air bubbles in a bitter cold snap or in a large vessel—one of the elongated, straw-like cells through which water flows in a flowering plant—may instead sever the column of water that runs the length of the tree. Freezing consequently poses a lower risk to trees with narrower vessels. Flowering plants that have evolved slenderer vessels are more likely to flourish in freezing environments. For this reason, tree communities tend to be dominated by plant families and genera that are specialists of either the tropics—low-elevation areas at low latitudes, where temperatures are generally above freezing year-round and don't fluctuate from month to month as much as rainfall does—or of freezing environments. It is hard for a plant species or clade to be successful in both environments.

It has proven easier for plant families and genera to move into biomes where they can thrive than to evolve into new biomes. In other words, tree lineages are quicker to take large steps across the landscape to find a suitable environment than to make evolutionary jumps from one environment to another. This phenomenon is called phylogenetic niche conservatism: close relatives often occupy similar roles in the ecosystem. The Red Oak and White Oak Groups migrated south through the Americas nearly to the equator, and they diversified in the mountains of Mexico, Central America, and Colombia. Despite their successes and the fact that South America is the heart of the world's tree diversity, oaks barely spilled into the lowland tropics and produced only one South American species. This may be due in large part to ecological and evolutionary priority: oaks are relative newcomers to the south, arriving in Mexico mostly likely no earlier than the Miocene and South America an estimated 430,000 years ago. But a lot of the evolutionary niche conservatism we see in oaks can't be explained by priority alone. Ring-Cupped Oaks (*Q.* sect. *Cyclobalanopsis*) are mostly restricted to East Asian evergreen broad-leaved forests, where they initially evolved. Yet Ring-Cupped Oaks were living in western Europe until the Pliocene. There is a good chance they were living in the western United States as well. They had their foot in the door, but as climates changed, the Ring-Cupped Oaks did not overcome their long-evolved ecological preferences.

The environments that favor different biological communities—forests, ephemeral ponds, prairies, bogs—tend to bring together species and lineages that are related and also share particular attributes. Yet evolution often surprises us. When we are looking for sedges, we usually head to the wetlands. But there are sedge species that thrive in dry, sandy oak savannas. We look for cacti in dry exposed soils, but there are also epiphytic cacti in periodically flooded forests of the Amazon. Oaks exhibit a balance between phylogenetic niche conservatism and evolutionary novelty. The largest and most ecologically diverse sections of the genus—the Red Oaks (*Quercus* sect. *Lobatae*) and the White Oaks (*Q.* sect. *Quercus*)— have largely evolved into the same regions of the Americas. They did so at approximately the same time, solving similar ecological problems each in their own way: almost everywhere in the Americas where you find a member of the Red Oak Group, you also find a member of the White Oak Group. This is evolutionary flexibility at its best. But you don't have to look long to notice that White Oaks have diversified into environments where the Red Oaks have not. White Oaks moved across the ocean to Eurasia, radiating to produce about twenty-five species while the Red Oaks were retreating from western Europe, finally expiring in the Late Pliocene. White Oaks staked a claim in the southern Rocky Mountains with Gambel's oak (*Q. gambelii*). A whole clade of White Oaks spread from northern Mexico into Arizona, and one of these populations made its way to southern Arizona to become *Q. engelmannii*.

What accounts for the disproportionate successes of the White Oaks? White Oak vessels are narrower on average than those of the Red Oaks, which may help White Oaks survive in drier environments. This same adaptation helps White Oaks adapt to freezing environments and thus may have equipped them for their Miocene trip back to Eurasia via one of the northern land bridges. Moreover, White Oaks have a particularly strong adaptation for keeping diseases at bay. After about a year of growth, the early-wood vessels of most species in the White Oak Group fill with balloon-like structures extruded by the cells that flank the vessels. These growths are called tyloses, and they grow into the vessels of oaks and many other tree genera. Tyloses form a barrier to water and diseases. In the Red Oak Group, by whatever accident of evolutionary history and perhaps physiological constraint, tyloses form more slowly. Thus when White Oaks are infected with oak wilt fungus, the rapid production of tyloses helps to restrict the disease to the branch it initially afflicted. The slower growth of tyloses in the Red Oaks gives diseases and fungi more time to spread. Moreover, White Oaks exhibit overall higher resistance to fungal infection: if you inoculate blocks of oak wood with fungi and let them rot for a few months, a species from the Red Oak Group will, on average, lose more mass to decay than the White Oak species. While both the

Red Oak and White Oak group have radiated into a similar range of habitats, the evolutionary legacies of the White Oaks may have set them up for success in a wider range of environments.

Evolutionary legacies may hamper future moves, but they also open doors. Evolution is, to paraphrase Robert Frost, a game played with the net up: beauty and diversity evolve under constraint. Biodiversity is an outcome of opportunities provided as well as limitations imposed by evolution on the Tree of Life.

I stood on a roadside near Miquihuana, Mexico, one afternoon, trying to convince my friend and colleague Socorro González-Elizondo that the oak branch I held was from the same species as the branch she was holding. Both had leaves as rough as elephant skin on the upper side. The leaf undersides were felted with a tangle of yellow-rusty hairs. I was wrong, however. One of our plants was *Q. miquihuanensis*, a Red Oak. The other was *Q. greggii*, a White Oak. The plants weren't even close relatives.

Oaks are champions of convergence, independent evolution of similar traits in distantly related lineages. Some cases of convergence have produced oaks so similar that even the most seasoned botanists have found them difficult to tell them apart. Per Axel Rydberg, the first curator of the Herbarium of New York Botanical Gardens, sometimes confused *Q. palmeri* with *Q. dumosa*, species of the Intermediate Oaks (*Q.* sect. *Protobalanus*) and White Oaks, respectively. The eminent plant ecologist Frederick Clements initially considered Emory oak (*Q. emoryi*), a Red Oak species from Mexico and the southwestern United States, to be a member of the Southern Live Oak Group (*Q.* sect. *Virentes*) due to its superficial similarity to *Q. fusiformis*. Interior live oak (*Q. wislizeni*) of the Red Oak Group and coastal sage scrub oak (*Q. dumosa*) of the White Oak Group are so similar that William Trelease gave a form of *Q. wislizeni* the name *Q. dumosa* forma *populifolia*. And Trelease knew his oaks if anyone did: his 1924 book *The American Oaks* is still the most complete treatment of oaks for the Americas.

These misclassifications and misidentifications are analogous to wrongly identifying an odd-looking baboon or proboscis monkey as a kind of human; the African and Asian monkeys are separated from humans by tens of millions of years of evolution, comparable to the separation between the Red, White, and Southern Live Oak Groups. Oaks have been evolving solutions to ecological problems for tens of millions of years, and different lineages have repeatedly happened upon similar solutions. The Red Oak, White Oak, and Cork Oak Groups (*Q.* sect. *Lobatae*, sect. *Quercus*, and sect. *Cerris*, respectively) have each evolved species with lobed leaves in response to drought, freezing, or the exigencies of packing leaves into a tiny bud for extended wintertime survival. Unrelated species of the White Oak, Red Oak, Holly Oak (*Q.* sect. *Ilex*), Intermediate Oak (*Q.* sect *Protobalanus*), and Cork Oak Groups have converged on small, thick leaves and the shrubby habit as adaptations to drought. These traits are adaptive in seasonally dry conditions today, yet the ancestors of these sections evolved in environments ranging from the temperate forests of North America to Mediterranean climates on both continents or tropical forests of East Asia.

Convergence shapes the interactions that make the natural world. Dolphins eat fish in part because they convergently evolved to live in the ocean. Bats and nighthawks eat insects on the wing; all three groups independently evolved wings. Natural selection hones species in response to environment and interactions with other organisms and also prepares species for ecological interactions. Then ecology kicks in: the environment filters species into habitats and climates where they can survive, and competition and cooperation among all the species who find themselves in a place influence the composition of biotic communities. These ecological interactions in turn shape the evolution of the species. Evolution and ecological interactions pass the biodiversity baton back and forth, generating and rearranging biodiversity.

In 1950, Philippe Bourdeau, who was studying oak-hickory forests of the eastern North American Piedmont, showed that both the White Oak and Red Oak Groups grow across a full range of drought and soil conditions. Species from each clade grew together at both ends of the moisture gradient rather than one clade dominating in one area, one clade in another. Almost twenty years later, the ecologist Robert Whittaker found that in forests from coast to coast, you are more likely to find a Red Oak and White Oak species growing together than two White Oak species growing together.

These data were reminiscent of a point made by Charles Darwin, who noticed that distantly related species seem to prefer each other's company. Close relatives, Darwin wrote, "from having nearly the same structure, constitution, and habits, generally come into the severest competition with each other" and consequently "tend to exterminate [one another]." Recall Darwin's trifold illustration in *The Origin of Species*. Each branch of Darwin's Tree of Life carries the legacy of natural selection fitting lineages to their environments. As lineages evolve and clades diversify, the individuals that make up their many species compete and collaborate. Traits evolving on the Tree of Life tend to make recently diverged species more similar to each other than more distantly related species because closely related species have had relatively little time to evolve differences. As an outcome, Darwin reasoned, the closest relatives should compete more viciously. They should tend to exclude one another from the forests.

But not always. In 2004, ecologist Jeannine Cavender-Bares demonstrated that in fact, Darwin's account is almost the opposite of what we find in North American oaks. As a PhD student, Cavender-Bares was interested in figuring out how leaf life span evolved and shapes plant communities. She headed to North Central Florida to scout out possible plant groups for her work. She was quickly impressed at the sheer diversity of oaks. Cavender-Bares found seventeen oak species growing within forty miles of Gainesville: nine species from the Red Oak Group, five from the White Oak Group, and three from the Southern Live Oak Group. They differed in leaf life span—evergreen to deciduous—as well as growth rate, ability to resprout after fire, vessel diameter, and numerous other traits. They also arrayed themselves along a habitat gradient from dry, sandy soils where fires regularly burn to wet clay soils that rarely see fire. Oaks offered the diversity of lineages, species, and ecology needed to draw connections between evolutionary history, trait evolution, and the assembly of species into communities.

Cavender-Bares collected data on tree bark thickness and wood density; leaf life span, thickness, size, and chlorophyll content; transpiration of water and tree vulnerability to drought; and growth rates, height at maturity, and trees' ability to resprout from underground runners (rhizomes). Where possible, she replicated these measurements in a greenhouse experiment. She used DNA sequences to estimate the portions of the oak tree of life that included her seventeen species. Cavender-Bares also recorded the number and size of oaks growing together in each plot she surveyed and all the trees they grew with. Along with the plant species composition of each plot, she collected data on the environment: soil texture, moisture, nutrients, and how frequently fires burn. These data characterized where oaks grow and what combinations of oaks grow together.

Putting these data all together made it possible to test how traits and habitat preferences evolve on the oak tree of life, and how these in turn shape the communities of trees we find in the field. Cavender-Bares first showed that many attributes of the northern Floridian oaks are strongly influenced by evolutionary history. Variation in leaf density, transpiration rate, life span, and chlorophyll content, as well as wood density, are explained largely by whether a species is from the Red Oak, White Oak, or Southern Live Oak Group. These traits seem likely to drive phylogenetic niche conservatism, and it also seems like they could set oaks up to behave as Darwin predicted: close relatives should compete more strongly than distant relatives, making it more likely to find Red and White Oaks growing together.

But in fact, many of the attributes that most strongly influence who lives and dies in a given environment evolved convergently from different ancestors. Bark thickness and the ability to resprout from underground runners (rhizomes), which are adaptations to fire; trees' maximum height and the rate at which they grow thicker; the ability to endure drought without disruption to the water column running from the roots to the leaves; and seed mass all evolved independently in the White Oaks and the Red Oaks. These convergent characteristics drive distantly related oaks into communities together. And species that were similar in these convergent traits tended to be clustered within the Florida plots together. Thus the environment in which an oak had evolved explained some of its ecologically most important traits better than who its ancestors were.

In other words, Cavender-Bares showed that Darwin's hypothesis for why close relatives are not found close to each other was simply not correct for the Floridian oaks. Instead of close relatives pushing each other out of the forest through competition, distant relatives thrived in forests together because of convergently evolved traits that made them well suited to similar environments. Cavender-Bares and her colleagues later showed that the same basic dynamic between evolution and ecology was at play in more than 112,000 forest plots across the continental United States.

The error I made on the roadside outside Miquihuana points to one of the keys to high diversity in oaks: convergence filters oaks into communities together. You can see this across the country. In Arizona and New Mexico, Emory oak (*Q. emoryi*), a member of the Red Oak Group, grows beside Arizona oak (*Q. arizonica*), a species of the White Oak Group. These two species arose from lineages that each made their way northward in parallel along the mountains of Mexico. Both have small, tough leaves with scalloped margins. Both grow to similar heights beside scaly-barked alligator junipers. You can quickly learn to recognize the smoother bark and bumpier acorn cap scales that distinguish Arizona oak as a White Oak, but seeing them together is jarring if you are from the north: at first they look more similar to each other than a bur oak does to its closest relatives.

If you climb the Chiricahua Mountains in southern Arizona, Emory oak grows between 1,400 and 2,200 meters and is gradually replaced by silverleaf oak (*Q. hypoleucoides*), another Red Oak; silverleaf oak grows from 1,800 to 2,800 meters. These two species grow with different White Oaks at every elevation, so you can almost always find species from two clades adjacent to one other. At lower elevations in the Siskiyou Mountains of northern California and southern Oregon, you can find deer oak (*Q. sadleriana*) of sect. *Ponticae* growing and sometimes hybridizing with *Q. garryana* of the White Oak Group. Growing nearby, you might find California black oak (*Q. kelloggii*) of the Red Oak Group and canyon live oak (*Q. chrysolepis*) of the Intermediate Oak Group, along with two closely related oak genera unique to the West Coast: *Notholithocarpus*, a genus of one species that is sister to the oaks, and the spiny-husked American chinkapins (*Chrysolepis*—the genus, not the oak species). The most recent ancestor shared by these four oak sections and the two closely related genera with which they grow lived about 60 million years ago.

You can find a similar pattern of different oak clades growing together in Eurasia as well. In northeastern Spain, you can walk through forests dominated by Kermes and Holm oaks (*Q. coccifera* and *Q. ilex*) of the Holly Oak Group, cork oak (*Q. suber*) of the Cork Oak Group, and pubescent oak (*Q. pubescens*) of the White Oak Group. These four Mediterranean species trace their ancestry back to around 50 million years ago. A hike on the popular trails of Wuyi-shan and nearby forests of East China will take you past members of the Ring-Cupped Oak, Holly Oak, Cork Oak, and White Oak Groups and two closely related Beech Family genera, the stone oaks (*Lithocarpus*) and the Eurasian chinkapins (*Castanopsis*).

There are, of course, lonely oaks in the world. Bur oak extends farther onto the Great Plains than any other oak species. Pedunculate oak (*Q. robur*) grows nearly to the Arctic Circle. Roble Colombiano (*Q. humboldtii*) is the sole oak of numerous Colombian forest types, where it is used for fuel, tools, and fenceposts. Given oaks' high species diversity, wide ecological range, and capacity to thrive where many other trees languish, it is not surprising to find oak species growing in solitude. You will find the same in any plant genus. What is striking about oaks is that because convergent evolution enables distantly related species to assemble together into communities, the oak diversity we encounter on a single walk through many watersheds, forest preserves, or woodlots is a skeleton of the oak evolutionary diversity of the entire continent.

Oaks repeatedly evolved into complementary and adjacent niches on their way southward to Sumatra and Java in the Eastern Hemisphere and to Central America and Colombia in the Western Hemisphere. Convergence and its converse—trait divergence among close relatives—hone every trait that fits oaks to their environments: acorns, leaf shapes, bark thickness and

Oak gall on *Quercus sapotifolia*: *Amphibolips* cf. *oaxacae*. The same gall is shown from the outside (upper) and in cross-section (middle). The gall wasp larva is illustrated in the lowest panel. Oaxaca, Mexico, April 12, 2023.

wood structure, growth rates, the ability to resprout after fire, and a wide range of other attributes. Each clade is a semi-independent experiment in natural selection, differentiating from ancestors to fill broad ecological space and become the progenitors of lineages we recognize today.

Convergence helps explain why distant relatives favor similar habitats, but it doesn't fully explain the diversity of oak communities. Why do I enter a forest in my region and find red oak, white oak, and bur oak growing so easily together, when any one of them might take over the whole thing? Ecological trade-offs are part of the answer, as they help explain why oaks are habitat specialists. Cavender-Bares's work showed that the oaks of Florida choose between playing it safe or growing fast. The driest, most fire-prone sites are populated by communities of oaks with thick, fire-protective bark. These are plants that grow cautiously, investing their energy in defenses against drought and fire. Trees of the wettest sites grow more rapidly, attain the greatest height at maturity, and have relatively thin bark. They live fast and grow as quickly as they can. Because self-defense takes energy that might have been put into growth, the oaks of North Central Florida exhibit a trade-off between growth and stress-tolerance. This finding is essential to explaining habitat specificity in oaks: trade-offs keep any one species from dominating across the entire landscape.

Trade-offs are not unique to Florida. In the West Texas sky islands, for example, drought-tolerant species such as Emory oak (*Q. emoryi*), Gambel's oak (*Q. gambelii*), and gray oak (*Q. grisea*), which are subject to relatively frequent fires, invest early in thick bark to protect young plants. Trees of more mesic species—chinkapin oak (*Q. muehlenbergii*) of moist low-elevation sites and Chisos oak (*Q. gravesii*) of higher elevation canyons, for example—hold off putting on thick bark until they are older and seem to grow more quickly when they are young. One species seems to play the game equally well whatever the fire frequency: silverleaf oak (*Q. hypoleucoides*) has thin bark when it is young and a propensity to resprout if burned, but it can resist fire when it grows older and its bark grows thick.

Oak diversity may also be shaped by facilitation, if different oak lineages make the soils or other aspects of the environment more suitable for one another. In the mountains around Morelia, Mexico, a dozen oak species grow commonly. Six of these are from the White Oak Group and six are from the Red Oak Group. The White Oak species drop their leaves while they are still full of nutrients. The Red Oak species pull more of their nutrients back into the tree before their leaves fall. The litter beneath different tree species is, as a consequence, potentially suited for different soil bacteria, fungi, algae, and other microorganisms. Microbes living in the soil beneath *Q. castanea*, a Mexican species of the Red Oak Group, produce lots of enzymes for breaking down leaves, presumably because the litter is so nutrient-poor. Beneath *Q. deserticola*, a White Oak species that grows with *Q. castanea*, microbes can throw more of their energy into growing instead of into making such a quantity of digestive enzymes. These single-species effects percolate up to multispecies oak communities: leaf litter beneath forest plots dominated by the White Oak Group is particularly high in nitrogen, but soil richness is highest beneath mixed-species communities with both Red and White Oak Group species. While this research is young, it suggests that the nutrient-rich White Oak leaf litter may help set the stage for oak diversity in this system by feeding the soil microbial community.

Another possible explanation for the co-occurrence of distant relatives is that distantly related species are often susceptible to different diseases and pests. They are therefore less likely to pass diseases on to one another. Tropical biologists Daniel Janzen and Joseph Connell hypothesized in 1970 and 1971 that sharing of diseases and pests among close relatives could increase tree diversity within forests. Janzen and Connell independently observed that while tropical forest seeds tend to fall close to their mother tree, adult trees are relatively evenly distributed. Tree species are not as clumped as you might expect based on their patterns of seed dispersal. Janzen and Connell hypothesized that plant parasites, pests, pathogens, and predators that prefer one species over others may prey opportunistically on nearby plants of their favored species. The outcome would look like social distancing among individuals of a single species, and higher intermixing of different species.

Expand this idea to clades—suppose that pests or diseases are clade-specific rather than species-specific—and Janzen-Connell effects could help explain the co-occurrence of distantly related oaks. For instance, some pathogens favor one oak group over another. Oak wilt is a fungal disease that is often lethal to species in the Red Oak Group and, to a lesser extent, the Southern Live Oaks. It does not generally kill White Oak Group species, which are better able to compartmentalize it and thus prevent its spread. Bur oak blight, a disease caused by the fungus *Tubakia iowensis*, primarily afflicts bur oaks. At least two other species of *Tubakia* can afflict multiple White Oak Group species, while nearby northern red oak appears to be unaffected. Bur oak blight may spread all the more rapidly if spring in the Midwest continues to be cool and wet, as it has been in the past few years. If *Tubakia* shifted to a wide range of species of the White Oak Group, the impacts on the composition of eastern North American forests could be profound.

Insect herbivores also tend to favor one oak group over another. In a detailed field study at the Shields Oak Grove at University of California, Davis, ecologist Ian Pearse demonstrated that the single best predictor of how much herbivory nonnative oaks suffered was how closely related they were to the local California valley oak (*Q. lobata*). He measured a number of traits that should influence what species an insect prefers: leaf life span; the timing of leaf emergence; the thickness, toughness, or water content of leaves; and several defensive chemicals in the leaves. In addition, he used a molecular fingerprinting method to estimate the oak tree of life. Remarkably, the evolutionary relatedness of species was more useful for predicting leaf damage by insects than any combination of leaf traits. Much of this effect was due simply to whether the oak host was in the White Oak Group (like *Q. lobata*) or not. In a companion study in botanical gardens distributed across the United States, Pearse found that the diversity and relatedness of native oaks to the nonnative species planted at a site predicted how much insect damage the nonnative species suffered. It appears that chewing and mining insects are influenced by oak relatedness: many are able to switch species, but they seek out close relatives, cueing in on attributes we may not even detect. Pearse's work does not prove that the Janzen-Connell mechanism shapes oak communities, but it does suggest that shared insect herbivores may help shape oak biodiversity site by site.

Distantly related oaks may also coexist by separating themselves in time instead of space. The acorns of species in the White Oak Group ripen in the same year the flowers that produced them are pollinated. They also germinate immediately. Many species of the Red Oak Group are slower: their acorns ripen the year after pollination, then they wait through a period of dormancy to germinate. Thus the pollinations of a given spring determine what trees may germinate in fall of the same year for most species in the White Oak Group but in spring two years later for many in the Red Oak Group. Spreading the wealth across years gives White Oaks a head start over Red Oaks in some years and Red Oaks the lead in others. If survival and competition among seedlings is an important part of oak coexistence, this balance may help distantly related oaks share the same communities.

There are exceptions to these coexistence patterns, but even some of the exceptions appear to highlight the ecological complementarity of oak species. In California and Arizona, for example, it is more common to find pairs of oak species that both have annual acorns or both have biennial acorns than pairs that differ in this aspect of reproductive timing. But in these same communities, the dominant oaks often include at least one species with evergreen leaves and one with deciduous leaves, which may be an equally effective way of divvying up resources. And in the New Jersey Pine Barrens, the common bear oak (*Q. ilicifolia*), a member of the Red Oak Group, often co-occurs with black oak (*Q. velutina*) or blackjack oak (*Q. marilandica*), which are also members of the Red Oak Group. Bear oak is, however, one of only two shrubby oak species in the northeastern United States, and it may be that its short stature facilitates co-occurrence with closely related tree species.

Oak evolutionary relationships and adaptation on the tree of life shape traits and ecological interactions. These interactions then sort oak species into communities, influencing who grows with whom and the natural selection that will shape the evolution of oak biodiversity in the future. Oak communities, all the species that live in those communities, and evolution of species on the Tree of Life share responsibility for the biodiversity we experience on a walk through the forest. No single process is in charge.

More than 1,000 insect species are known to feed on oaks. This is certainly an underestimate, and many more insects likely depend on oaks for some part of their livelihood. In one study, 541 insect species were collected from a single red oak seed orchard. Among native plants, no one beats oaks for diversity of leaf-eating caterpillars. Mites make their homes in the patches of hairs in axils of the leaf veins on the undersurfaces of California coast live oak (*Q. agrifolia*) leaves. The mites are predatory and may protect the leaf from herbivores in return for the protection they receive from the tree. The communities of insects, fungi, and other organisms that depend on oaks make every oak tree—every leaf and acorn, in fact—an ecosystem.

Larvae of weevils, moths, and midges often colonize developing acorns and feed together on the nutrient-rich cotyledons. The only sign we typically see of this community-within-the-forest is a hole drilled in the side of the nut, the trail of a weevil that has eaten its fill and then escaped. Often ants join the party after the fact. *Temnothorax* ants enter acorns through holes left by weevils that chewed their way out. The ants sometimes pack the edges of the entrance hole with grains of sand so that only they and their kin can pass through. They hunt for food on the forest floor and carry it back to the acorn. They lay eggs in the acorn and tend their young through all seasons, transporting their brood upward and downward to moderate the

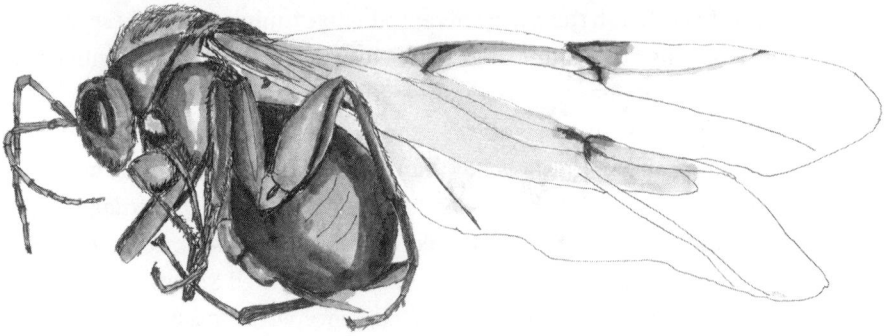

Gall wasp: *Disholcaspis quercusmamma*, emerged from a gall on *Quercus macro-carpa*. Collected along the Delaware and Hudson Rail Trail, southwestern Vermont, October 11, 2022. Illustrated from a photograph provided by Leah Samuels.

babies' temperatures as needed, just as other ant species do within an ant mound. They move their young away from mold and microbes that colonize the bottom of the acorn shell. The nutshell serves as nursery and home.

Fungi colonize acorns as well. Often an acorn's first endophytic fungi—*endo* for inside, *phytic* for plant—are a gift passed on from the flowers or leaves of the mother tree. Effectively, the mother tree infects her own offspring, creating a rich community of fungi in the developing cotyledons. Some of these endophytic fungi that have been living inside the mother tree's leaves are particular about which oak species they colonize, so the diversity of oak species in the forest shapes the fungal diversity of the forest canopy. Other fungi enter the acorn only after it falls, when they are transmitted from the leaf litter. Endophytes that are particular to oaks have more highly evolved tolerance to the defensive chemicals in oak leaves—phenolics—than non–oak specialist fungi do, and they may play a role in breaking down oak leaf litter. The fungi in the fallen leaves are thus as important to the forest as the fungi in the still-growing leaves.

As acorns and leaves decompose, nutrients crawl away inside mature insects or seep into the soil. There, they drain past or feed mycorrhizal fungi—amanites, boletes, chanterelles, and many others—that grow in intimate connection with oak roots. Mycorrhizal fungi pull carbon from the oaks in the form of sugars that the trees produce through photosynthesis. In exchange, the fungi forage broadly, acting as an extension to the trees' root systems and passing along phosphorus, nitrogen, other minerals, and water. Mycorrhizal fungi have been working with plants for about 450 million years, long before flowers evolved. Many mycorrhizal fungal species are generalists, shared among oaks and maples, elms, crabapples, even grasses and sedges. Shared mycorrhizal connections may help shunt resources between plants, though their importance in this respect compared to root grafts and the flow of nutrients through water moving in the soil is not known. Not all fungi are shared, however. Some mycorrhizal fungi prefer the roots of one oak species in the forest, or they may thrive in the chemical brew of the leaf litter dropped by a particular oak species or group of species. The outcome is that the oak tree of life shapes the fungal community belowground perhaps just as much as it shapes the biodiversity of the forest above ground: the community of trees, insects, birds, mammals, shrubs, mosses, fungi, understory herbs, and everything else that makes the forest.

Oaks have evolved to work with mycorrhizal fungi. When fungal species colonize the roots of pedunculate oak (*Q. robur*), the oak evaluates whether the fungi are likely to be mycorrhizal and thus most likely beneficial. If it deems that they are, the oak activates genes that build cell walls; these genes help produce roots that can take advantage of the fungi. At the same time, the oak dampens the activity of genes that produce proteins to defend the tree against pathogens. In so doing, the tree enables the mycorrhizal fungi to envelop the root without fear of attack by the tree.

Individual interactions between oaks and their fungi scale up to mold the species diversity of the whole forest. In one particularly massive study, researchers collected seeds of fifty-five different tree species from 550 forests across the northeastern and northwestern United States. They planted the seeds into pots in a greenhouse, incorporating soil or roots collected from beneath trees of their own species or a different randomly selected species. Mixing soils and roots from different trees into the potting soil seeded the experimental pots with fungi and other microorganisms. Tree species such as oaks and pines, which associate with ectomycorrhizal fungi—mycorrhizal fungi that form a coat or mantle around the root—grew better when they were planted into soil inoculated with roots or soil of their own species. In other words, parents sharing fungi with their offspring may give oaks and their relatives (as well as pines and their relatives) an advantage. But close relatives who are not the same species might have the opposite effect: in a separate greenhouse study on two pairs of sister oak species, growing in live soil of one's own species improved seedling growth, while growing in the soil of the sister species decreased seedling growth and survival. While more research is needed to validate how general these findings are, particularly for

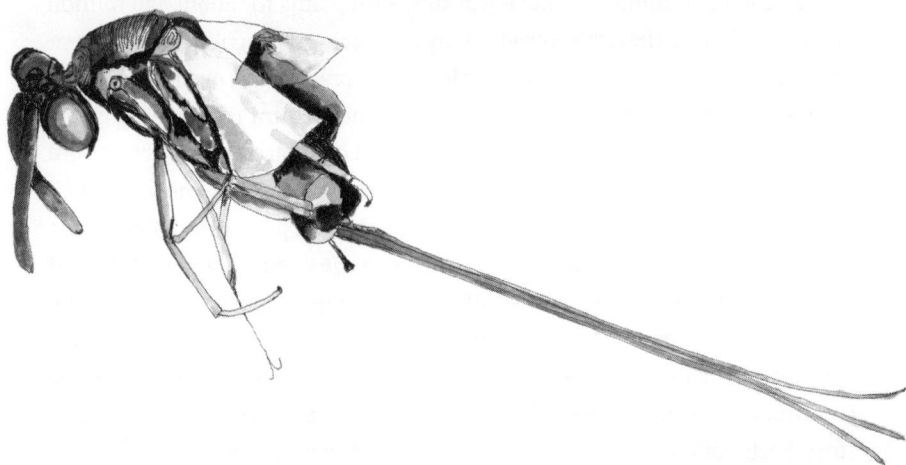

Gall wasp parasitoid: *Torymus* sp., emerged from *Disholcaspis quercusmamma* gall on *Q. alba*. Collected along the Delaware and Hudson Rail Trail, southwestern Vermont, October 11, 2022. Illustrated from a photograph provided by Leah Samuels.

the sister-species study, these two studies in combination suggest that fungi help diversify oak communities.

Oak gall wasps, Cynipids, number approximately 1,400 species world-wide. Seventy percent of Cynipids live exclusively on oaks; the others live on other Beech Family genera (e.g., *Castanea*, *Castanopsis*, *Chrysolepis*, or *Notholithocarpus*), southern beeches (*Nothofagus*), or genera outside the Fagaceae and Nothofagaceae. Gall wasps—like all kinds of gall-forming insects, which include sawflies, midges, and thrips—lay their eggs inside plant tissues. Chemicals on the surface of the egg cause the plant to grow oddly, manipulating how the tree expresses its own genes. The result is a gall built at the insect's behest. Galls may look like bullets, pom-poms, clusters of berries, pimples, ceramic beads, ping-pong balls, volcanoes, kernels of corn, bouquets of leaves, blisters, fuzzy donuts, or just thickenings of the stem. Galls are nurseries, siphoning resources from growing plant tissues to feed the larva developing inside. Oak galls are often enriched in tannic acids to protect the larva from fungi; in fact, these concentrated tannins were a central ingredient of ink used from the Middle Ages through the nineteenth century.

Galls have been forming on plants of various kinds for at least 385 million years. Cynipid wasps evolved 75–45 million years ago, if not earlier. If the older estimate is correct, Cynipids specialized on other Fagaceae before the oaks evolved. They really took off, though, when oaks started to diversify. Most gall wasps live exclusively on one oak species or clade, and gall wasp speciation—at least in the Americas—is often associated with transitions in which species or plant organ is used by the larvae. Thus, as a gall wasp population makes exploratory shifts from acorns to leaves, from stems to roots, or from species to species, its descendants can become isolated from the other populations of its species. These new populations may become re-productively separated and ecologically and genetically distinct enough to form new species. In the Americas, gall wasps of the Red Oak, Intermediate Oak, and White Oak Groups are mostly distinct from one another, though some gall wasp species or genera may grow on oaks from multiple groups. In Eurasia, gall wasps that live on the Cork Oak Group are distinct from gall wasps that live on the Holly Oak Group, and these two are distinct from the gall wasps of the White Oak Group. And within the Eurasian White Oaks, a group of gall wasps arose that uses not one but two oak clades: the Cork Oak Group and White Oak Group are alternate hosts for at least two gall wasp genera, who depend on having members of both lineages growing in close proximity to complete their life cycle.

Gall wasps, in turn, provide food for parasites, who attack the growing larvae. Woodpeckers and other birds also eat gall wasps, and some fungi live on the galls themselves. Galls also sometimes serve as homes for late-coming wasps, who burrow in and either curl up to grow in an unused corner or kick out the larva who was there first. The gall wasps protect themselves as best they can from these parasites. Some of the galls produce nectar to attract ants, which may in turn prey on insects who would eat the galls or oak leaves. At least one genus of gall wasps—the genus *Kokkocynips* of Mexico and eastern North America—produces galls with a nutritious lump on the side, full of fatty acids resembling the insect blood that ants adore. When these galls tumble from the tree, ants are as quick to pick them up and carry them home as they do the similarly equipped seeds of bloodroot, wild ginger, trillium, and many other forest understory herbs. The ants feed on the gob of food and, inadvertently, protect the wasp growing inside the gall, in whom they have no particular interest.

Cynipid wasps depend on oaks, and in turn the oak tree of life struc-tures the cynipid wasp tree of life. A network of organisms assembles around the galls. In the same way, oaks and fungi, oaks and squirrels, and oaks and the leaf-tying insects that chew their leaves and make a shelter for additional insects to chew with them are a network of intersecting evolutionary histories.

What is the importance of oak species? Wouldn't it be enough to have just as many oak individuals as we have today, but all of one species? In some ways, perhaps. The diversity of butterflies, bees, wasps, and understory plants in some forests is more strongly affected by oak abundance than by the total tree diversity of the surrounding forests. For those organisms, the number of oak species may be less important than the number or density of oak individuals in the forest. Oaks provide an estimated $22 billion in net value to the United States each year through their effects on cooling the climate and improving air quality. That's another effect of having so many oak individuals in the world, irrespective of diversity. Oaks' effects on the world are due in part to their sheer number and cumulative biomass.

But abundance and mass are only part of the story. Across the Tree of Life, diversity begets ecological impact. Forests with a greater diversity of tree species pull more carbon from the atmosphere and lock it up in their roots, in the soil, and in the plant material other organisms depend on for food and shelter. Tree diversity in forested communities promotes decomposition and the accumulation of litter needed by invertebrate, fungal, and microbial species. Each of the eight sections of *Quercus* bears a unique evolutionary legacy, which gives each clade a unique role in the world. Red Oaks and White Oaks have been evolving together in North America since the Eocene, working out a way to pack more trees onto the continent. Cork Oaks, Holly Oaks, and White Oaks united in western Europe in the Miocene, divvying up and sharing communities as they filled the continent. If a forest loses an oak species, the other oaks in the community won't simply expand to fill all the roles that species played. The combined effects of biomass, species diversity, and evolutionary diversity give oaks an oversized impact in the ecosystems they occupy.

Oak species also matter to the ongoing evolution of the genus itself. Species serve as experimental grounds for alleles. New alleles arise within a population. Those alleles are then tested against the environmental and biological conditions unique to a particular set of forests, woodlands, savannas, hillsides of chaparral, hammocks, or other communities. Some forests select for trees that can tolerate cold winters and strong fires, others for trees that can resist a particular suite of pathogens. The resulting progeny then disperse to places where natural selection acts on them in new ways, producing trees and populations that may one day become new species or extend the reach of an existing species. Each forest, by favoring some individuals over others, tweaks the gene pool. The species' niche evolves as pollen and acorns travel among communities and the resulting offspring survive or perish.

Occasionally—oaks being oaks—alleles escape from one species into another. Introgressed alleles are tested anew in the genomic and environmental context of the species in which they find themselves. More often than not, they are weeded out, ill-suited to their new environment and genomic context. Alternatively, alleles may persist after introgression if they are lucky or natural selection favors them in their new species. Thus the balance between variation, introgression, and the particularity of each species' niche forms what geneticist Sewall Wright described as the "trial and error" that allows partially isolated populations to explore adaptive landscapes.

Biotic communities, with their component species, form an interconnected network of evolutionary laboratories. Oaks give us unique insights into this network and into the evolutionary and ecological history of the Northern Hemisphere. Every oak is a leaf on the Tree of Life. It sprouts from a clade that invaded the continents more than 50 million years ago. It connects by way of one lineage after another to oaks across the world from you, and to oaks right around the corner. Every oak is also part of a species, the outcome of adaptation, migration, and hybridization both ancient and ongoing. It is a player in the evolution of its population and every population intersecting in the biotic community in which it grows. And every oak grows, reproduces, and competes in a biotic community. It fights exotic diseases and insects. A succession of soil biotic communities develops in the gaps between its roots, shaping and shaped by the plants and animals that live with the oak. It makes a home for a multitude of other organisms.

A biotic community is both a snapshot in time and an evolutionary winnow. The lives of the oaks transpire in these communities. Species originate here, and the Tree of Life originates here. Both in turn form the oak communities we know.

The Future of Oaks

Almost every lineage will go extinct, eventually. The diversity of humans who coexisted during the Pleistocene—*Homo erectus*, the oldest unambiguous member of our genus; *H. floresiensis*, an island-bound human who grew smaller over time and lived for perhaps 150,000 years before going extinct; the Neanderthals and Denisovans, with whom our own species, *H. sapiens*, both competed and mated as recently as 86,000–37,000 years ago—reminds us that we are the sole survivors of our genus. The extinctions of nonavian dinosaurs, diverse forms of mammals, and passenger pigeons show that entire lineages can be lost in a geological instant. Likewise, oaks won't always reign in the northern temperate forests.

How long will oaks survive? That's hard to say. But given how long they have been around—at least 56 million years—and how little time modern humans have been around—about 200,000 years—we might suppose that oaks as a genus will outlive the human species. Let's assume that we early twenty-first-century humans are observing a time drawn at random from the life span of modern humans and oaks. We'll assume, in other words, that we are no more or less likely than any other observers at any other time in Earth's history to be witnessing the adolescence or senescence of our species and of the oaks. If this assumption is correct, we can be 95% certain that we aren't seeing the first 2.5% or last 2.5% of the life of either our own species or the lineage of oaks. By this reasoning, we might give humans somewhere between 5,100 and 7.8 million more years. Oaks as a genus should have somewhere between 1.2 million and 2.1 billion years.

This kind of conjecture doesn't help me sleep all that well. Take the assumption that we are observing oaks at a point drawn at random from the life span of the genus *Quercus*. This doesn't seem likely, given what we have been doing to the Earth, especially since the Industrial Revolution. The spike in temperatures in the Paleocene-Eocene Thermal Maximum (PETM) was one of the most rapid temperature changes our world has ever seen. The PETM may not have caused a lot of terrestrial plant extinction, but it would have looked catastrophic to us had we been living through it. It shuffled species distributions and reconfigured ecosystems. The moment we're living in now is equally exceptional. Humans are driving atmospheric CO_2 upward at around ten times the rate of increase at the onset of the PETM. We may be rolling back tens of millions of years of cooling climates that favored oak diversification. Under one particularly dire scenario for greenhouse gas emissions, which assumes we make little effort to change the status quo, most of the modern range of oak species could reach the temperature of the early Eocene by the year 2200. In the short term, oaks may benefit, at least in some places: overall oak abundance could rise in the Upper Great Lakes and New England in the coming decades under expected climate change trends, due in large part to northward shifts in more southern species. Some of our most widespread species, however—northern red oaks, bur oaks, pin oaks—are projected to contract to smaller ranges and lower overall importance in our forests. Several oak species risk being extirpated from much of the southern United States by the end of the twenty-first century.

What, then, should we expect to see? How many oaks should we expect will go extinct in our lifetime? How many would go extinct if it were not for humans? The average rate of species extinction is hard to estimate: fossil data probably underestimate species diversity, and it can be difficult to distinguish the extinction of a fossil species from evolutionary change within that species. Phylogenetic inferences also present challenges for estimating extinction rates, but they give us a place to start. Using data from across the Tree of Life, one estimate for extinction rates over time—the background extinction rate, the one we should compare with today's rate of extinction, in human hands—ranges from 2.3 to 13.5 extinctions per million species per century. If we simply projected the worst-case estimate of the background extinction rate onto today's 425 or so oak species, we might expect one oak extinction every 17,400 years or so before humans got into the mix.

The latest conservation assessment of oaks suggests that 31% of the world's oaks are threatened with extinction. The estimated rate of extinction for plants worldwide in the 270 years since Linnaeus wrote *Species Plantarum* ranges anywhere from 50 to 1,000 times the background extinction rate for plants, and current rates may be as high as 1,000 to 10,000 times the background extinction rate. Pathogens moved around by humans are driving the demise of some of our most abundant tree species. We have witnessed, in the span of a human lifetime, the loss of American chestnut as an important component of eastern North American forests. A forest in my town that had abundant, large ashes fifteen years ago is now littered with ash corpses. Butternut is listed as endangered, and eastern hemlock as near-threatened. Each of these is a consequence of humans shuttling pathogens around the globe, and no amount of good hygiene is going to prevent every potential tree disease from spreading.

We are all witnesses to the decline of oak populations and the near-cessation of oak reproduction in a range of ecosystems across the Northern Hemisphere. Even our most widespread species are suffering. Populations of California valley oak (*Q. lobata*) have been declining for more than seventy years, due in part to cattle, deer, and rodents browsing and digging up seedlings. Seedlings of Garry oak (*Q. garryana*), a keystone species of savannas from British Columbia to California, are decimated in some areas of British Columbia by hungry European rabbits and competitive exotic grasses. Across eastern North America, eastern white oak (*Q. alba*) was one of the dominant species prior to European settlement, accounting in many forests for anywhere from 17% to 36% of the forest composition in the Northeast and a frequency in plots as high as 80% in forests of southern Illinois. Since European colonization, however, fire suppression has favored maples and other shade-tolerant species. White oak is barely replacing itself across large portions of its range.

It seems that the early twenty-first century is a special time indeed, both for us and for oaks. I don't expect oaks as a genus to go anywhere very fast, but we have good reason to expect that if we don't make changes in how we live, we may drive at least a few oak species extinct in the coming decades.

The history of oaks illustrates that species are begotten of extinction. Oaks have been the recipients of a long string of good fortune, survivors despite more than 50 million years of dramatic climate change. Born on the cusp of a spike in temperatures, then sweeping the continents as the tropical forests were extirpated from much of the Northern Hemisphere, they have survived by being in the right place at the right time, with the right set of tools. Oaks migrated rapidly as the glaciers receded, swapped genes with other survivors, and repeatedly solved the problems of freezing temperatures, drought, fire, and inundation. If we had been there to see them trundling across North America and Eurasia during the Eocene, we might have tried to stop them. The oaks were an exotic clade at one time.

Oaks are protean, transforming as they cover the landscape. They adapt to changes in climate and habitat. They vary from one population to the next. But they are not invincible. Not even Proteus was boundlessly protean. When Menelaus got hold of him, Proteus turned into a bearded lion, then a snake, a leopard, a boar, rushing water, an enormous tree. Menelaus held him tight until finally, exhausted, Proteus turned back to his original form and allowed Menelaus to question him.

Tree species shift and turn, adapt to almost everything the environment can throw at them. They track climates over time, evolving as they migrate. Then, sometimes without warning, they run up against a wall and go extinct. There are boundaries to adaptation. Hold Proteus long enough, and he stops changing form. Lean hard on an oak population or species, and it may not bend far enough to survive. Oaks as a lineage may survive longer than the human species, but many individual species will go down before us if we don't take action.

The good news is that we still have time. Admittedly, some ground has been lost for good: more than three hundred seed plant species are estimated to have gone extinct worldwide since the mid-1700s, fifty of these in the United States and Canada alone. We have driven greenhouse gas levels high enough that there is no fully rolling back climate change. But we know enough to slow climate change and undo some of what we've done to our planet. Research on oak propagation is making it possible to bank oak genes beyond the longevity of individual acorns, which cannot be stored in conventional seed banks. Botanical gardens are safeguarding genetic diversity of at-risk species by planting living conservation collections, providing the seeds needed to establish species beyond their current ranges when dispersal and adaptation cannot keep up. Scientists, restorationists, local land managers, and others are teaming up in sites across the Americas to address the reproduction crisis in *Q. brandegeei* (Baja California), *Q. insignis* (Costa Rica), *Q. garryana* (the Northwest), *Q. alba* (the eastern United States), and other oak species. Above all, people recognize that trees matter, and that there are concrete actions all of us can take.

In addition to all this, oaks will persist in part because of one of their superpowers, the one that at once impresses and vexes us: hybridization and gene flow. Oaks do not have the capacity to think ahead and make plans for the future. But as an oak species becomes less common, the probability of

Fox squirrel in snow, retrieving acorns. The Morton Arboretum, February 15, 2021.

it producing acorns that are of hybrid origin tends to increase. The resulting gene flow may be more of a lifeline than a death knell. Gene flow may provide needed genetic diversity to a species with dwindling populations. And alleles that are particularly well adapted in the rarer species may contribute to the success of the more widespread species. Neanderthals and Denisovans live in our genomes. Some oak species will go extinct but live on, in part, within others.

The oak tree of life bristles with dead-end branches. Each marks the death of a species. Each such loss scars the world. When you look more closely, though, you will find strands of gossamer trailing between the branches, genes moving between lineages. These are the future of a species beyond its last individual. Lineages are lost to the world over and over. These strands moving between species give me hope, nonetheless, that few species are ever lost utterly. Every individual impacts the world permanently, far beyond the boundaries of the forest where it was born.

Acknowledgments

This book grew from my interests in natural history, evolution, and stories. It would not exist without support, friendship, and mentorship of many who fostered these interests early on: my parents, Maureen Laustsen and Mark Hipp, and my grandparents, Ruth and Laurence Hipp; Devin Biggs, who first got me interested in plants; Dave Egan, Molly Fifield-Murray, Bill Jordan, Sylvia Marek, Ken Wood, Friends of Little Bluestem, and other friends and colleagues at the University of Wisconsin–Madison Arboretum; and teachers who impacted my thinking about evolution and plant biology as I was shifting into this field, including David Baum, Paul Berry, Ray Evert, Tom Givnish, Ken Sytsma, and Don Waller.

This book covers many fields beyond my areas of immediate research, and it has grown and benefited from conversations in the field, in hallways, at conferences, at lab meetings, in the classroom and lab, all interspersed between hours of reading and research. It would be hard to list everyone who has made a mark on this book, but I have learned much of what I know about oaks and evolutionary biology from research, conversations, and fieldwork with Stephanie Adams, Kieran Althaus, Silvia Alvarez Clare, Tricia Bethke, Adam Black, Catherine Bodénès, Johannes Bouchal, Marlin Bowles, Christian Burban, Roderick Cameron, Chuck Cannon, John Carlson, Antonio Castilla, Arturo Castro, Jeannine Cavender-Bares, Béatrice Chassé, Allen Coombes, Richard Cronn, Andy Crowl, Thomas Denk, Sara Desmond, Alexis Ducousso, Kurt Dreisilker, Michael Eason, Deren Eaton, Elizabeth Fitzek, Ryan Fuller, Oliver Gailing, Elliot Gardner, Mira Garner, M. Soccorro González-Elizondo, Jesús G. González-Gallegos, Antonio González-Rodríguez, Guido Grimm, Paul Gugger, Erwan Guichoux, Marlene Hahn, James Hitz, Sean Hoban, David Jablonski, Richard Jensen, Matthew Kaproth, Peter Kennedy, Tom Kimmerer, Wes Knapp, Austin Koontz, Antoine Kremer, Chai-Shian Kua, Desanka Lazic, Emma Leavens, Thibault Leroy, Xiaojuan Liu, Bela Loza, Paul Manos, Rúben Martin Sánchez, Heather McCarthy, John McVay, Meghan Midgley, Jake Miesbauer, Rebekah Mohn, Jorge Noriega, Ian Pearse, Rémy Petit, Kasey Pham, Christophe Plomion, Trevor Price, Rick Ree, Gabe Ribicoff, Senna Robeson, Maricela Rodriguez, Hernando Rodríguez-Correa, Christy Rollinson, Jeanne Romero-Severson, Leah Samuels, Lucy Schroeder, Dave Shepard, Marco Simeone, Victoria Sork, Emma Spence, Guy Sternberg, Nick Stoynoff, Jay Sturner, Susana Valencia-A., Baosheng Wang, Gary Watson, Jaime Weber, Murphy Westwood, Alan Whittemore, Lindsey Worcester, Yington "Amanda" Wu, Sofia Zorrilla Azcué, friends and colleagues of the International Oak Society, my lab group—Team Herbarium and the Systematics Lab of the Morton Arboretum—and friends and colleagues at University of Wisconsin–Madison Department of Botany, the Morton Arboretum, the Field Museum, and University of Chicago's Committee on Evolutionary Biology. I apologize if I have missed anyone. This book would be a lesser work had all these friends and colleagues not shared with me their love of and insights into evolution and plants, particularly oaks.

I am grateful to colleagues, friends, researchers, oak experts, and excellent readers who read parts or all of this book at one time or another, often more than once, and did their best to disabuse me of misunderstandings: Tali Babila, Jérôme Bartholomé, Catherine Bodénès, Louise Bodt, Johannes Bouchal, Chuck Cannon, Jeannine Cavender-Bares, Min Deng, Thomas Denk, Guido Grimm, Al Keuter, Antoine Kremer, Thibault Leroy, James Mallet, Paul Manos, Luke McCormack, Ian Pearse, Lily Peck, Kasey Pham, Trevor Price, Dylan Schwilk, José Ramírez-Valiente, Victoria Sork and her lab group, Michael Steele, Fernando Villanea, Jack Williams, and Team Herbarium and the Systematics Lab of the Morton Arboretum. Juli Pujade identified the gallwasp genus on *Q. sapotifolia* (chapter 7). Roderick Cameron, Rachel Davis, Maureen Laustsen, and Joshua Moses read early drafts of several chapters and helped in thinking through voice and scope. Joseph Calamia, Béatrice Chassé, Tracy Honn, Richard Powers, Alan Whittemore, and two anonymous reviewers read complete drafts of the manuscript and provided insights and improvements throughout. All have improved the book in important ways. Any errors, misstatements, or confusing passages are nonetheless entirely mine.

My editor Joseph Calamia has been enthusiastic and supportive throughout the project; I have enjoyed talking ideas through with him and feel fortunate to have been able to work with him. Nicole Balant's edits and insights have improved the manuscript greatly, and her enthusiasm for the text and missives from Midcoast Maine came at just the right time. Tamara Ghattas and the entire editorial and production team at University of Chicago Press have been excellent to work with at every stage of the process. I am grateful to them for bringing this book to light.

This book was written independent of my position at the Morton Arboretum, but it would not have been possible without the Arboretum's many years of support for my research and their mission to plant, protect, and understand trees. Insights discussed within this book have been gained in part through research funded by the U.S. National Science Foundation (Awards 114648, 2129281), a Fulbright Scholar Award through the Franco-American Commission (2014, INRAE-BioGeCo and Université de Bordeaux), a Fulbright Specialist Award (2023, Georg-August-Universität Göttingen), the Michigan Botanical Club Hanes Fund, the American Philosophical Society Research Grant Program, the U.S. National Arboretum, the United States Botanic Garden, and the Center for Tree Science of the Morton Arboretum.

Above all, I am grateful to my wife, Rachel Davis, and our sons, David and Louis Hipp. They have been supportive, patient, and good-humored when this project was straightforward and when it was not, and honest about what they found most interesting. I have been grateful for Rachel's insights and wisdom at every turn. Her view into oaks and oak communities as she illustrated the book breathed new life into the chapters and helped knit them together. Being able to share the natural world in this way, looking at it together from different angles and somehow making of the parts a shared whole, is one of my greatest joys.

Oak Names

Scientific Names to Common Names

Scientific names precede the colon, one of the common names follows; names in parentheses are the clade to which the named species, genus, or family belongs.

Castanea: Chestnut (Fagaceae)

Castanopsis: Eurasian chinkapin (Fagaceae)

Chrysolepis: American chinkapin (Fagaceae)

Fagaceae: Beech Family (Fagales)

Fagales: Beech Order (Rosids)

Fagus: beech (Fagaceae)

Juglandaceae: Walnut Family (Fagales)

Juglans nigra: black walnut (*Juglans*)

Nothofagaceae: Southern Beech Family (Fagales)

Notholithocarpus densiflorus: Tanoak (Fagaceae)

Quercus acutissima: sawtooth oak (sect. *Cerris*)

Quercus agrifolia: coast live oak (sect. *Lobatae*)

Quercus alba: eastern white oak (sect. *Quercus*)

Quercus arizonica: Arizona oak (sect. *Quercus*)

Quercus berberidifolia: California scrub oak (sect. *Quercus*)

Quercus bicolor: swamp white oak (sect. *Quercus*)

Quercus cerris: Turkish oak (sect. *Cerris*)

Quercus chrysolepis: canyon live oak (sect. *Protobalanus*)

Quercus coccifera: Kermes oak (sect. *Ilex*)

Quercus coccinea: scarlet oak (sect. *Lobatae*)

Quercus dumosa: coastal sage scrub oak (sect. *Quercus*)

Quercus ellipsoidalis: Hill's oak (sect. *Lobatae*)

Quercus emoryi: Emory oak (sect. *Lobatae*)

Quercus engelmannii: Engelmann's oak (sect. *Quercus*)

Quercus gambelii: Gambel's oak (sect. *Quercus*)

Quercus garryana: Garry oak (sect. *Quercus*)

Quercus gilva: Ichiigashi, red-bark oak (sect. *Cyclobalanopsis*)

Quercus gravesii: Chisos oak (sect. *Lobatae*)

Quercus grisea: gray oak (sect. *Quercus*)

Quercus humboldtii: Roble colombiano, or simply "roble," as it is the only oak in South America (sect. *Lobatae*)

Quercus hypoleucoides: silverleaf oak (sect. *Lobatae*)

Quercus ilex: Holm oak (sect. *Ilex*)

Quercus ilicifolia: bear oak (sect. *Lobatae*)

Quercus insignis (sect. *Quercus*)

Quercus kelloggii: California black oak (sect. *Lobatae*)

Quercus lobata: California valley oak (sect. *Quercus*)

Quercus lyrata: overcup oak (sect. *Quercus*)

Quercus macrocarpa: bur oak (sect. *Quercus*)

Quercus marilandica: blackjack oak (sect. *Lobatae*)

Quercus michauxii: swamp chestnut oak (sect. *Quercus*)

Quercus minima: dwarf live oak (sect. *Virentes*)

Quercus mongolica: Mongolian white oak (sect. *Quercus*)

Quercus montana: chestnut oak (sect. *Quercus*)

Quercus muehlenbergii: chinkapin oak (sect. *Quercus*)

Quercus oleoides (sect. *Virentes*)

Quercus palmeri: Palmer's Oak (sect. *Protobalanus*)

Quercus parvula var. *shrevei*: Shreve oak (sect. *Lobatae*)

Quercus petraea: sessile oak (sect. *Quercus*)

Quercus phellos: willow oak (sect. *Lobatae*)

Quercus pontica: Armenian oak (sect. *Ponticae*)

Quercus pubescens: pubescent oak (sect. *Quercus*)

Quercus pyrenaica: Pyrenean oak (sect. *Quercus*)

Quercus robur: pedunculate oak (sect. *Quercus*)

Quercus rubra: northern red oak (sect. *Lobatae*)

Quercus rugosa: netleaf oak (sect. *Quercus*)

Quercus sadleriana: deer oak (sect. *Ponticae*)

Quercus serrata: jolcham oak (sect. *Quercus*)

Quercus skinneri (sect. *Lobatae*)

Quercus stellata: post oak (sect. *Quercus*)

Quercus suber: cork oak (sect. *Cerris*)

Quercus tardifolia: lateleaf oak (sect. *Lobatae*)

Quercus variabilis: Chinese cork oak (sect. *Cerris*)

Quercus velutina: black oak (sect. *Lobatae*)

Quercus wislizeni: interior live oak (sect. *Lobatae*)

The Roburoid oaks: Eurasian white oak clade (sect. *Quercus*)

section Cerris: Cork Oak Group, Cork Oaks (subg. *Cerris*)

section Cyclobalanopsis: Ring-Cupped Oak Group, Ring-Cupped Oaks (subg. *Cerris*)

section Ilex: Holly Oak or Holm Oak Group, Holly Oaks or Holm Oaks (subg. *Cerris*)

section Lobatae: Red Oak Group, Red Oaks (subg. *Quercus*)

section Ponticae: Deer Oak Group, Deer Oaks (subg. *Quercus*)

section Protobalanus: Golden-Cup or Intermediate Oaks Group, Golden-Cup or Intermediate Oaks (subg. *Quercus*)

section Quercus: White Oak Group, White Oaks (subg. *Quercus*)

section Virentes: Southern Live Oak Group, Southern Live Oaks (subg. *Quercus*)

subgenus Cerris: Eurasian Oak Group, Eurasian Oaks (*Quercus*)

subgenus Quercus: American Oak Group, American Oaks (*Quercus*)

Common Names to Scientific Names

American chinkapin: *Chrysolepis* (Fagaceae)

American Oak Group, American Oaks: subgenus *Quercus* (*Quercus*)

Arizona oak: *Quercus arizonica* (sect. *Quercus*)

Armenian oak: *Quercus pontica* (sect. *Ponticae*)

bear oak: *Quercus ilicifolia* (sect. *Lobatae*)

beech: *Fagus* (Fagaceae)

Beech Family: Fagaceae (Fagales)

Beech Order: Fagales (Rosids)

blackjack oak: *Quercus marilandica* (sect. *Lobatae*)

black oak: *Quercus velutina* (sect. *Lobatae*)

black walnut: *Juglans nigra* (*Juglans*)

bur oak: *Quercus macrocarpa* (sect. *Quercus*)

California black oak: *Quercus kelloggii* (sect. *Lobatae*)

California scrub oak: *Quercus berberidifolia* (sect. *Quercus*)

California valley oak: *Quercus lobata* (sect. *Quercus*)

canyon live oak: *Quercus chrysolepis* (sect. *Protobalanus*)

Chestnut: *Castanea* (Fagaceae)

chestnut oak: *Quercus montana* (sect. *Quercus*)

Chinese cork oak: *Quercus variabilis* (sect. *Cerris*)

chinkapin oak: *Quercus muehlenbergii* (sect. *Quercus*)

Chisos oak: *Quercus gravesii* (sect. *Lobatae*)

coastal sage scrub oak: *Quercus dumosa* (sect. *Quercus*)

coast live oak: *Quercus agrifolia* (sect. *Lobatae*)

cork oak: *Quercus suber* (sect. *Cerris*)

Cork Oak Group, Cork Oaks: section *Cerris* (subg. *Cerris*)

deer oak: *Quercus sadleriana* (sect. *Ponticae*)

Deer Oak Group, Deer Oaks: section *Ponticae* (subg. *Quercus*)

dwarf live oak: *Quercus minima* (sect. *Virentes*)

eastern white oak: *Quercus alba* (sect. *Quercus*)

Emory oak: *Quercus emoryi* (sect. *Lobatae*)

Engelmann's oak: *Quercus engelmannii* (sect. *Quercus*)

Eurasian chinkapin: *Castanopsis* (Fagaceae)

Eurasian Oak Group, Eurasian Oaks: subgenus *Cerris* (*Quercus*)

Eurasian white oak clade: no formal name, often called the Roburoid oaks (sect. *Quercus*)

Gambel's oak: *Quercus gambelii* (sect. *Quercus*)

Garry oak: *Quercus garryana* (sect. *Quercus*)

Golden-Cup or Intermediate Oaks Group, Golden-Cup or Intermediate Oaks: section *Protobalanus* (subg. *Quercus*)

gray oak: *Quercus grisea* (sect. *Quercus*)

Hill's oak: *Quercus ellipsoidalis* (sect. *Lobatae*)

Holly Oak or Holm Oak Group, Holly Oaks or Holm Oaks: section *Ilex* (subg. *Cerris*)

Holm oak: *Quercus ilex* (sect. *Ilex*)

Ichiigashi, red-bark oak: *Quercus gilva* (sect. *Cyclobalanopsis*)

interior live oak: *Quercus wislizeni* (sect. *Lobatae*)

jolcham oak: *Quercus serrata* (sect. *Quercus*)

Kermes oak: *Quercus coccifera* (sect. *Ilex*)

lateleaf oak: *Quercus tardifolia* (sect. *Lobatae*)

Mongolian white oak: *Quercus mongolica* (sect. *Quercus*)

netleaf oak: *Quercus rugosa* (sect. *Quercus*)

northern red oak: *Quercus rubra* (sect. *Lobatae*)

overcup oak: *Quercus lyrata* (sect. *Quercus*)

Palmer's Oak: *Quercus palmeri* (sect. *Protobalanus*)

pedunculate oak: *Quercus robur* (sect. *Quercus*)

post oak: *Quercus stellata* (sect. *Quercus*)

pubescent oak: *Quercus pubescens* (sect. *Quercus*)

Pyrenean oak: *Quercus pyrenaica* (sect. *Quercus*)

Quercus insignis (sect. *Quercus*)

Quercus oleoides (sect. *Virentes*)

Quercus skinneri (sect. *Lobatae*)

Red Oak Group, Red Oaks: section *Lobatae* (subg. *Quercus*)

Ring-Cupped Oak Group, Ring-Cupped Oaks: section *Cyclobalanopsis* (subg. *Cerris*)

Roble colombiano, or simply "roble," as it is the only oak in South America: *Quercus humboldtii* (sect. *Lobatae*)

sawtooth oak: *Quercus acutissima* (sect. *Cerris*)

scarlet oak: *Quercus coccinea* (sect. *Lobatae*)

sessile oak: *Quercus petraea* (sect. *Quercus*)

Shreve oak: *Quercus parvula* var. *shrevei* (sect. *Lobatae*)

silverleaf oak: *Quercus hypoleucoides* (sect. *Lobatae*)

Southern Beech Family: Nothofagaceae (Fagales)

Southern Live Oak Group, Southern Live Oaks: section *Virentes* (subg. *Quercus*)

swamp chestnut oak: *Quercus michauxii* (sect. *Quercus*)

swamp white oak: *Quercus bicolor* (sect. *Quercus*)

Tanoak: *Notholithocarpus densiflorus* (Fagaceae)

Turkish oak: *Quercus cerris* (sect. *Cerris*)

Walnut Family: Juglandaceae (Fagales)

White Oak Group, White Oaks: section *Quercus* (subg. *Quercus*)

willow oak: *Quercus phellos* (sect. *Lobatae*)

Notes

References and some auxiliary information are organized by chapter, then by section within each chapter. Sections separated by white space in the text are indicated here in the notes by page number and the first sentence of the section. Within each section, references are provided mostly in the order in which they appear in the main text, with main topics indicated in boldface and subtopics in some cases indicated by underlines. Only key references are provided, but they should suffice to provide the interested reader with access to some of the most relevant literature for essentially all topics covered in the book.

Chapter One

P. 7 *"Spring in the upper Midwest begins downhill from the oaks."*

Root phenology studies are available for a few oaks and other temperate forest species, especially in North America (McCormack et al. 2015; Reich, Teskey, and Hinckley 1980); but the results mentioned for *Q. rubra* have not yet been published (McCormack, pers. comm., May 9, 2023).

Floral and pollination phenology discussion draws on personal experience and published sources (Cecich 1996, 1997; Stairs 1964; Ducousso, Michaud, and Lumaret 1993). Environmental conditions and timing for pollen dispersal additionally draw from Sharp and Chisman (1961). Acorn development is drawn from Sharp and Sprague (1967). Counts of the number of pollen grains on a stigma come from pollen supplementation studies of *Q. lobata* (Pearse et al. 2015). Number of pollen tubes growing together through the style is from Abadie et al. (2012), who documented reproductive barriers between *Q. robur* and *Q. petraea*.

The growth of self-pollen and interspecific pollen is based on studies of *Q. ilex* (Yacine and Bouras 1997) and *Q. suber* (Boavida, Varela, and Feijó 1999). Variation in pollen growth coordination: Boavida, Silva, and Feijó (2001, especially fig. 8). The week-long viability of the stigma is from a study of *Q. rubra* and *Q. velutina* (Cecich and Haenchen 1995).

Pollen migration distances are from *Q. macrocarpa* (Dow and Ashley 1998); *Q. petraea*–*Q. pyrenaica* (Valbuena-Carabana et al. 2005); *Q. ilex* (Hampe, Pemonge, and Petit 2013); and *Q. robur* (Buschbom, Yanbaev, and Degen 2011; Moracho et al. 2016). The estimate of 100 kilometers (Kremer et al. 2010, 107) is based on pollen viability models. Pollen grain length: averaged from lengths reported in Tekleva, Polevova, and Naryshkina (2023, table 1).

Delayed fertilization and pollen growth. As reported in oaks: Deng et al. (2022); Cecich (1997); Joseph Williams, Boecklen, and Howard (2001); Sogo and Tobe (2006); Satake and Kelly (2021). Delayed fertilization in conifers: Gelbart and von Aderkas (2002). Deng et al. (2022) argue that the arrest of pollen tubes at the style joint in *Q. acutissima* may allow pollen grains to catch up with one another; only two to four pollen tubes commence growing after the pause. This paper also demonstrates porogamy (fertilization via the micropyle), counter to a previous study suggesting chalazogamy (fertilization via the chalaza). Phylogenetic distribution of biennial and annual acorn production: Denk et al. (2017). Variation in acorn development time in *Q. suber*: Díaz-Fernández, Climent, and Gil (2004).

Two-seeded acorns. The data reported are from Illinois *Q. macrocarpa* (Garrison and Augspurger 1983); the authors report that circa 2% of acorns were two-seeded (a high estimate for bur oaks, given my own observations), and their seedlings grew 80% as tall as seedlings from single-seeded acorns, with leaves about 80% as long.

Fertilization of multiple ovules is based on *Q. gambelii*, *Q. alba*, and *Q. velutina* (Mogensen 1975) and *Q. cerris* (Boavida, Silva, and Feijó 2001; Boavida, Varela, and Feijó 1999). The embryo description comes largely from Mogensen's (1970) account of embryo development in *Q. arizonica*.

P. 14 *"The innovation of packing a to-go meal into one's cotyledons evolved many times across the flowering plant tree of life . . ."*

Cotyledon effects on acorn growth may persist for twelve to fifteen years (Cieslar 1923, cited in Kleinschmit 1993). Acorn nutrition content: Mason (1992, app. 1). Use of oaks by 180 birds and mammals: Van Dersal (1940). Lyme disease: Ostfeld et al. (2006). Acorn-rodent population connection: Wolff (1996).

Weevil life history is given for the genus *Curculio*; Steele (2021, fig. 12.3) provides a summary and states that adult weevils will move from tree to tree through the canopy if they fail to find acorns but likely don't fly very often (Steele, pers.

comm., March 9, 2023). Covariance of rostrum length with seed size: J. Hughes and Vogler (2004). Seed mycobiome: Fort et al. (2021).

Scatter-hoarding and squirrels. The effect on herbivory of burial by deer and mice: Steele (2021). Desiccation and insect herbivory are from a study of *Q. schottkyana* (Xia, Tan, et al. 2016). Codistribution of oaks and scatter-hoarding rodents: Steele (2021, 192–93). Paw-maneuver and head-shaking behaviors of gray squirrels: Steele (2021, 88, 118); Preston and Jacobs (2009). Squirrels' ability to detect the maturity stage of acorns from their scent: Steele (2021, 45); Sundaram et al. (2020) investigated the breakdown of the wax layer. Squirrels' memory: Jacobs and Liman (1991); Delgado and Jacobs (2017). Seed dormancy in *Q. macrocarpa*: Bonner and Karrfalt (2008, 928–38).

Jays. The name "jay" is used here for several unrelated Corvid genera (S. De Kort and Clayton 2005). Steele (2021, 214, for numbers of jays worldwide who scatter-hoard acorns); Bossema (1979); Pesendorfer, Sillett, et al. (2016); and Darley-Hill and Johnson (1981; includes the *Q. palustris* study) provide broad reviews. Eight thousand acorns dispersed in a couple of days: Steele (2021, 216). Distances cited: 4–5 kilometers (W. Johnson and Paterson, Wisconsin, cited in Darley-Hill and Johnson 1981); 3–4 kilometers for rooks (Källander 2007); studies using direct observation or radio tracking (Gómez 2003; Pons and Pausas 2007; Källander 2007) found the vast majority of dispersals to be under 1 kilometer. Weevil-avoidance by jays is from Mexican jay studies in *Q. emoryi*, which also showed jays favoring tannin-rich acorns (Hubbard and McPherson 1997). Holm oak study: Gómez (2003). Planting cost estimates: Pearse et al. (2021).

P. 20 *"Oaks strike a balance between giving their dispersers what they want . . . and making sure some of their acorns make it to adulthood."*

Tannins. General review: Constabel, Yoshida, and Waker (2014). Human uses: Baldwin and Booth (2022). Tannin effects and acclimation: Koenig and Heck (1988; jays, acorn woodpeckers); Chung-MacCoubrey, Hagerman, and Kirkpatrick (1997; gray squirrels); Shimada et al. (2006; mice). Acorn tannin gradients: Steele et al. (1993); Steele (2021, 111–13). Variety of methods used historically by Native Americans for treating acorns: Driver (1952).

Acorn woodpeckers. Codistribution of acorn woodpeckers and oaks: P. Moore (1999); Koenig and Haydock (1999); Freeman and Mason (2015; Colombia only). Food preferences of Colombian acorn woodpeckers: Oikawa and Pulgarín-R. (2019). Life history and ecology: Koenig and Mumme (1987, esp. chaps. 1–4, 14).

Mechanical/developmental protection. Acorn fruit wall thickness in *Q. variabilis*: Yi and Yang (2010). Two other studies suggest a trade-off between mechanical and chemical defenses (X. Chen, Cannon, and Conklin-Brittan 2012; Y. Wang et al. 2023), but with a strong phylogenetic component not accounted for. Acorn depredation and growth post–cotyledon removal: Steele (2021, 43, chap. 6); Yi et al. (2019). *Quercus insignis* germination trials: García-Hernández et al. (2023). Acorn survival: Korstian (1927).

P. 22 *"Acorn defenses are clever, but in a typical year, 70%–90% or more of the acorn crop may be devoured . . ."*

General background on masting. Jack-in-the-pulpit sex: Bierzychudek (1984). Continental seed production in white spruce: LaMontagne et al. (2020). Global analysis of masting: Pearse et al. (2020). *Quercus schottkyana* seed size: Xia, Harrower, et al. (2016).

Explanations of masting in oaks draw on recent reviews (e.g., Koenig et al. 2015; Pesendorfer, Koenig, et al. 2016; Pearse, Koenig, and Kelly 2016; Pearse et al. 2020; Koenig 2021) and individual studies (pollen limitation vs. flower production: Fleurot et al. 2023; summer droughts: Sork, Bramble, and Sexton 1993). Periodical cicadas: Koenig et al. (2022); White and Strehl (1978); Karban (1980); Koenig and Liebhold (2003).

P. 25 *"The geographic ranges of oak species track changing climates, fires, human migrations, competition from other species, and changes in how humans manage forests and savannas."*

Contemporary migration rates. *Quercus ilex*: Delzon et al. (2013); *Q. petraea*: Truffaut et al. (2017); *Q. rubra*: Sittaro et al. (2017); *Q. mongolica*: Tang et al. (2023).

Glacial dynamics. Last Glacial Maximum (LGM): P. Clark et al. (2009). On 100,000-year cycles following the mid-Pleistocene Transition: Barker et al. (2022); P. Hughes and Gibbard (2018); Augustin et al. (2004). Glacial thickness: Willeit et al. (2019).

Migration rates post-LGM.

EUROPE. Petit, Brewer, et al. (2002); Brewer et al. (2002, 2005); Kremer et al. (2010); additional references in Kremer and Hipp (2020, particularly fig. 1).

EASTERN UNITED STATES. H. Delcourt and Delcourt (1984); P. Delcourt and Delcourt (1987, 246–56); Jackson et al. (2000); Ordonez and Williams (2013, especially suppl. fig. 5); Lumibao, Hoban, and McLachlan (2017); John Williams et al. (2004); McLachlan, Clark, and Manos (2005); Pielou (2008). In eastern North America, ice-marginal refugia may be essential to understanding the rate of tree migration (see previous references and Bemmels, Knowles, and Dick 2019; Magni et al. 2005; Snell and Cowling 2015). The importance of northern refugia in Europe is less certain (cf. Giesecke 2016; Robin et al. 2016).

CALIFORNIA/MEDITERRANEAN. Grivet et al. (2006); de Heredia et al. (2007); Dodd and Kashani (2003).

EAST ASIA. Cao et al. (2015) include pollen records primarily for eastern China; I broaden their report to include Southeast Asia as a possibility. Multiple refugia for East Asian oaks: Y. Li, Zhang, and Fang (2019); X. Chen, Cannon, and Conklin-Brittan (2012); Hao, Cheng, and Song (2022).

LITTLE ICE AGE. Cultural impacts, especially in Europe: Blom (2019). Temperature fluctuations: Ahmed et al. (2013).

P. 29 *"In 1899, Clement Reid, a largely self-trained geologist and naturalist, published a book entitled* The Origin of the British Flora.*"*

Reid's paradox. "Paradox" described: J. Clark et al. (1998); Reid (1895; 1899, 25).

Dispersal. Dispersal by corvids: Steele (2021); Bossema (1979); Pesendorfer, Sillett, et al. (2016); W. Johnson and Webb (1989). Dispersal distances: J. Moore et al. (2007). Generation times: W. Johnson and Webb (1989, table 2); Kleinschmit (1993, 180). Importance of long-distance dispersal events: Le Corre et al. (1997). Passenger pigeons: Bucher (1992); Hung et al. (2014); Webb (1986).

Humans. Acorn use by humans: Logan (2005, 29–76); Chassé (2016). Masting trees clustered around Native American settlements: Munos et al. (2014). Earliest documented evidence of humans in North America: M. Bennett et al. (2021). *Quercus rubra* in Eurasia: Merceron et al. (2017). Gall wasps traveling with oaks: Stone et al. (2007).

P. 32 *"After their travels across seasons and landscapes . . . a few acorns germinate."*

Germination-to-seedling success. Basal emergence of the radicle refers to the Kerrii group of section *Cyclobalanopsis* (Deng, Zhou, and Li 2013; X.-Q. Sun et al. 2021) and *Q. alnifolia* of section *Ilex* (Cinar-Yilmaz and Akkemik 2007). Dormancy: X.-Q. Sun et al. (2021, esp. table 1); section *Virentes* dormancy: Lewis (1911). Seedling growth and photosynthate gradients: P. Johnson, Shifley, and Rogers (2009, 92–100). The approximate transitions from flowering to seedling provided here are based on: pistillate flower to acorn transition rates circa 7% (Pearse et al. 2015); acorn to first-year seedling transition rates circa 25%, ignoring predation prior to dispersal (Borchert et al. 1989); seedling survival to the second or third year circa 1% (F. Davis, Tyler, and Mahall 2011).

Gene flow "map." I coarsely estimate "billions of oaks" worldwide, starting with Crowther et al.'s (2015) estimate that temperate broad-leaved forests contain circa 360 billion individual trees. If the northern temperate forest is half this amount and oaks constitute even 1% of the total stem count, this gives us over a billion oaks. Directional (anisotropic) pollen movement: Austerlitz et al. (2007); Dutech et al. (2005); Moracho et al. (2016). Billions of pollen grains per tree: Gómez-Casero et al. (2004) estimate 4.20 to 55.0 billion pollen grains per tree in 4,500 to 100,000 catkins per tree (four oak species, Córdoba). The same study estimates pollen viability to drop from an average of 80% to 50% over the course of a month. Elevation of 3,000 meters: Kleinschmit (1993, 180), who also cites earlier authors reporting greater than 550,000 pollen grains per catkin.

Chapter Two

P. 39 *"Find an oak close to your home."*

Individual tree variation. Leaf shape variation on a single tree: personal observations and several studies (Bruschi, Grossoni, and Bussotti 2003; Baranski 1975, 77; Blue and Jensen 1988; Kusi and Karsai 2020; Rubio De Casas et al. 2007). Functional plasticity: fine roots of *Q. alba* produce fewer ellagitannins, a chemical protection against fungi, in dry soils. *Quercus rubra* does not show the same plasticity (Suseela et al. 2020). *Quercus ilex* fine root plasticity: Encinas-Valero et al. (2022).

Oak longevity and vegetative reproduction. *Quercus macrocarpa*: the tree discussed was cored July 2018, inner ring aged to 1772; the pith at breast height was estimated to date to the 1760s (Christy Rollinson, pers. comm., June 1, 2023). *Quercus alba*: Eastern Oldlist (n.d.). *Quercus petraea*: Piovesan et al. (2020). *Quercus palmeri*: May et al. (2009). Holocene temperature change: Osman et al. (2021). *Q. petraea* resprouting: Alberto et al. (2010). Rhizome-forming oak species: Muller (1951). *Quercus hinckleyi* clones: Backs et al. (2015). Cannon, Piovesan, and Munné-Bosch (2022) demonstrate the potential importance of long-lived trees as genetic bridges.

"Variation is the fountainhead of evolution" is from Dobzhansky (1965).

P. 42 *"Modern biodiversity science traces back to* The Origin of Species, *published by Charles Darwin in 1859."*

Darwin. The full title of the first five editions was actually *On the Origin of Species by Means of Natural Selection, or the Preservation of Favoured Races in the Struggle for Life*, which is a mouthful. In the 1872 edition, which I reference often in this book as Darwin's final edition, Darwin or his editors lopped off the initial "On." In this book, I mostly use *The Origin of Species*. Darwin's "rhinoceros-sized rodent" turned out to be an extinct notoungulate, *Toxodon* (M. Buckley 2015). Darwin's finches: Sato et al. 1999. The account of Darwin's early years is from his autobiography (C. Darwin 1958). Darwin and his children tracking bees: Allen (1977, 150–51). "I would like to know grasses and sedges—and care. Then my least journey into the world would be a field trip, a series of happy recognitions": Dillard (1974).

Artificial selection. "Domestic dogs" (Frantz et al. 2016) in this paragraph excludes the more recently domesticated "wolfdogs." Darwin's view of pigeon varieties as deriving from a single species was based on morphology and anatomy but is supported by molecular data (Stringham et al. 2012).

Natural selection. "He who can read Sir Charles Lyell's grand work on the Principles of Geology, which the future historian will recognise as having produced a revolution in natural science, and yet does not admit how vast have been the past periods of time, may at once close this volume": C. Darwin (1872, 266). Estimates when Darwin was writing pushed Earth's age to 66,000 times older than the inferred Genesis age of circa 6,000 years, but this (rightly) struck Darwin as too little time; he hypothesized that natural selection might have been substantially stronger in the early years of the Earth (C. Darwin 1872, 286).

Individual variation in oaks. C. Darwin, January 14, 1863, to Alphonse de Candolle (Darwin Correspondence Project n.d.); and C. Darwin (1872, 40, chap. 4). "I look at the term species": C. Darwin (1872, 42).

P. 46 *"The Earth bakes in one place and is*
blanketed with snow in another."

Common gardens overview. General history of provenance trials: Callaham (1964). Oak provenance trials through 1993: Kleinschmit (1993). Reviews of historical common garden work: Núñez-Farfán and Schlichting (2001); Lowry (2012); Endersby (2018); Hagen (1983); Clausen et al. (1940, chap. 11, 394).

Oak common gardens. Variation in oaks in a variety of traits: Kleinschmit (1993); Kremer and Hipp (2020, 995–96). "Heritability" here refers to the ratio of among-family phenotypic variance to total phenotypic variance, typically measured in common gardens; for an example of oak trait heritability measured in a natural community of *Q. robur* and *Q. petraea* using genomic relatedness and parentage over two generations to estimate additive variance, see Alexandre, Truffaut, Ducousso, et al. (2020). Interestingly, while both species exhibited high heritable variation in numerous traits that adapt them to their environment, sessile oak exhibited higher variance in overall fitness (Alexandre, Truffaut, Klein, et al. 2020). This may not be surprising: sessile oak is expanding into pedunculate oak territory across wide geographic ranges and even within single stands (Truffaut et al. 2017). *Quercus alba* common garden: Huang et al. (2016); O'Connor and Coggeshall (2011). Differences between source populations in this work explained less than 20% of height variation after twenty-three years of growth. Flowering phenology impacts on acorn production: Koenig et al. (2021).

Reciprocal transplant experiments. *Quercus rubra*: Sork, Stowe, and Hochwender (1993). "One of the most massive common garden experiments": Sáenz-Romero et al. (2017). Additional studies of transfer distances in oaks (Nagamitsu and Shuri 2021; George et al. 2020; Browne et al. 2019) have found qualitatively similar results.

"Even without reciprocal transplant experiments." Bud-break phenology:

Wright et al. (2021) for *Q. lobata*; Firmat et al. (2017) for *Q. petraea*. *Quercus suber* drought adaptations: Morillas et al. (2023). *Quercus suber* specific leaf area (SLA): Ramírez-Valiente et al. (particularly 2010; also 2017, 2020). The finding that trees with lower SLA grew more rapidly is also found in *Quercus oleoides* (Ramírez-Valiente et al. 2017). Water-use efficiency of *Q. petraea*: Rabarijaona et al. (2022).

***Quercus oleoides*, section *Virentes*, and the evolution of plasticity.** A series of studies by Cavender-Bares and colleagues in section *Virentes* are foundational to our understanding of climate adaptation in oaks. A few key studies on this work focusing on freezing-tolerance, cold-acclimation, and drought-tolerance: Cavender-Bares (2019); Koehler, Center, and Cavender-Bares (2012); Cavender-Bares and Ramírez-Valiente (2017); Ramírez-Valiente and Cavender-Bares (2017). Climatic distribution of *Q. oleoides*: Cavender-Bares et al. (2015). Local adaptation, *Q. oleoides*: Deacon and Cavender-Bares (2015). *Quercus coccifera*: Balaguer et al. (2001). Countergradient plasticity, *Q. petraea*: Caignard et al. (2021). Comparison of field to common garden variation, *Q. petraea*: Bresson et al. (2011). *Quercus douglasii*: Papper (2021).

Variance increasing as geographic distance increases. Kleinschmit (1993, 176) illustrates how variation increases as the distance between populations increases. Petit and Hampe (2006) describe this phenomenon in their article, "Some Evolutionary Consequences of Being a Tree." Opedal et al. (2023) demonstrate that variation within populations scaled by the mean trait value—evolvability, a term coined for the coefficient of variation for traits (e.g., Hansen, Pélabon, and Houle 2011)—predicts average divergence among populations within plant species; their study is based on common garden studies conducted in twenty-four herbaceous genera.

P. 53 *"The things we can observe and measure on
an organism are called its phenotype."*

Background: genotypes.

GENOMES REFERENCED. *Quercus acutissima* (758 Mb, 31,490 protein-coding genes); *Q. lobata* (811 Mb assigned to 12 chromosomes; 39,373 mapped protein-coding genes); *Q. mongolica* (775 Mb assigned to 12 chromosomes, 33,489 protein-coding genes mapped of 36,553 total); *Q. rubra* (739 Mb mapped to 12 chromosomes; 31,783 protein-coding genes mapped of 33,333 total); and *Q. gilva* (549 Mb mapped to 12 chromosomes, 36,442 predicted protein-coding genes) (Fu et al. 2022; Sork et al. 2022; Ai et al. 2022; X. Zhou et al. 2022). I use 35,000 genes in this book for simplicity. The older *Q. robur* genome (Plomion et al. 2018) was not included in gene count estimates.

GENOTYPE. *Genotype* and *genome* are employed variously, and sometimes synonymously (Mahner and Kary 1997). In this book, I use *genotype* to describe the genetic variations that make the phenotype; I use *genome* to refer to the entire body of DNA within an individual, or the collective genomes of individuals comprising a species or lineage.

MUTATIONS. 234-year-old tree mutation rates are from the "Napolean oak": Schmid-Siegert et al. (2017). Mouse coloration: Hoekstra et al. (2006). Callipyge phenotype in sheep: Freking et al. (2002). Effects of single-nucleotide polymorphisms (SNPs) on human disease phenotypes: Shastry (2009).

Powdery mildew and QTL studies. Family effects: in wild forests, Ekholm et al. (2017); in a common garden, Desprez-Loustau et al. (2014). Number of powdery mildew species worldwide: Chater and Woods (2019). Effects of powdery mildew on plant performance: Hajji, Dreyer, and Marçais (2009); Copolovici et al. (2014); Marçais and Desprez-Loustau (2014); Bert et al. (2016). Molecular evidence of East Asian *Erysiphe alphitoides* origin: Takamatsu et al. (2007). Movement onto other oak species, eucalyptus, and mango: Cho et al. (2018); Desprez-Loustau et al. (2017). The QTL study that forms the heart of this story is by Bartholomé et al. (2020). Broader list of oak QTL studies: Gailing et al. (2022, sect. 4.3).

Genotype-environment association studies. *Quercus ellipsoidalis–Q. rubra*: Lind-Riehl, Sullivan, and Gailing (2014); *Quercus lobata*: Gugger et al. (2021); *Quercus rugosa*: Martins et al. (2018). List of environmental associations: Gao et al. (2021); also see Du et al. (2020); Pina-Martins et al. (2019). Identification of seven genes associated with same environmental factor across three European species: Rellstab et al. (2016).

P. 57 *"Natural selection is most effective when there is high genetic variation within populations."*

Genetic variation in woody plants. Contrast between macroevolutionary rates and microevolutionary rates in trees: Petit and Hampe (2006). Relative nucleotide substitution rates: Smith and Donoghue (2008) for woody and herbaceous plant lineages; Lanfear et al. (2013) for correlation of nucleotide substitution rates with plant height. Genetic diversity across plants: Hamrick, Godt, and Sherman-Broyles (1992); Hamrick and Godt (1996); H. De Kort et al. (2021).

Colbert's oak forests, background. Forests visited: Forêt Domaniale de la Petite Charnie; Forêt Domaniale de Réno-Valdieu. On 30,000 euros for an old oak: original price estimate from a field discussion with Antoine Kremer's colleagues in the Office National des Forêts (ONF) at Forêt de Réno-Valdieu, 2014. Kremer subsequently recontacted his colleagues in ONF in 2023, and they indicated that in the oak wood–selling season of 2022, the highest price reached in France was 3,200 euros per cubic meter. These were from sessile oak (*Q. petraea*) trees in Forêt de Réno-Valdieu. Kremer also reported that at an international auction in the same season, the highest price was 4,300 euros per cubic meter (Antoine

Kremer, pers. comm., May 4, 2023). Even-aged silvicultural practices in France: Alexandre, Truffaut, Klein, et al. (2020, suppl. 1). Population densities: Saleh et al. (2022); Antone Kremer (2021).

Genomic study of allele frequency changes during and after the Little Ice Age. Little Ice Age effects on European society: Blom (2019); Degroot (2018, 2019). Removed from the text because it's speculative was a fascinating argument by Burckle and Grissino-Mayer (2003) that the unique sound of Stradivari's violins is due in part to the slow growth of the spruces he used to make them. Temperature estimates: Luterbacher et al. (2004) document three particularly cold periods when temperatures dropped to approximately one degree Celsius lower than the twentieth-century average for decades at a time: one beginning around 1650, one around 1770, and one around 1850. The genomic study reported here is Saleh et al. (2022); a follow-up study investigated phenotypic variation in a common garden established from acorns collected at these sites (Caignard et al. 2023).

P. 61 *"In the 20,000 years since the Last Glacial Maximum, bur oaks have expanded to their current distribution in eastern North America . . ."*

Oak adaptation lagging behind climate change. Climatic niche models suggest that *Q. robur* may be suited for climates as far as 70 degrees N latitude in Europe (Ülker, Tavşanoğlu, and Perktaş 2018). *Quercus petraea* maladaptation: Sáenz-Romero et al. (2017). *Quercus lobata* lagged responses to climate: Browne et al. (2019). Accessibility of new environments with climate change: Mauri et al. (2022). Genes common across European white oaks, risk of nonadaptedness (RONA) analysis: Rellstab et al. (2016). *Quercus acutissima* landscape genomics: Yuan et al. (2023). Climate scenarios referenced: RCP8.5 and the less extreme RCP4.5.

Nature versus nurture. Leonard Darwin's (1913) point has also been attributed to the psychologist Donald Hebb, but this may be apocryphal (Serpell 2013).

Species. The image of the "single, hypervariable species" was suggested by Silvertown (2005). Species as genotypic and phenotypic clusters: Mallet (2020).

Chapter Three

P. 65 *"The swamp white oak in front of our house*
is hemmed in by pavement."

Excavations of oak roots: Watson et al. (2014). "The suburban residue of a mixed oak forest": Maple Grove Forest Preserve, Dupage Co., Illinois (Hipp 2021b). The "king of kings": Peattie (1991, 195). Oak biomass and species diversity: Cavender-Bares (2019, 2016). Hybrid epithets: *Oak Name Checklist*, accessed January 27, 2024.

P. 67 *"Approximately 600,000 years ago, a population of our distant relatives migrated from Africa into Europe."*

Human evolution. This narrative is summarized from several studies and reviews (including Bergström et al. 2021; Gellis and Foley 2023; Holliday 2006; Macciardi and Martini 2022; Rosas, Bastir, and García-Tabernero 2022; Stringer and Crété 2022).

Biological species concept (BSC). The BSC was initially articulated by Edward Poulton (1904) and Theodosius Dobzhansky (1935): "A species is a group of individuals fully fertile inter se, but barred from interbreeding with other similar groups by its physiological properties . . ." (Dobzhansky 1935, 353). Ernst Mayr (1942) named and popularized the concept. Short historical perspectives on the BSC: Mallet (2003); Grant (1981, 45–47). The interaction between Anderson and Mayr in the year leading up to their paired Jesup lectures at Columbia University is particularly instructive (Kleinman 2013); these lectures formed a groundwork for Mayr's 1942 book (beyond Mayr's extensive researches before these lectures). Aristotle on reproduction by mules: *De Generatione Animalium* II.8, trans. Platt (Aristotle [350 BC] 2010, ll. 11–29). Molecular phylogeny of horses, mules, and their relatives: Jónsson et al. (2014). The "view commonly entertained by naturalists": C. Darwin (1872, 235).

Darwin on hybridization between species. Gärtner and Kölreuter were among the most-cited authors in *The Origin of Species* (de Lima Navarro and de Amorim Machado 2020) but "arrived at diametrically opposite conclusions in regard to some of the very same forms" (C. Darwin 1872, 237, chap. 9). Darwin's species definition: C. Darwin (1874), as discussed in Mallet (1995).

P. 70 *"Many potential hybrid combinations are never found in the wild . . ."*

Hybrids observed in oaks. Limits of oak hybridization: Wiegand (1935); Palmer (1948); Cottam, Tucker, and Santamour (1982). According to Engelmann (1878, 398–400): "White-oaks and Black-oaks are too distinct to be crossed. Among the White-oaks hybrids seem to be much rarer than among the Black-oaks, or it may be that they are more difficult to discover." He then lists three putative white oak hybrids (all involving the widespread *Quercus alba*) and seven putative black/red oak lineage hybrids, adding that "hybrid Black-oaks are much more numerous, or, to speak more correctly, more have thus far been noticed, perhaps because their leaf-forms are more various, and thus the intermediate ones are more easily recognized." Summary of Red Oak hybrids: Rauschendorfer, Rooney, and Külheim (2022).

California oaks. California black oak fossil co-occurrence: Mensing (2005). California black oak genetic distinctness: Hauser et al. (2017). Divergence time between *Q. engelmannii* and the California White Oaks: Hipp et al. (2020). Hybridization between the California scrub White Oaks and *Q. engelmannii*: Kim et al. (2018).

Bartram's oak. Phylogenomic study and background: Crowl, Bruno, et al. (2020). Bartram Oak as a hybrid: Engelmann (1878); MacDougal (1907); Hollick (1919). Bartram Oak as *not* a hybrid: S. Buckley (1861); L. Gale (1856).

> P. 72 *"Plant biologists working in the 1930s and 1940s . . .*
> *enriched our understanding of species . . ."*

Anderson. Anderson (1928, 310): "We may think of each species as a net, with the knots representing individuals." Introducing the term "introgression": Anderson and Hubricht (1938).

Studies following Anderson. *Quercus marilandica* and *Q. ilicifolia*: Stebbins et al. (1947). Palmer's study of hybrid oaks: (Palmer 1948). Ecological control of hybridization: Muller (1952). *Quercus gambelii* and *Q. macrocarpa*: Tucker and Maze (1966); Maze (1968). *Quercus undulata*: Tucker (1961, 1963, 1970, 1971); Tucker, Cottam, and Drobnick (1961). "If the oaks could intermix": L. Gale (1856).

> P. 75 *"The conflict between evidence for hybridization and gene*
> *flow on one hand and species coherence on the other came to a*
> *head with a trio of papers published in 1975 and 1976."*

White oak syngameon. Hardin (1975), which also drew on work by Baranski (1975). James Mallet (pers. comm., March 13, 2023) alerted me to the fact that Lotsy coined the term *syngameon* in a paper originally written in French in 1917, but that many citations of the term reference the (1925) English translation of his article (Lotsy 1925; 1917). White Oaks as something other than a biological species: Burger (1975). White Oaks as an ecological species: Van Valen (1976).

Syngameon, **use of the term.** It's worth noting that Lotsy (1917), who coined the term *syngameon*, had in mind something at least a little different. He did not believe in genetic diversity within species, and he introduced the term to explain any crossing between genetically distinct populations. Dobzhansky (1937, 311)—whose *Genetics and the Origin of Species* is one of the foundational texts of twentieth-century evolutionary biology, which helped in teasing apart the origin of species from the evolution of variation by natural selection as two intertwined but distinct domains of evolutionary study—viewed *syngameon* as a near-synonym for *biological species*, namely, a reproductively interconnected set of populations. In this sense, it would perhaps have become a redundant term had it not been co-opted in the 1950s and onward to its current use. Not everyone uses the term *syngameon* in the same way; Barton (2020), for example, defines it thus: "We can envisage biological species as collections of locally adapted populations, connected by a trickle of gene flow, yet able to coexist in sympatry, if sufficiently reproductively isolated; this has been termed a 'syngameon.'" Buck and Flores-Rentería (2022) provide a brief history of the term.

P. 78 *"Hardin, Burger, and Van Valen were writing a decade before oak molecular data would become available."*

Isozymes. Manos and Fairbrothers (1987); Guttman and Weigt (1989); Ducousso, Michaud, and Lumaret (1993); Hokanson et al. (1993); Bacilieri et al. (1996). Note that my use of *isozymes* in this chapter is a simplification. Many studies distinguish between variants encoded by different genes—isozymes in the strict sense—and alleles at a single gene—allozymes. Not all studies draw the distinction clearly, however, and the difference is not necessary for this discussion.

Chloroplasts. The key study for this section is Whittemore and Schaal (1991).

CHLOROPLAST INHERITANCE. Survey in angiosperms: Corriveau and Coleman (1988); exceptions to maternal inheritance of chloroplasts are scattered across the plant phylogeny, and environmental variation can shift within individual species (e.g., in tobacco, where cold conditions can lead to 2%–3% paternal transmission of plastids; Chung et al. 2023). Maternal inheritance of oak chloroplasts: Dumolin-Lapegue, Demesure, and Petit (1995).

EUROPEAN OAK STUDIES. Petit, Kremer, and Wagner (1993); Ferris et al. (1993); Dumolin-Lapegue et al. (1997); Dumolin-Lapegue, Kremer, and Petit (1999); Petit, Csaikl, et al. (2002). Migration of oaks by pollen swamping: Petit et al. (2003).

P. 80 *"Whittemore and Schaal had, importantly, sampled a second genome alongside the chloroplast: the nuclear genome . . ."*

Dispersal distance of acorns by jays: Petit et al. (2003). Pollen movement: Ashley (2021). Species coherence for nuclear vs. chloroplast genome: Petit and Excoffier (2009). Isozyme study separating *Q. robur* from *Q. petraea*: Zanetto, Roussel, and Kremer (1994). Microsatellite study: Muir, Fleming, and Schlötterer (2000).

Exemplar oak species studies. Some emphasize gene flow between species (e.g., González-Rodríguez et al. 2004; Dodd and Afzal-Rafii 2004; Moran, Willis, and Clark 2012; Craft and Ashley 2006), while others emphasize species distinctions in spite of gene flow (e.g., Hipp and Weber 2008; Owusu et al. 2015; Ramos-Ortiz et al. 2016; Cavender-Bares and Pahlich 2009). The figures reported here are based on a sample of studies from 2002 through 2023 that all use an explicit model-based population structure analysis (as implemented in STRUCTURE, FastStructure, Admixture, or analogous software) to estimate admixture proportions from multilocus nuclear data (Burge et al. 2019; Cavender-Bares and Pahlich 2009; Craft, Ashley, and Koenig 2002; R. Fu et al. 2022; Y. Li et al. 2021;

X. Li 2021; Nagamitsu et al. 2019; Ortego, Gugger, and Sork 2015; O'Donnell, Fitz-Gibbon, and Sork 2021; Pérez-Pedraza et al. 2021; Reutimann et al. 2023; Shi et al. 2023; Valbuena-Carabana et al. 2005; Yücedağ, Müller, and Gailing 2021).
Quercus ellipsoidalis **and** *Q. velutina*: Hipp and Weber (2008); Hipp (2010).

P. 83 *"Swamp white oaks, bur oaks, and eastern white oaks live within a few meters of each other . . ."*

Reproductive isolation. Isolation mechanisms in *Q. petraea* and *Q. robur* (Rushton 1993; Abadie et al. 2012). Self-pollination effects on flower abortion in *Q. ilex*: Yacine and Bouras (1997). *Quercus ellipsoidalis* and *Q. coccinea*: Hipp and Weber (2008); Shepard (2009, 1993); Owusu et al. (2015).

"All species have traveled along a passage." I am following Darwin in using the analogy of a passage here: "Certainly no clear line of demarcation has as yet been drawn between species and sub-species—that is, the forms which in the opinion of some naturalists come very near to, but do not quite arrive at, the rank of species: or, again, between sub-species and well-marked varieties, or between lesser varieties and individual differences. These differences blend into each other by an insensible series; and a series impresses the mind with the idea of an actual passage" (C. Darwin 1872, 41).

Species distinctions in oaks. "Worst-case scenario": Coyne and Orr (2004, 43). Remington Hills Flora, *Q. ×morehus*: Condit (1944). Species of North America north of Mexico: Manos and Hipp (2021). Mexico: Valencia-Avalos (2020, 2004). Selected recent taxonomic studies supporting the genetic distinctiveness of Mexican oak species: McCauley, Cortés-Palomec, and Oyama (2019); Morales-Saldaña et al. (2022); Pérez-Pedraza et al. (2021).

Chapter Four

P. 89 *"The world in the mid-Cretaceous, about 100 million years ago, was hot and growing hotter."*

Cretaceous world. General Cretaceous history: A. Gale (2000); Graham (2011, chap. 5). Atmospheric CO_2 levels: Bice et al. (2006) (1,000–1,500 ppmv for the mid-Cretaceous); Cui, Schubert, and Jahren (2020) (contemporary value of ca. 400 ppmv and mean of 200–400 over the length of the Pleistocene). Connection between volcanism and atmospheric CO_2 in the mid-Cretaceous: Weissert and Erba (2004). Antarctic glaciation: Ladant and Donnadieu (2016). Position of continents and mountain heights: Scotese (2021). Temperature profile for the Cretaceous: Scotese et al. (2021). Timing of mammal diversification: Upham, Esselstyn, and Jetz (2019); Carlisle et al. (2023); Brusatte (2022).

Early angiosperms. The stem age of angiosperms may be more than 340 million years old (Magallón et al. 2015; Barba-Montoya et al. 2018), but first fossils date to about 130 million years ago (Herendeen et al. 2017). Early Cretaceous angiosperm abundance, diversity, and physiognomy: Friis, Crane, and Pedersen. (2011b, chap. 19, esp. sect. 19.5). Timing of angiosperm families and orders: Magallón et al. (2015); Barba-Montoya et al. (2018).

P. 90 *"Yggdrasil, the Norse Tree of Life, stands at the center of the world."*

Yggdrasil: Gaiman (2017). Inferred life history of LUCA: Weiss et al. (2018). Clade analogies: Baum and Smith (2013). Bird origins: Prum et al. (2015). Taxonomy of *Homo*: Gellis and Foley (2023, table 17.1).

P. 94 *"I did not know the difference between a pine and a spruce the year before I took my first plant taxonomy course."*

Linnaeus. Linnaeus's natural system: J. Larson (1967). Linnaeus's plant collectors: Müller-Wille (2007). Kinsey and Fernald: Peter del Tredici (2006). Proportion of the Tree of Life that are bacteria: Larsen et al. (2017). "If you do not know the names": Linné (2003, 169).

P. 98 *"Until the late twentieth century, members of the Beech Family, Fagaceae, were classified as belonging to a group comprising most of the plant families that have aments."*

Amentiferae. History and systematics of the "Amentiferae": Thorne (1973); Stern (1973); Walker and Doyle (1975); Cronquist (1965). Concerted convergence in *Trillium* and *Arisaema*: Givnish et al. (2005). Function of "baldness" in vultures: Ward et al. (2008). Bird phylogeny: Hackett et al. (2008).

Fagales. Original concept: Engler (1892) cited in Constance (1955). The story of Chase et al. (1993) is also told in Silvertown (2005). Manos et al. (1993) used restriction site fragment length polymorphism (RFLP); the follow-up by Manos and Steele (1997) used chloroplast sequences. Fagales as we recognize it today was designated as in APG I (Angiosperm Phylogeny Group 1998). Species counts and some aspects of broader classification are from Angiosperm Phylogeny Web (APWeb) (Stevens 2001).

P. 103 *"Follow the Fagales tree of life back to its root, and you'll find fossils scattered along the way."*

Fossils versus DNA. "Fossilized flower petals": (Herendeen et al. 2017). Million-year-old DNA sequences are from mammoths (van der Valk et al. 2021); plants have yielded DNA back to about 8,400 years ago (Wagner et al. 2018; Pont et al. 2019).

Normapolles. Overview: Friis, Crane, and Pedersen (2011a, chap. 14, sect. 14.6.6);

Friis, Pedersen, and Schönenberger (2006); Polette and Batten (2017); Sims et al. (1999); T. Taylor (2009). Map of the Normapolles province: Baskin and Baskin (2016). Two preprints that analyze Normapolles fossils along with nuclear sequence data suggest that Normapolles may fall closer to the root of the Fagales tree (Siniscalchi et al. 2023; Yang et al. 2023), but additional study is needed. *Rhoiptelea*: Cheng-Yih and Kubitzki (1993); Bouchal et al. (2014).

P. 106 *"The Normapolles give us some ideas of what early Fagales looked like."*

Unnamed Fagales. Circa 120 charcoalified flowers and fruits from Martha's Vineyard, United States, earliest Campanian (D. Taylor, Hu, and Tiffney 2012). Cladistic analysis shows that the small flowers, a dry indehiscent fruit, a single-whorled perianth, and a possible cupule group these fossils with Fagales; while bisexual flowers, floral nectaries, ovules that are mature at anthesis, and two stamen characters separate them from the Fagales.

Archaefagacea. Circa 100 flower and fruit fossils, circa 89 million years ago [mya], Kamikitaba, Japan (Takahashi et al. 2008). The fossils are considered not assignable to Fagales by Gandolfo et al. (2018) for lack of a cupule and the presence of three-seeded fruits. The flowers are bisexual with six tepals each and apparently six stamens—though the preservation is so poor it is hard to be sure—each with a narrow filament up to half a millimeter long and pollen like that of today's chestnuts (*Castanea*). Friis, Crane, and Pedersen (2011a, 338) consider the fossil to be representative of "a possible stem-group lineage of Fagales." Fagaceae genera indicated by Denk to be similar to *Archaefagacea* are the old Castaneoideae, which includes the genera *Castanea, Castanopsis,* and all three of the closest *Quercus* relatives: *Notholithocarpus, Lithocarpus,* and *Chrysolepis.*

Antiquacupula: Sims et al. (1998).

Protofagacea: Herendeen, Crane, and Drinnan (1995). *Protofagacea* nutlets had Nothofagaceae-like characters—nutlets three to a cupule, with one flattened nutlet sandwiched between two triangular nutlets—along with Fagaceae-like characters: hairs on the inner surface of the fruit walls, short styles with expanded stigmas, male flowers with vestigial pistils, pistillate flowers with vestigial stamens. Male flowers were borne on short stalks, like Nothofagaceae and *Fagus,* and packed into spikes. They did not have elongated aments.

Soepadmoa: Gandolfo et al. (2018).

Fagales morphology. Cupule development: Fey and Endress (1983). Insect pollination: extrafloral nectaries are reported for *Quercus alba* and *Q. mongolica* (Weber and Keeler 2013). Pollination mode summarized by genus: Larson-Johnson (2016). Insect pollination in *Castanea*: Petit and Larue (2022). Scenarios for the evolution of wind pollination from insect pollination inspired in part by Culley et al. (2002).

Fagales biogeography. Southeast Asian hypothesis: Hill (1992); the difficulty of dispersing from Southeast Asia to Australia was pointed out by Guido Grimm (pers. comm., January 24, 2023). Paleogeographic maps: Scotese (2021, fig. 6); Scotese, Boucot, and Xu (2014, map 23). The most recent common ancestor of Fagales as a widespread syngameon parallels a hypothesis for modern human evolution (Stringer and Crété 2022) and is compatible with deep introgression in *Quercus* and Fagaceae (Cardoni et al. 2022; McVay, Hipp, and Manos 2017; B.-F. Zhou et al. 2022). Possible hybrid origin of Juglandaceae: Y. Ding et al. (2023). The role played by selected genes shuttling around was suggested by Morjan and Rieseberg (2004).

P. 109 *"Temperatures began dropping about 93 million years ago."*

Boreotropics: Wolfe (1975). Fagales topology: sources include R.-Q. Li et al. (2004); Xiang et al. (2014); Xing et al. (2014); Larson-Johnson (2016); Yang et al. (2023; 2021).

Nothofagaceae. Seed dispersal: Navarro-Cerrillo et al. (2020). Biogeography: Hill (1992); Manos (1997); Sanmartín and Ronquist (2004); Knapp et al. (2005); Sauquet et al. (2012). Modern distribution: Veblen, Hill, and Read (1996).

Fagaceae. *Fagus*: Denk and Grimm (2009); Renner et al. (2016); Cardoni et al. (2022); Schulze and Grimm (2022). Jiang et al. (2022) suggest younger divergence times than Renner et al. (2016); the latter uses a much denser collection of fossils analyzed using tip dating (a fossilized birth-death model) and is given preference in this narrative, but it may be biased toward older divergence times. Fossil distribution: Forman (1964); Crepet and Nixon (1989a, "Earliest megafossil evidence"). Climate and vegetation: Wolfe (1987); Crepet and Nixon (1989b, "Extinct transitional Fagaceae").

Core Fagales. Juglandaceae biogeography: Zhang et al. (2022); the oldest mesofossil, a circa 90–85 million–year-old fruiting structure similar to modern-day *Rhoiptelea*, the first-diverging genus in the Juglandaceae, with pollen found in Late Cretaceous sediments of the Czech Republic (Heřmanová, Kvaček, and Friis 2011). Myricaceae: Herbert (2005). Casuarinaceae: Steane, Wilson, and Hill (2003). Ticodendraceae: Description: Gómez-Laurito and Gómez P. (1989); Red List Assessment: Rivers et al. (2019); fossils: Manchester (2011).

P. 114 *"The next step in the evolution of the Fagaceae was a figurative explosion of species, which may have been coincident with a literal explosion."*

Meteor impact. End-of-the-Cretaceous events largely follow Brusatte (2022, 167; 2018, chap. 9). Fossilized ejecta deposited in Haiti: Kring (1995). Effects of meteorite impacts in Montana: DePalma et al. (2019). Deccan Traps, India: Schoene et al. (2015). Effects on forest tree community: Wappler et al. (2009). Effects on angiosperm lineages and species: Thompson and Ramírez-Barahona (2023). Effects on northern South America: Wing et al. (2009). Impact effect on evolution and spread of deciduousness: Wolfe (1987).

Nut evolution. Seed dispersal and Fagales diversity: Bouchenak-Khelladi et al. (2015); Xing et al. (2014).

Castanea. Phylogeny: Lang, Kubisiak, and Huang (2007); W. Zhou and Xiang (2022). The oldest reported fossil from the group, *Castanea intermedia*, is from the Rocky Mountains, but it is ambiguous, as no fruits are reported, and the leaves of *Castanea* and relatives can be confused with other genera (Manchester 2014). Distribution of American chestnut: Faison and Foster (2014).

Castanopsis. Clarno Nut Beds: Manchester (1994). *Castanopsis* fossil history: Sadowski et al. (2018). The *Castanopsis* South American dispersal history was presented and debated in Wilf et al. (2019b, 2019a) and Denk et al. (2019). Molecular evidence for ancient shared history with *Lithocarpus*: Cannon and Manos (2003).

Fruit evolution. Cupule evolution: Oh and Manos (2008); the phylogeny and some evolutionary inferences are outdated but include good background on and tests of evolutionary hypotheses and structure of the cupule. Development: Fey and Endress (1983). *Lithocarpus* fruit diversity: X. Chen and Kohyama (2021). *Notholithocarpus*: Manos et al. (2008).

P. 120 *"By the end of the Paleocene, 56 million years ago, all the families and nearly all the genera of the Fagales were in place."*

Western Inland Seaway evolution: Slattery et al. (2013); Scotese (2021). *Rhoiptelea* and Juglandaceae: Bouchal et al. (2014); Cheng-Yih and Kubitzki (1993); Zhang et al. (2022).

Chapter Five

P. 123 *"If we could head back in time 56 million years . . . we would be hard-pressed to find any oaks."*

Animal fossils of Ellesmere Island: Dawson et al. (1976). Plant diversity is based on pollen assemblages (G. Harrington et al. 2012); woody plants represented 52% of early Eocene angiosperm species (Luo et al. 2023).

Paleocene-Eocene Thermal Maximum (PETM).

CAUSES. Volcanic activity was likely one of the triggers for the PETM (Jones et al. 2019; Gutjahr et al. 2017; Svensen et al. 2004; Kender et al. 2021). Giant's Causeway (Northern Ireland) is a remnant of this event (Storey, Duncan, and Swisher 2007), as are land exposures in western Scotland, East Greenland,

and Denmark (Tali Babila, pers. comm., May 3, 2023). Potential role of melting Antarctic permafrost: Kender et al. (2021); these same feedback loops are at play today (Chadburn et al. 2017; Flanagan 2021). Humans currently release about ten gigatons of extra carbon into the atmosphere per year, a rate estimated to be 9–10 times more rapid than the early PETM (Babila et al. 2022; Friedlingstein et al. 2022; Zeebe, Ridgwell, and Zachos 2016); the PETM represents the sharpest increase in atmospheric carbon over the 66 million years prior to the past century (Zeebe, Ridgwell, and Zachos 2016).

REBOUNDING FROM THE PETM. End processes of the PETM are uncertain, but photosynthesis and erosion played a role (McInerney and Wing 2011; Penman and Zachos 2018). Rapid carbon sequestration at the end of the PETM: Bowen and Zachos (2010). PETM recovery time: McInerney and Wing (2011).

BIOLOGICAL EFFECTS OF THE PETM. North America: McInerney and Wing (2011); Wing and Currano (2013). South American vegetation: Jaramillo et al. (2010). Foraminifera: McInerney and Wing (2011). Mammal migrations: Brusatte (2022, 212–19); Clyde and Gingerich (1998). Insect herbivory: Currano et al. (2008).

St. Pankraz oak pollen. Original study: Hofmann, Mohamad, and Egger (2011); site context relative to other PETM sites: Egger, Heilmann-Clausen, and Schmitz (2009).

Early oak fossils. Early oak fossils referenced: Barrón et al. (2017); Grímsson et al. (2016, 2015); Liu, Song, and Jin (2020). Miocene fossils in Iceland: Denk et al. (2010). Boreotropic concept: Wolfe (1975, 1977); also reviewed by Baskin and Baskin (2016). In Alaska at this time, the polar broad-leaved deciduous forest was limited to 70°–75°N, placing it above the Arctic Circle (Graham 1999, p. 177).

P. 128 *"The oaks were not born at a particular moment or in a particular place."*

Stem of the oak clade. Three other possible synapomorphies for *Quercus*—styles that are a sizable portion of the entire ovary at pollination, hemianatropous ovules (ovules tipped on their side relative to their point of connection), and a stalkless connection between the ovule and the placenta of the ovary—are postulated by Deng et al. (2008), but not discussed here because I am not aware of comparable developmental work in *Notholithocarpus*.

Oak classification and timing. The split between the Eurasian and American oak clades bracketed by calibrated nuclear phylogenies, estimating it between circa 56 and 50 mya (Hipp et al. 2020; B.-F. Zhou et al. 2022). The oak classification outlined is from Denk et al. (2017).

"Our classifications will come to be, as far as they can be so made, genealogies": C. Darwin (1872, 427).

P. 131 *"Earth began to cool following the Early Eocene*
Climatic Optimum."

Eocene cooling was protracted and gradual, and the causes are not fully known. *Azolla* bloom: Speelman et al. (2009). Oceanic currents: Bijl et al. (2013). Lauretano et al. (2021) demonstrate the role of decreasing atmospheric CO_2 from the middle Eocene to early Oligocene and demonstrate that growth of the Antarctic ice sheet cannot on its own explain the observed cooling. Role of erosion in the carbon cycle: Tipper et al. (2021).

Chasing them southward were oaks and the other species of the deciduous forests. *Eucommia* was present as well, now known from a single species in China. Genera and interpretation: Graham (1999a, 197).

Continental connections.

BERING LAND BRIDGE. Paleomaps: Wen, Nie, and Ickert-Bond (2016). Timing of the peopling of the Americas: Raff (2021).

TURGAI SEA: Akhmetiev et al. (2012); Palcu and Krijgsman (2021).

NORTH ATLANTIC LAND BRIDGE (NALB). History of connectivity between the continents: Milne and Abbot (2002). Mapped northern and southern NALB routes: Tiffney (1985, esp. map 3); Brikiatis (2014) also provides timing for both the NALB and Bering Land Bridge. McKenna (1975) presents fossil evidence for severing of connectivity between mammalian biota of North America and Europe by way of the NALB at around 49 Ma. Graham (1999a, 191) presents a more updated picture of mammal movement, indicating that the late Eocene and earliest Oligocene "marks the end of extensive land-mammal interchange with Europe across the North Atlantic Land Bridge." More recent research demonstrates a succession of migrations between North America and western Europe via the NALB until the Pliocene (Denk, Grímsson, and Zetter 2010; Denk et al. 2011a, 2011b).

P. 134 *"The mostly Eurasian oaks—subgenus* Cerris—
appear to have arisen in East Asia . . ."

Early fossils. Eurasian sections *Cerris* and *Ilex* are known from the Shkotovo Basin in far eastern Russia, 56–48 mya (Denk et al. 2023, suppl. table 2). Fossil distribution of section *Cyclobalanopsis*: Jia et al. (2015); Xu et al. (2016); Barrón et al. (2017). Geography and chronology of the Tibetan Plateau and Himalayan uplift: C. Wang et al. (2008). Evidence for *Cyclobalanopsis* persisting in Pliocene Europe: Vieira et al. (2023), updating a previous report (Sadowski, Schmidt, and Denk 2020).

Phylogenetic histories. *Cyclobalanopsis*: Deng et al. (2018). *Cerris* and *Ilex* (in part): Denk et al. (2023). The section *Ilex* history is largely from Jiang et al. (2019), but the timing is adjusted based on Denk et al. (2023). *Quercus ilex* cultural significance and human uses: Shirone, Vessella, and Varela (2019); Drori (2018, 48). Sharing of chloroplast genomes in Eurasia: "Some Iberian Cork oaks were found to carry a chloroplast genome unique to the western Mediterranean Holm oaks (there have been reports of inter-sectional hybrids in the region, but conclusive genetic evidence has yet to be produced), while in the Aegean region, some holly oaks have been found to carry the plastomes typical for the cork oaks" (G. Grimm, pers. comm., March 9, 2023; see also Simeone et al. 2016, 2018).

P. 136 *"North America crept westward as the Eurasian oaks diversified."*

Early fossil evidence. Axel Heiberg Island White Oak: McIntyre (1991); McIver and Basinger (1999). Sect. *Protobalanus* in Eocene Greenland: Grímsson et al. (2015). Sect. *Lobatae* pollen in British Columbia: Grímsson et al. (2016).

Shifts in North American climate and vegetation. Overview: Graham (1999, 194, 197). Western Inland Sea: Graham (1999, 188–189; 2011, 26); Slattery et al. (2013; see especially Figs. 9–16). Forest-to-grassland transition. Fossil soils in Oregon, the Great Plains, and Pakistan: Retellack (1997; 2001); grass abundance and photosynthetic pathway evolution: Keeley and Rundel (2005); Edwards and Smith (2010).

Western North America. Diversification within the Rocky Mountains: Wing et al. (1987); Bouchal et al. (2014); Mensing (2015). Red and White Oak fossils of Oligocene Texas: Daghlian and Crepet (1983). *Virentes* is particularly complex, including a radiation in the southeastern United States and species in Cuba, Mexico, Central America, and the Baja Peninsula (Cavender-Bares et al. 2015). Section *Protobalanus*: Ortego, Gugger, and Sork (2018).

Species counts. The species counts presented are rough numbers, even for such small groups as the California oaks, in part because some varieties in each of the groups are genetically distinctive. At this point it is not clear whether *Q. gambelii* of the southern Rocky Mountains should be considered a California white oak, as its genome is a mix of *Q. macrocarpa* and *Q.* sect. *Dumosae* (McVay, Hauser, Hipp, and Manos 2017). California White Oaks: Fitz-Gibbon et al. (2017). California Red Oaks: Hauser et al. (2017). Overall numbers for the North American oaks may seem low to some readers: there are more than ninety species of oaks treated in the *Flora of North America* (Nixon 1997). However, I do not include the Texas and Mexican/C. American/SW U.S. lineages for either the white oaks or the red oaks here, as those lineages did not originate until the Oligocene or Miocene and were secondary radiations from eastern North American ancestors (Hipp et al. 2020, 2018). Thus the species I include in this count correspond to white oak subsections *Prinoideae*, *Albae*, and *Stellatae*, and red oak subsections *Palustres*, *Coccineae*, and *Phellos* (Manos and Hipp 2021).

P. 140 *"Much of the world became dramatically colder 34 million years ago, at the beginning of the Oligocene."*

Climate and geography. Global temperatures: Westerhold et al. (2020). Temperature changes in northern Europe: Hren et al. (2013). Temperature heterogeneity: Zanazzi et al. (2007). Global synthesis of global changes at the Eocene-Oligocene transition (EOT): Hutchinson et al. (2021). Decreasing atmospheric CO_2 as the primary driver in the Southern Hemisphere: Lauretano et al. (2021). The Paratethys Sea covered much of what is now the Eurasian Steppe through the Oligocene, becoming an isolated and massive lake at about 12 mya. It drained away dramatically in the Miocene (Palcu et al. 2021).

Extinctions and migrations. Extinctions and biodiversity turnover reviewed: Prothero (1994). Effects of the EOT on North American biota: Zanazzi et al. (2007). Mollusk extinction: Ivany, Patterson, and Lohmann (2000). Floristic and vegetation EOT trends: Pound and Salzmann (2017); Hutchinson et al. (2021). Spread and dominance of temperate forests: Martinetto et al. (2020).

P. 142 *"In October 2022, one of my graduate students, Kieran Althaus, and I flew to northeastern Mexico to collect oaks with twelve of our colleagues."*

Intro. This opening is adapted from Hipp et al. (2023). The trip was supported in part by National Geographic Society Grant #NGS-73961R-20 to my colleague Antonio González-Rodríguez and would have been impossible without the work of the group—primarily Antonio, Socorro González-Elizondo, Hernando Rodríguez-Correa, Allen Coombes, and their respective students and staff—in organizing it. The group also included regional and taxonomic experts Lucio Caamaño, Jacinto Treviño Carreón, and Arturo Mora-Olivo.

Eastern North American–Eastern Mexican disjunctions. Reviews: Martin and Harrell (1957); Graham (1999b); Hipp et al. (2023); Stull (2023). The estimate of 100 species and lineages comes from Stull's review. Additional examples. *Tilia*: McCarthy and Mason-Gamer (2016); *Ulmus*: Whittemore et al. (2021); *Viburnum*: Donoghue et al. (2022); *Pinus*: Jin et al. (2021); *Acer saccharum* complex: Vargas-Rodriguez et al. (2020, 2015).

Oaks move into Mexico. *Quercus tardifolia*: Renault (2021, 2022). *Quercus* fossils. Earliest putative fossils are from Chiapas, in Chavez and Rzedowski (1993), cited in Graham (1999b, 35) with some hesitation ("Some confirmation of these intriguing identifications is necessary to assess this enigmatic flora") and not indicated in his figures. However, Late Miocene fossils are also found in Guatemala and Panama (Graham 1999b, fig. 4). Miocene timing aligns with molecular dates (Hipp et al. 2018, 2020, 2023). Initial Mexican oak diversification in the mountains may be somewhat complicated by the fact that some of the earliest oak fossils in Mexico are from lower-elevation sites (Graham 1999b).

Mexican oak diversification. Oak differentiation along environmental gradients in Mexico: Spellenberg, Bacon, and González-Elizondo (1998); Figueroa-Rangel and Olvera-Vargas (2022). Diversification rates: Hipp et al. (2018). Oak diversity: Valencia-Avalos (2004); this is likely an underestimate, as numerous species have been described subsequently, and molecular work has split some wide-ranging species into a morphologically, ecologically, and genomically distinct entities (Morales-Saldaña et al. 2022; see also McCauley, Cortés-Palomec, and Oyama 2019).

Particular species. *Quercus oleoides*: Cavender-Bares et al. (2015). *Quercus insignis*: García-de la Cruz et al. (2014); Jerome (2018). *Quercus humboldtii*: Avella Muñoz and Rangel Churio (2014); Zorrilla-Azcué et al. (2021); Ortego et al. (2023). My use of "lonely oak" is from Zorrilla-Azcué et al. *Quercus engelmannii*: Mensing (2005); O'Donnell et al. (2021).

P. 145 *"Around the time oaks were entering Mexico,
an ancestor of three eastern North American oak
species . . . was heading toward Eurasia."*

North Atlantic Land Bridge (NALB). From Denk et al. (2010, 285): "From Middle Miocene sediments of Alaska, Heer (1869) reported lobed leaves of *Quercus furuhjelmii* Heer[,] . . . [which] Wolfe and Tanai (1980) compared . . . to modern East Asian and North American oaks with 'chestnut'-like laminas. Closer inspection shows that the fossil species resembles East Asian white oaks such as *Q. dentata* Thunb. in Murray, *Q. fabri* Hance, and *Q. mongolica* Fisch. ex Ledeb." There appear to have been pulses of White and Red Oak migration between North America and Europe via the NALB until at least 5 mya (Denk, Grímsson, and Zetter 2010). The chloroplast topology referenced here: Pham et al. (2017).

White oak success. White Oak Group habitat description: Guido Grimm (pers. comm., March 10, 2023). Vessel diameter: Cavender-Bares and Holbrook (2001). Hybridization with *Q. pontica*: McVay et al. (2017); Crowl et al. (2020b); B.-F. Zhou et al. (2022). Distribution of the White Oak Group: Denk et al. (2017, fig. 2.3). Temperate forest biome distribution: Martinetto et al. (2020, fig. 2.1a).

P. 147 *"Today's oaks look quite organized."*

Baltic amber: Sadowski et al. (2020). Portugal distribution and other western European distributions from the Miocene through the present day: Vieira et al. (2023). Section *Cyclobalanopsis*: Liu et al. (2019, though see the critical evaluation of some fossils in Sadowski et al. 2020; Vieira et al. 2023).

Chapter Six

P. 151 *"Miles Davis walked into Columbia Studio B on August 19, 1969, with an engineer, twelve musicians, and his producer, Teo Macero."*

This history comes largely from Tingen (2017; 2001, chap. 5) and Szwed (2004). "There were some low moments, some starts and stops": Szwed (2004, 295). "Dash of Jack DeJohnette": Lenny White, a nineteen-year-old drummer who played on the album (Tingen 2017). The timing for "Pharaoh's Dance" follows the 1999 remix and annotations in Tingen (2001, 312).

P. 152 *"When I watch a robin flipping oak leaves over on the forest floor, I experience the cumulative effect of about 35,000 protein-coding genes in the oak genome . . ."*

Repetitive regions in the oak genome: Kapoor et al. (2023). "The number of particles in the visible universe": Eddington's estimate of the number of baryons in the visible universe (Vopson 2021). Limits to recombination: cf. Anderson's (1949, 34) "recombination spindle."

P. 154 *"The spool of tape in the Columbia Recording studio, with all the cuts and splices, is not the music itself."*

Sturtevant's (1913) first linkage-mapping project: Mukherjee (2016). First rudiments of an oak genomic map: Zanetto et al. (1996), two chromosomes only. First linkage map of all 12 chromosomes: Barreneche et al. (1998). Oak chiasmata: Nativadade in Igens-Moller (1955). *Quercus rubra* RAD-seq linkage map: Konar et al. (2017). Marker transmission from parent to offspring: Bodénès et al. (2016). Linkage maps linked to traits: Scotti-Saintagne et al. (2004b, 2004a). Carbon isotope composition QTL: Brendel et al. (2008). Brief review of oak QTL studies: Gailing et al. (2022, sect. 4.3).

P. 158 *"If a linkage map is a notation of the points where the raw recording has been cut and spliced back together, then an assembled reference genome is, at its best, the completed album."*

Assembling a reference genome. Telomere-to-telomere human genome, save for the Y-chromosome: Nurk et al. (2022). "Genes become duplicated when": Panchy et al. (2016); Grover and Wendel (2010). Complexities of plant genomes: Schatz et al. (2012); Paterson et al. (2010); Lynch and Conery (2003); Yanqing Sun et al. (2022). Sequencing coverage: it is common to use many more sequences than

I present in the hypothetical (and fanciful) example of "sequencing" "Pharaoh's Dance"; the *Q. mongolica* genome was based on more than 16 million PacBio sequencing reads averaging nearly 14,000 base pairs each, with about 2.6 times as many base pairs of shorter Illumina and Hi-C reads (Ai et al. 2022, tables S3, S7).

Quercus robur **genome**: Plomion et al. (2018; 2016). Transcriptome resource developed from "3P" and others: Ueno et al. (2010). SNP-based linkage map: Bodénès et al. (2016). It is worth noting that this first reference genome was based on early genome-sequencing technology, and it yielded a genome that had, like the human genome, numerous gaps, more than 36,000 of them by one estimate. To create a completed genome assembly, the researchers tethered fragments ("contigs") of the oak genome to a high-density oak linkage map and a previously published peach genome assembly, which is similar at fine scales to the oak genome. This practice is no longer the norm, but relating genomes to linkage maps is still a good practice, a kind of sanity-check.

New oak reference genomes. I am only referencing chromosome-level, annotated reference genomes (Sork et al. 2022; Plomion et al. 2018; e.g., Ai et al. 2022; Ye Sun et al. 2021; Mishra et al. 2022; Shirasawa et al. 2021; R. Fu et al. 2022; X. Zhou et al. 2022; Han et al. 2022; Yuan et al. 2023; Kapoor et al. 2023; Rey et al. 2023). Genomic similarity between oaks and grasses: Sork et al. (2022).

P. 161 *"You and I are modern humans . . ."*

Human introgression. The percent of the modern human genome that comes from Neanderthal introgression varies among different populations, with populations of East Asian origin generally having the most, followed by European, followed by North African and then Subsaharan African populations (Ahlquist et al. 2021; L. Chen et al. 2020; Peyrégne, Slon, and Kelso 2023). Denisovans do not have a Linnaean (scientific) name, as they are known primarily from ancient DNA sequences (Brown et al. 2022). Tissue types affected by Neanderthal introgression: Ahlquist et al. (2021). Facial features: Bonfante et al. (2021). Altitude adaptation: Huerta-Sánchez et al. (2014). Skin color, freckling, and lipid metabolism: Racimo et al. (2015). Blood coagulation, adaptive introgression: Simonti et al. (2016). Disease-related genes, introgression from now-extinct human lineages: Benton et al. (2021); Dannemann and Kelso (2017). Ust'-Ishim: Q. Fu et al. (2014); Moorjani et al. (2016). "How important is introgressive hybridization?": Anderson (1949, 102).

Oak introgression. *Quercus grisea* and *Q. gambelii*: Howard et al. (1997); Swenson, Fair, and Heikoop (2008). *Quercus ellipsoidalis* and *Q. rubra*: Khodwekar and Gailing (2017); Lind-Riehl, Sullivan, and Gailing (2014). California White Oak transcriptomes: Oney-Birol et al. (2018). California Red Oaks: Dodd and Afzal-Rafii (2004); Dodd, Kashani, and Azal-Rafii (2002). Shrub white oaks of California: Burge et al. (2019). Engelmann oak adaptive introgression: O'Donnell et al. (2021); O'Donnell (2023). European White Oaks: Leroy et al. (2020a, 2017, 2020b). *Quercus variabilis* and *Q. acutissima*: R. Fu et al. (2022).

P. 166 *"Some genes track branches of the Tree of Life."*

Chloroplast genomes. Oak chloroplast genome size: X. Li et al. (2018); the oak mitochondrial genome is 2.5 times as large with half as many genes (Bi et al. 2019). Maternal inheritance of oak chloroplasts: Dumolin-Lapegue, Demesure, and Petit (1995). Even given the opportunity, chloroplasts rarely recombine (according to experimental work in *Oenothera*: Chiu and Sears 1985).

Selected milestones in oak molecular phylogenics. First robust molecular phylogeny of *Quercus*: Manos et al. (1999). The turn to single-copy nuclear genes: Oh and Manos (2008); Hubert et al. (2014). Application of nuclear ribosomal DNA repeats to dissect the effects of gene flow: Denk and Grimm (2010); Cardoni et al. (2022); Piredda et al. (2021); Simeone et al. (2018). Use of a DNA fingerprinting technique, amplified fragment length polymorphisms (AFLPs) to resolve fine-scale relationships among closely related species: Pearse and Hipp (2009).

Eurasian White Oaks. Initial work: Manos et al. (1999); Oh and Manos (2008); Denk and Grimm (2010); Hubert et al. (2014). Morphological hypothesis from the 1980s: Axelrod (1983). Initial RAD-seq phylogeny: Hipp et al. (2014). Eurasian White Oak phylogeny: McVay, Hipp, and Manos (2017); Crowl et al. (2020b); B.-F. Zhou et al. (2022, fig. S10a).

Introgression in other oak clades. Southern Live Oaks (sect. *Virentes*): Eaton et al. (2015); Cavender-Bares et al. (2015). California White Oaks (subsect. *Dumosae*): Kim et al. (2018); O'Donnell et al. (2021); Oney-Birol et al. (2018); Ortego et al. (2014). *Quercus gambelii*: McVay et al. (2017). Golden-Cupped Oaks (sect. *Protobalanus*): Ortego, Gugger, and Sork (2018).

P. 170 *"Intrinsic reproductive isolation has evolved between at least some pairs of related oaks."*

Genome rearrangement and speciation. Sunflowers: numerous papers by Loren Rieseberg and collaborators (e.g., Rieseberg 2001; Rieseberg, Van Fossen, and Desrochers 1995). Sedges and other holocentric organisms: Escudero et al. (2016; 2023); Lucek, Augustijnen, and Escudero (2022).

Genome collinearity. *Quercus gilva*, *Q. lobata*, and *Q. mongolica*: Ai et al. (2022); X. Zhou et al. (2022). *Quercus rubra*, *Q. lobata*, *Q. mongolica*: Kapoor et al. (2023).

The syngameon. Evidence of multispecies introgression in other tree groups: Cannon and Lerdau (2022, 2015); Caron et al. (2019); D. Larson et al. (2021). "What we know as *Bitches Brew* could have been assembled twenty different ways": Bob

Belden, in Szwed (2004). "We argue here that the syngameon is more than the sum of the pairwise interactions between species": Cannon and Petit (2020, 979).

Chapter Seven

P. 175 *"Woodcocks migrate through the upper Midwest in the latter half of October."*

Late October phenology: pers. obs. (https://botanistsfieldnotes.com/2019/10/20/a -hesitant-turn-toward-fall/; accessed September 19, 2023).

P. 176 *"Give natural selection a few populations of bacteria and 2 billion years, and it may hand you back an orangutan and an oak tree."*

Evolutionary hangers-on. Whales: Brusatte (2022, chap. 9, "Extreme mammals"); Coyne (2009, 50). Cavefish: Krishnan and Rohner (2017). Wisdom teeth and other vestigial organs: Dhawan, Yedavalli, and Massoud (2023); Spinney et al. (2008). Variation evolving through the retooling of old solutions: Shubin, Tabin, and Carroll (2009).

Phylogenetic niche conservatism.

ANGIOSPERMS AND TREES. Importance of phylogenetic niche conservatism: Donoghue (2008). Evolution of freezing-tolerance: Zanne et al. (2014). Phylogenetic conservatism in trees of the Americas: Segovia et al. (2020). Attributes of the tropics: adapted from Segovia et al. (2020), as based in turn on Feeley and Stroud (2018). Embolism mechanism: Charra-Vaskou et al. (2023). *Selenicereus wittii*, an epiphytic cactus with water-dispersed seeds, endemic to inundated forests of the Amazon: Barthlott et al. (1997). Fagales niche conservatism: Segovia et al. (2020); Folk et al. (2023). Priority effects are discussed in the context of evolutionary legacy effects in Cavender-Bares et al. (2016).

IN OAKS. Holly oaks (sect. *Ilex*): Alonso-Forn et al. (2023). Last gasps of *Castanopsis*, *Trigonobalanus*, and sections *Lobatae* and *Cyclobalanopsis* in western Europe: Vieira et al. (2023). Rot resistance in Red versus White Oaks: Scheffer, Englerth, and Duncan (1949). Vessel diameter in White versus Red Oaks: Cavender-Bares and Holbrook (2001). Tyloses in Red versus White Oaks: Fallon et al. (2020). Comparative oak wood anatomy: Nixon (2009). Seed strategies in Red versus White Oaks: Steele et al. (2004); this book, chapter 1. Seed selection in mice: Ancillotto, Sozio, and Mortelliti (2015). Robert Frost on writing free verse as "playing tennis with the net down": 1933 February 10, *Scranton (Pennsylvania) Times*, "Frost Reads Poems at Century Club Meeting," February 10, 1933, p. 22, col. 6. https://quoteinvestigator.com/2021/05 /24/poem-tennis.

P. 180 *"I stood on a roadside near Miquihuana, Mexico . . ."*

Oak convergence. Oak examples: Tucker (1974); Trelease (1924, 117, cited in Tucker 1974). "African and Asian monkeys": Family Cercopithecidae; divergence times from Grabowski and Jungers (2017, fig. 1a). Arguments for evolution of lobedness: Edwards et al. (2016). Sclerophylly: Gil-Pelegrín and his lab and collaborators (e.g., Gil-Pelegrín et al. 2017; Alonso-Forn et al. 2020; Sancho-Knapik et al. 2021). Convergence on drought-resistance in the Mexican oaks: Aguilar-Romero et al. (2017).

Initial observations of oak phylogenetic overdispersion. Initially reported in the eastern North American Piedmont (Bourdeau 1954); expanded by Whittaker (1969) as quoted in Mohler (1990, 247).

Connecting the Tree of Life to oak communities. Foundational papers on the Florida oaks: Cavender-Bares (2019); Cavender-Bares, Ackerly, Baum, and Bazzaz (2004); Cavender-Bares, Kitajima, and Bazzaz (2004); Cavender-Bares, Keen, and Miles (2006). Continental U.S. oaks, phylogenetic effects: Cavender-Bares et al. (2018). Biographical perspective on Cavender-Bares's work: Sridhar (2021).

Western United States and Eurasia. Oaks of the Chiricahua Mountains: Fallon and Cavender-Bares (2018). Siskiyou Mountains: These species are from a composite of several transects (Whittaker 1960, table 12, transect no. 3); *Q. garryana* is not listed by Whittaker for the composite transect, but I include it in this list because it is known to hybridize with *Q. sadleriana*. Northeast Spain: Pascual, Molinas, and Verdaguer (2002). *Quercus humboldtii*: Hooghiemstra, Cleef, and Flantua (2022). Additional examples from the oak community: Keator and Bazell (1998, 199–237).

Darwin on phylogenetic overdispersion. "From having nearly the same structure, constitution, and habits": C. Darwin (1872, 59, 86).

P. 186 *"Convergence helps explain why distant relatives favor similar habitats, but it doesn't fully explain the diversity of oak communities."*

Trade-offs.

OAKS OF THE WEST TEXAS SKY ISLANDS: Schwilk, Gaetani, and Poulos (2013). *Quercus hypoleucoides* seeming to grow more quickly when young: Schwilk (pers. comm., October 30, 2023) notes that while no growth rates have been published, resprout heights measured after fires suggest that the species of mesic sites do, in fact, grow more quickly when they are young.

QUERCUS ILEX: Amimi et al. (2023).

SECTION *VIRENTES*: Koehler, Center, and Cavender-Bares (2012); population effect in this common garden study was less than the species effect, but significant in the widespread *Q. oleoides* and *Q. virginiana*.

TRADE-OFFS ARE, IN MANY CASES, MORE COMPLEX. Within the Americas, evergreen oak species exhibit a clear trade-off between relative growth rate and drought-tolerance, with species that invest most highly in protecting themselves against drought having the lowest ability to take quick advantage of water when it comes; deciduous species do not exhibit the same trade-off (Kaproth et al. 2023). And in a cross-species study of oaks ranging from temperate to Mediterranean climates in western Europe, on average, species from mesic climates simply started growing earlier, not more rapidly. Drought-tolerance did not come at an obvious cost (Ramírez-Valiente et al. 2020). But habitat specificity and convergence are still clear in both of these systems.

Mexico, nutrient cycling: González-Rodríguez et al. (2019); Chávez-Vergara et al. (2015). A potential facilitation effect has been reported based on fire frequency data in oak savannas of Minnesota, in which the growth and survivorship of Hill's oak—a member of the Red Oak Group—are improved when it grows with bur oak, a member of the White Oak Group, more than when it grows with other members of its own species. Bur oak did not benefit as much from growing with Hill's oak, though individual bur oaks survived longer on average when growing with Hill's oak than when growing alone in the prairie. These effects appear not to be driven by soil, but perhaps are driven by the high fire tolerance of bur oak (M. Davis and Condit 2022).

Janzen-Connell effects. Janzen-Connell effects have been demonstrated in tropical forests, temperate forests, temperate grasslands, and coral reefs (Terborgh 2020).
OAK WILT: Appel (2008). Juzwik et al. (2008) showed that oak wilt can also spread to Eurasian White Oaks, but this would have little bearing on the phylogenetic structure of North American plant communities.
BUR OAK BLIGHT (*TUBAKIA IOWENSIS*). Spread on bur oak and swamp white oak: T. Harrington, McNew, and Yun (2012). Host distribution, eastern North American *Tubakia*: West (2015). Species specificity of other *Tubakia*: Matsumura, Morinaga, and Fukuda (2022). "Bur oak blight may spread all the more rapidly": Swanston et al. (2018).

Insects faithful to oak clades: Pearse and Hipp (2009, 2012, 2014).
Co-occurrence of distantly related oaks across North America: Mohler (1990); Cavender-Bares et al. (2018). Counterexamples to the Red Oak–White Oak pattern are mostly from Mohler (1990).

P. 189 *"More than 1,000 insect species are known to feed on oaks."*

Insects. Insects on oaks: Trieff (2002, 11). Caterpillars: Narango, Tallamy, and Shropshire (2020), analyzed by genus, not by species. Diversity of butterflies, bees, wasps as a function of oak abundance: Andreas et al. (2023), who found a landscape-scale effect from urban forests in Prague, at 20 to 40 m; at the plot level, oak relative abundance had no statistically significant effect. "Patches of hairs": Coltharp et al. (2021). *Temnothorax* (acorn ants): Karlik et al. (2016); Mitrus et al. (2021); Tallamy (2021, 24–26); Giannetti et al. (2022).

Endophytes. Transfer from leaves to acorns: U'Ren and Zimmerman (2021); species specificity: Moricca et al. (2012); tolerance to phenolics: Nickerson, Moore, and U'Ren (2023); effects on leaf decomposition: Weatherhead et al. (2022).

Mycorrhizae.

FOSSIL HISTORY. Oldest fossils demonstrating mycorrhizal associations in plants dates to 407 mya, but spores of mycorrhizal fungi are known from fossils dating to 50 Ma earlier (Tedersoo, Bahram, and Zobel 2020; Brundrett and Tedersoo 2018). Fungi colonized the land an estimated 250 million years before plants did (Lutzoni et al. 2018).

MYCORRHIZAL NETWORKS. Both interspecific fungal connections and the movement of water and nutrients are documented between oaks and other species connected by mycorrhizal fungi (e.g., Egerton-Warburton, Querejeta, and Allen 2007; Meding 2007; Toju et al. 2013). However, the data are still inconclusive regarding the extent and importance of common mycorrhizal connections (the "wood-wide web") (Karst, Jones, and Hoeksema 2023).

EFFECTS OF LEAF LITTER on mycorrhizal fungal communities: Aponte et al. (2010) (*Quercus suber* and *Q. canariensis*). Fungus-specific gene activity within trees: Bouffaud et al. (2020).

ECM/AM DICHOTOMY: Phillips, Brzostek, and Midgley (2013).

TRANSPLANT EXPERIMENT, FIFTY-FIVE SPECIES: J. Bennett et al. (2017).

GREENHOUSE EXPERIMENT: Wu et al. (2022) tested the sister species *Q. michauxii* and *Q. alba* (white oaks) and *Q. shumardii* and *Q. acerifolia* (red oaks). The results bear additional study, as they performed the experiment in only one direction (all soil was from either *Q. shumardii* or *Q. alba*) and used acorns from a single maternal tree of each species.

P. 192 *"Oak gall wasps, Cynipids, number*
approximately 1,400 species worldwide."

Gall wasp biology: Stone et al. (2002); Egan et al. (2018); Stone and Schönrogge (2003). Evidence that gall wasps hijack genes similar to those found in the nitrogen-fixing root nodules of legumes: Hearn et al. (2019). Gall ink: Corregidor et al. (2019).

Galls in deep history. Fossil gall, Middle Devonian liverwort: Labandeira (2021). Estimated ages: bracketed by earliest fossil evidence of cynipids (Ronquist 1999; Stone et al. 2009) and a calibrated molecular phylogeny (Blaimer et al. 2020). Evidence that gall wasps may have specialized on other Fagaceae before *Quercus*: *Dryocosmos kuriphilis*, which falls near the edge of the oak gall phylogeny in two of the most recent genomic phylogenetic analyses (Blaimer et al. 2020; Ward et al. 2022), forms galls on chestnuts, not oaks.

Gall interactions. Nectar production: Nicholls, Melika, and Stone (2017). Ant-dispersal of galls: Warren II et al. (2022). Leaf-tying insects: H. Wang et al. (2023)

P. 194 *"What is the importance of oak species?"*

Ecological arguments. Much has been written on the connection between diversity and productivity; for studies from which these particular points are drawn, see Tang et al. (2022); J. Ding et al. (2021). Biomass and diversity, United States and Mexico: Cavender-Bares (2019, 2016). Net monetary value of oaks: Cavender-Bares et al. (2022). Oaks fall second in net monetary value to pines (*Pinus*); which provide an estimated $7.4 billion per year in wood products (in comparison to $557 million from oaks). Value to mammals and birds: Van Dersal (1940).

Evolutionary arguments. This argument about the creative role of species in the syngameon is particularly clearly articulated by Stebbins (1950, 278), who analogizes "closely related, incompletely isolated species" to Wright's (1940) model of adaptive gene flow between populations of a single species.

Epilogue

P. 197 *"Almost every lineage will go extinct, eventually."*

Human evolution. *Homo floresiensis*: Sutikna et al. (2016). Timing of the gene flow between *Homo sapiens* and *H. neanderthalensis* is covered in chapter 6 of this book.

How long will oaks survive? This calculation is from Jim Holt's essay "The Riemann Zeta Conjecture and the Laughter of the Primes" (Holt 2018, chap. 4); following Gott (1993). Assuming our observations are drawn from a random moment in Earth's history and we are correct about the age of the things we are observing, then the 95% confidence interval around their remaining age is 39 × the observed age for the upper bound, 1/39 × the observed age for the lower.

Climate change. The scenario alluded to is RCP 8.5, which assumes "high population and relatively slow income growth with modest rates of technological change and energy intensity improvements" (Riahi et al. 2011, 33). Projections compared to paleoclimates: Burke et al. (2018). Rangewide projections, North American trees: Overpeck et al. (1991); Iverson and Prasad (1998); Iverson et al. (2019); projections for the southern United States: W. Wang et al. (2019). Projections account for neither the potential for evolutionary change, nor the risk of nonnative insects and emerging pathogens.

Extinction rate estimates. Background extinction rate: De Vos et al. (2015). The back-of-the-envelope calculation I am using to extrapolate from De Vos et al. to oaks is 0.135 extinctions per million species per year × 425 oak species

= 0.000057375 oak species per year. Oak species at risk: Carrero et al. (2020). Recent extinction rates in plants—comparisons with the background rate: Humphreys et al. (2019). Current rate of extinction: Pimm and Joppa (2015).

Oak risks. Oak reproduction crisis: Gottschalk and Wargo (1997). *Quercus lobata*: F. Davis, Tyler, and Mahall (2011). *Quercus garryana*: MacDougall, Duwyn, and Jones (2010). *Quercus alba*: Abrams (2003).

P. 199 *"The history of oaks illustrates that species are begotten of extinction."*

Proteus and Menelaus: Homer (1919 book IV, lines 380–480).

Literature Cited

Abadie, P., G. Roussel, B. Dencausse, C. Bonnet, E. Bertocchi, J.-M. Louvet, A. Kremer, and P. Garnier-Géré. 2012. "Strength, Diversity and Plasticity of Postmating Reproductive Barriers between Two Hybridizing Oak Species (*Quercus robur* L. and *Quercus petraea* (Matt) Liebl.)." *Journal of Evolutionary Biology* 25 (1): 157–73. https://doi.org/10.1111/j.1420-9101.2011.02414.x. .

Abrams, Marc D. 2003. "Where Has All the White Oak Gone?" *BioScience* 53 (10): 927–39. https://doi.org/10.1641/0006-3568(2003)053[0927:WHATWO]2.0.CO;2.

Aguilar-Romero, Rafael, Fernando Pineda-Garcia, Horacio Paz, Antonio González-Rodríguez, and Ken Oyama. 2017. "Differentiation in the Water-Use Strategies among Oak Species from Central Mexico." *Tree Physiology* 37 (7): 915–25. https://doi.org/10.1093/treephys/tpx033.

Ahlquist, K. D., Mayra M. Bañuelos, Alyssa Funk, Jiaying Lai, Stephen Rong, Fernando A. Villanea, and Kelsey E. Witt. 2021. "Our Tangled Family Tree: New Genomic Methods Offer Insight into the Legacy of Archaic Admixture." *Genome Biology and Evolution* 13 (7): evab115. https://doi.org/10.1093/gbe/evab115.

Ahmed, Moinuddin, Kevin J. Anchukaitis, Asfawossen Asrat, Hemant P. Borgaonkar, Martina Braida, Brendan M. Buckley, Ulf Büntgen, et al. [PAGES 2k Consortium]. 2013. "Continental-Scale Temperature Variability during the Past Two Millennia." *Nature Geoscience* 6 (5): 339–46. https://doi.org/10.1038/ngeo1797.

Ai, Wanfeng, Yanqun Liu, Mei, Xiaolin Zhang, Enguang Tan, Hanzhang Liu, Xiaoyi Han, et al. 2022. "A Chromosome-Scale Genome Assembly of the Mongolian Oak (*Quercus mongolica*)." *Molecular Ecology Resources* 22 (6): 2396–2410. https://doi.org/10.1111/1755-0998.13616.

Akhmetiev, Mikhail A., Nina I. Zaporozhets, Vladimir N. Benyamovskiy, Galina N.

Aleksandrova, Alina I. Iakovleva, and Tatiana V. Oreshkina. 2012. "The Paleogene History of the Western Siberian Seaway—a Connection of the Peri-Tethys to the Arctic Ocean." *Austrian Journal of Earth Science* 105:50–67.

Alberto, F., J. Niort, J. Derory, O. Lepais, R. Vitalis, D. Galop, and A. Kremer. 2010. "Population Differentiation of Sessile Oak at the Altitudinal Front of Migration in the French Pyrenees." *Molecular Ecology* 19 (13): 2626–39. https://doi.org/10.1111/j.1365-294X.2010.04631.x.

Alexandre, Hermine, Laura Truffaut, Alexis Ducousso, Jean-Marc Louvet, Gérard Nepveu, José M. Torres-Ruiz, Frédéric Lagane, et al. 2020. "In Situ Estimation of Genetic Variation of Functional and Ecological Traits in *Quercus petraea* and *Q. robur*." *Tree Genetics and Genomes* 16 (2): 32. https://doi.org/10.1007/s11295-019-1407-9.

Alexandre, Hermine, Laura Truffaut, Etienne Klein, Alexis Ducousso, Emilie Chancerel, Isabelle Lesur, Benjamin Dencausse, et al. 2020. "How Does Contemporary Selection Shape Oak Phenotypes?" *Evolutionary Applications* 13 (10): 2772–90. https://doi.org/10.1111/eva.13082.

Allan, Mea. 1977. *Darwin and His Flowers: The Key to Natural Selection*. London: Faber and Faber.

Alonso-Forn, David, Domingo Sancho-Knapik, María Dolores Fariñas, Miquel Nadal, Rubén Martín-Sánchez, Juan Pedro Ferrio, Víctor Resco de Dios, et al. 2023. "Disentangling Leaf Structural and Material Properties in Relationship to Their Anatomical and Chemical Compositional Traits in Oaks (*Quercus* L.)." *Annals of Botany* (February): mcad030. https://doi.org/10.1093/aob/mcad030.

Alonso-Forn, David, Domingo Sancho-Knapik, Juan Pedro Ferrio, José Javier Peguero-Pina, Amauri Bueno, Yusuke Onoda, Jeannine Cavender-Bares, et al. 2020. "Revisiting the Functional Basis of Sclerophylly within the Leaf Economics Spectrum of Oaks: Different Roads to Rome." *Current Forestry Reports* 6 (December): 260–81. https://doi.org/10.1007/s40725-020-00122-7.

Amimi, Nabil, Hana Ghouil, Rim Zitouna-Chebbi, Thierry Joët, and Youssef Ammari. 2023. "Intraspecific Variation of *Quercus ilex* L. Seed Morphophysiological Traits in Tunisia Reveals a Trade-Off between Seed Germination and Shoot Emergence Rates along a Thermal Gradient." *Annals of Forest Science* 80 (1): 12. https://doi.org/10.1186/s13595-023-01179-7.

Ancillotto, Leonardo, Giulia Sozio, and Alessio Mortelliti. 2015. "Acorns Were Good until Tannins Were Found: Factors Affecting Seed-Selection in the Hazel Dormouse (*Muscardinus avellanarius*)." *Mammalian Biology* 80 (2): 135–40. https://doi.org/10.1016/j.mambio.2014.05.004.

Anderson, Edgar. 1928. "The Problem of Species in the Northern Blue Flags, *Iris versicolor* L. and *Iris virginica* L." *Annals of the Missouri Botanical Garden* 15 (3): 241–332. https://doi.org/10.2307/2394087.

———. 1949. *Introgressive Hybridization*. New York: Wiley.

Anderson, Edgar, and Leslie Hubricht. 1938. "Hybridization in *Tradescantia*. III. The Evidence for Introgressive Hybridization." *American Journal of Botany* 25 (6): 396–402. https://doi.org/10.2307/2436413.

Andreas, Michal, Romana Prausová, Tereza Brestovanská, Lucie Hostinská, Markéta Kalábová, Petr Bogusch, Josef P. Halda, et al. 2023. "Tree Species-Rich Open Oak Woodlands within Scattered Urban Landscapes Promote Biodiversity." *Urban Forestry and Urban Greening* (March): 127914. https://doi.org/10.1016/j.ufug.2023.127914.

Angiosperm Phylogeny Group. 1998. "An Ordinal Classification for the Families of Flowering Plants." Annals of the Missouri Botanical Garden 85 (4): 531. https://doi.org/10.2307/2992015.

Aponte, Cristina, Luis V. García, Teodoro Marañón, and Monique Gardes. 2010. "Indirect Host Effect on Ectomycorrhizal Fungi: Leaf Fall and Litter Quality Explain Changes in Fungal Communities on the Roots of Co-Occurring Mediterranean Oaks." *Soil Biology and Biochemistry* 42 (5): 788–96. https://doi.org/10.1016/j.soilbio.2010.01.014.

Appel, David N. 2008. "Oak Wilt Biology, Impact, and Host/Pathogen Relationships: A Texas Perspective." In *The Proceedings of the 2nd National Oak Wilt Symposium*, edited by Ronald F. Billings and David N. Appel, 43–54. International Society of Arboriculture—Texas Chapter. https://texasoakwilt.org/assets/studies/NOWS/conference_assets/conferencepapers/MacDonaldandDouble.pdf.

Aristotle. 1910. *De Generatione Animalium*. Edited by W. D. Ross and J. A. Smith. Translated by Arthur Platt. Oxford: Clarendon Press. Original work ca. 350 BCE.

Ashley, Mary V. 2021. "Answers Blowing in the Wind: A Quarter Century of Genetic Studies of Pollination in Oaks." *Forests* 12 (5): 575. https://doi.org/10.3390/f12050575.

Augustin, Laurent, Carlo Barbante, Piers R. F. Barnes, Jean Marc Barnola, Matthias Bigler, Emiliano Castellano, Olivier Cattani, et al. [EPICA Community Members]. 2004. "Eight Glacial Cycles from an Antarctic Ice Core." *Nature* 429 (6992): 623–28. https://doi.org/10.1038/nature02599.

Austerlitz, F., C. Dutech, P. E. Smouse, F. Davis, and V. L. Sork. 2007. "Estimating Anisotropic Pollen Dispersal: A Case Study in *Quercus lobata*." *Heredity* 99 (2): 193–204. https://doi.org/10.1038/sj.hdy.6800983.

Avella Muñoz, Andrés, and Jesús Orlando Rangel Churio. 2014. "Oak Forests Types of *Quercus humboldtii* in the Guantiva-La Rusia-Iguaque Corridor (Santander-Boyacá, Colombia): Their Conservation and Sustainable Use." *Colombia Forestal* 17 (1): 100–116.

Axelrod, D. I. 1983. "Biogeography of Oaks in the Arcto-Tertiary Province." *Annals of the Missouri Botanical Garden* 70:629–57.

Babila, Tali L., Donald E. Penman, Christopher D. Standish, Monika Doubrawa, Timothy J. Bralower, Marci M. Robinson, Jean M. Self-Trail, et al. 2022. "Surface Ocean Warming and Acidification Driven by Rapid Carbon Release Precedes Paleocene-Eocene Thermal Maximum." *Science Advances* 8 (11): eabg1025. https://doi.org/10.1126/sciadv.abg1025.

Bacilieri, Roberto, Alexis Ducousso, Rémy J. Petit, and Antoine Kremer. 1996. "Mating System and Asymmetric Hybridization in a Mixed Stand of European Oaks." *Evolution* 50 (2): 900–908. https://doi.org/10.1111/j.1558-5646.1996.tb03898.x.

Backs, Janet Rizner, Martin Terry, Mollie Klein, and Mary V. Ashley. 2015. "Genetic Analysis of a Rare Isolated Species: A Tough Little West Texas Oak, *Quercus hinckleyi* C.H. Mull." *Journal of the Torrey Botanical Society* 142 (4): 302–13. https://doi.org/10.3159/TORREY-D-14-0009.

Balaguer, L., E. Martínez-Ferri, F. Valladares, M. E. Pérez-Corona, F. J. Baquedano, F. J. Castillo, and E. Manrique. 2001. "Population Divergence in the Plasticity of the Response of *Quercus coccifera* to the Light Environment." *Functional Ecology* 15 (1): 124–35.

Baldwin, Andrew, and Brian W. Booth. 2022. "Biomedical Applications of Tannic Acid." *Journal of Biomaterials Applications* 36 (8): 1503–23. https://doi.org/10.1177/08853282211058099.

Baranski, Michael J. 1975. *An Analysis of Variation within White Oak* (Quercus alba L.). *Tech. Bul. No. 236*. Raleigh: North Carolina Agricultural Experiment Station.

Barba-Montoya, Jose, Mario dos Reis, Harald Schneider, Philip C. J. Donoghue, and Ziheng Yang. 2018. "Constraining Uncertainty in the Timescale of Angiosperm Evolution and the Veracity of a Cretaceous Terrestrial Revolution." *New Phytologist* 218 (2): 819–34. https://doi.org/10.1111/nph.15011.

Barker, Stephen, Aidan Starr, Jeroen van der Lubbe, Alice Doughty, Gregor Knorr, Stephen Conn, Sian Lordsmith, et al. 2022. "Persistent Influence of Precession on Northern Ice Sheet Variability since the Early Pleistocene." *Science* 376 (6596): 961–67. https://doi.org/10.1126/science.abm4033.

Barreneche, T., C. Bodénès, C. Lexer, J. F. Trontin, S. Fluch, R. Streiff, C. Plomion, et al. 1998. "A Genetic Linkage Map of *Quercus robur* L. (Pedunculate Oak) Based on RAPD, SCAR, Microsatellite, Minisatellite, Isozyme and 5S rDNA Markers." *Theoretical and Applied Genetics* 97:1090–1103.

Barrón, Eduardo, Anna Averyanova, Zlatko Kvaček, Arata Momohara, Kathleen B. Pigg, Svetlana Popova, José María Postigo-Mijarra, et al. 2017. "The Fossil History of *Quercus*." In *Oaks Physiological Ecology. Exploring the Functional Diversity of Genus Quercus L.*, edited by Eustaquio Gil-Pelegrín, José Javier Peguero-Pina, and Domingo Sancho-Knapik, 39–105. Tree Physiology. Cham, Germany: Springer International. https://doi.org/10.1007/978-3-319-69099-5_3.

Barthlott, Wilhelm, Stefan Porembski, Manfred Kluge, Jörn Hopke, and Loki Schmidt. 1997. "*Selenicereus Wittii* (Cactaceae): An Epiphyte Adapted to Amazonian Igapó Inundation Forests." *Plant Systematics and Evolution* 206 (1): 175–85. https://doi.org/10.1007/BF00987947.

Bartholomé, Jérôme, Benjamin Brachi, Benoit Marçais, Amira Mougou-Hamdane, Catherine Bodénès, Christophe Plomion, Cécile Robin, et al. 2020. "The Genetics of Exapted Resistance to Two Exotic Pathogens in Pedunculate Oak." *New Phytologist* 226 (4): 1088–1103. https://doi.org/10.1111/nph.16319.

Barton, Nicholas H. 2020. "On the Completion of Speciation." *Philosophical Transactions of the Royal Society B: Biological Sciences* 375 (1806): 20190530. https://doi.org/10.1098/rstb.2019.0530.

Baskin, Jerry, and Carol Baskin. 2016. "Origins and Relationships of the Mixed Mesophytic Forest of Oregon–Idaho, China, and Kentucky: Review and Synthesis 1." *Annals of the Missouri Botanical Garden* 101 (April): 525–52. https://doi.org/10.3417/2014017.

Baum, David A., and Stacey D. Smith. 2013. *Tree Thinking: An Introduction to Phylogenetic Biology.* Greenwood Village, CO: Roberts and Co.

Bemmels, Jordan B., L. Lacey Knowles, and Christopher W. Dick. 2019. "Genomic Evidence of Survival near Ice Sheet Margins for Some, but Not All, North American Trees." *Proceedings of the National Academy of Sciences* 116 (17): 8431–36. https://doi.org/10.1073/pnas.1901656116.

Bennett, Jonathan A., Hafiz Maherali, Kurt O. Reinhart, Ylva Lekberg, Miranda M. Hart, and John Klironomos. 2017. "Plant-Soil Feedbacks and Mycorrhizal Type Influence Temperate Forest Population Dynamics." *Science* 355 (6321): 181–84. https://doi.org/10.1126/science.aai8212.

Bennett, Matthew R., David Bustos, Jeffrey S. Pigati, Kathleen B. Springer, Thomas M. Urban, Vance T. Holliday, Sally C. Reynolds, et al. 2021. "Evidence of Humans in North America during the Last Glacial Maximum." *Science* 373 (6562): 1528–31. https://doi.org/10.1126/science.abg7586.

Benton, Mary Lauren, Abin Abraham, Abigail L. LaBella, Patrick Abbot, Antonis Rokas, and John A. Capra. 2021. "The Influence of Evolutionary History on Human Health and Disease." *Nature Reviews Genetics* 22 (5): 269–83. https://doi.org/10.1038/s41576-020-00305-9.

Bergström, Anders, Chris Stringer, Mateja Hajdinjak, Eleanor M. L. Scerri, and Pontus Skoglund. 2021. "Origins of Modern Human Ancestry." *Nature* 590 (7845): 229–37. https://doi.org/10.1038/s41586-021-03244-5.

Bert, Didier, Jean-Baptiste Lasnier, Xavier Capdevielle, Aline Dugravot, and Marie-Laure Desprez-Loustau. 2016. "Powdery Mildew Decreases the Radial Growth of Oak Trees with Cumulative and Delayed Effects over Years." *PLOS ONE* 11 (5): e0155344. https://doi.org/10.1371/journal.pone.0155344.

Bi, Quanxin, Dongxing Li, Yang Zhao, Mengke Wang, Yingchao Li, Xiaojuan Liu, Libing Wang, et al. 2019. "Complete Mitochondrial Genome of *Quercus variabilis* (Fagales, Fagaceae)." *Mitochondrial DNA Part B* 4 (2): 3927–28. https://doi.org/10.1080/23802359.2019.1687027.

Bice, Karen L., Daniel Birgel, Philip A. Meyers, Kristina A. Dahl, Kai-Uwe Hinrichs, and Richard D. Norris. 2006. "A Multiple Proxy and Model Study of Cretaceous Upper Ocean Temperatures and Atmospheric CO_2 Concentrations." *Paleoceanography* 21 (2): PA2002. https://doi.org/10.1029/2005PA001203.

Bierzychudek, Paulette. 1984. "Determinants of Gender in Jack-in-the-Pulpit: The Influence of Plant Size and Reproductive History." *Oecologia* 65 (1): 14–18. https://doi.org/10.1007/BF00384456.

Bijl, Peter K., James A. P. Bendle, Steven M. Bohaty, Jörg Pross, Stefan Schouten, Lisa Tauxe, Catherine E. Stickley, et al. 2013. "Eocene Cooling Linked to Early Flow across the Tasmanian Gateway." *Proceedings of the National Academy of Sciences* 110 (24): 9645–50. https://doi.org/10.1073/pnas.1220872110.

Blaimer, Bonnie B., Dietrich Gotzek, Seán G. Brady, and Matthew L. Buffington. 2020. "Comprehensive Phylogenomic Analyses Re-Write the Evolution of Parasitism within Cynipoid Wasps." *BMC Evolutionary Biology* 20 (1): 155. https://doi.org/10.1186/s12862-020-01716-2.

Blom, Philipp. 2019. *Nature's Mutiny: How the Little Ice Age of the Long Seventeenth Century Transformed the West and Shaped the Present.* New York: Liveright.

Blue, Marguerite P., and Richard J. Jensen. 1988. "Positional and Seasonal Variation in Oak (*Quercus*, Fagaceae) Leaf Morphology." *American Journal of Botany* 75:939–47.

Boavida, Leonor C., J. P. Silva, and J. A. Feijó. 2001. "Sexual Reproduction in the Cork Oak (*Quercus suber* L). II. Crossing Intra- and Interspecific Barriers." *Sexual Plant Reproduction* 14 (3): 143–52. https://doi.org/10.1007/s004970100100.

Boavida, Leonor C., M. Carolina Varela, and J. A. Feijó. 1999. "Sexual Reproduction in the Cork Oak (*Quercus suber* L.). I. The Progamic Phase." *Sexual Plant Reproduction* 11 (6): 347–53. https://doi.org/10.1007/s004970050162.

Bodénès, Catherine, Emilie Chancerel, François Ehrenmann, Antoine Kremer, and Christophe Plomion. 2016. "High-Density Linkage Mapping and Distribution of Segregation Distortion Regions in the Oak Genome." *DNA Research: An International Journal for Rapid Publication of Reports on Genes and Genomes* 23 (2): 115–24. https://doi.org/10.1093/dnares/dsw001.

Bonfante, Betty, Pierre Faux, Nicolas Navarro, Javier Mendoza-Revilla, Morgane Dubied, Charlotte Montillot, Emma Wentworth, et al. 2021. "A GWAS in Latin Americans Identifies Novel Face Shape Loci, Implicating VPS13B and a Denisovan Introgressed Region in Facial Variation." *Science Advances* 7 (6): eabc6160. https://doi.org/10.1126/sciadv.abc6160.

Bonner, Franklin T., and Robert P. Karrfalt, eds. 2008. *The Woody Plant Seed Manual.* Agriculture Handbook 727. Washington, DC: U.S. Department of Agriculture, Forest Service.

Borchert, Mark I., Frank W. Davis, Joel Michaelsen, and Lyn Dee Oyler. 1989. "Interactions of Factors Affecting Seedling Recruitment of Blue Oak (*Quercus douglasii*) in California." *Ecology* 70 (2): 389–404. https://doi.org/10.2307/1937544.

Bossema, I. 1979. "Jays and Oaks: An Eco-Ethological Study of a Symbiosis." *Behaviour* 70 (1–2): 1–117.

Bouchal, Johannes, Reinhard Zetter, Friðgeir Grímsson, and Thomas Denk. 2014. "Evolutionary Trends and Ecological Differentiation in Early Cenozoic Fagaceae of Western North America." *American Journal of Botany* 101 (8): 1332–49. https://doi.org/10.3732/ajb.1400118.

Bouchenak-Khelladi, Yanis, Renske E. Onstein, Yaowu Xing, Orlando Schwery, and H. Peter Linder. 2015. "On the Complexity of Triggering Evolutionary Radiations." *New Phytologist* 207 (2): 313–26. https://doi.org/10.1111/nph.13331.

Bouffaud, Marie-Lara, Sylvie Herrmann, Mika T. Tarkka, Markus Bönn, Lasse Feldhahn, and François Buscot. 2020. "Oak Displays Common Local but Specific Distant Gene Regulation Responses to Different Mycorrhizal Fungi." *BMC Genomics* 21 (1): 399. https://doi.org/10.1186/s12864-020-06806-5.

Bourdeau, Philippe. 1954. "Oak Seedling Ecology Determining Segregation of Species in Piedmont Oak-Hickory Forests." *Ecological Monographs* 24 (3): 297–320. https://doi.org/10.2307/1948467.

Bowen, Gabriel J., and James C. Zachos. 2010. "Rapid Carbon Sequestration at the Termination of the Palaeocene–Eocene Thermal Maximum." *Nature Geoscience* 3 (12): 866–69. https://doi.org/10.1038/ngeo1014.

Brendel, Oliver, Didier Le Thiec, Caroline Scotti-Saintagne, Catherine Bodénès, Antoine Kremer, and Jean-Marc Guehl. 2008. "Quantitative Trait Loci Controlling Water Use Efficiency and Related Traits in *Quercus robur* L." *Tree Genetics and Genomes* 4 (2): 263–78. https://doi.org/10.1007/s11295-007-0107-z.

Bresson, Caroline C., Yann Vitasse, Antoine Kremer, and Sylvain Delzon. 2011. "To What Extent Is Altitudinal Variation of Functional Traits Driven by Genetic Adaptation in European Oak and Beech?" *Tree Physiology* 31 (11): 1164–74. https://doi.org/10.1093/treephys/tpr084.

Brewer, Simon, Christelle Hélyalleaume, Rachid Cheddadi, Jacques-Louis de Beaulieu, Jeanne-Marine Laurent, and Joseph Le Cuziat. 2005. "Postglacial History of Atlantic Oakwoods: Context, Dynamics and Controlling Factors." *Botanical Journal of Scotland* 57 (1–2): 41–57. https://doi.org/10.1080/03746600508685084.

Brewer, Simon, R. Cheddadi, J. L. de Beaulieu, and M. Reille. 2002. "The Spread of Deciduous *Quercus* throughout Europe since the Last Glacial Period." *Forest Ecology and Management* 156 (1): 27–48. https://doi.org/10.1016/S0378-1127(01)00646-6.

Brikiatis, Leonidas. 2014. "The De Geer, Thulean and Beringia Routes: Key Concepts for Understanding Early Cenozoic Biogeography." *Journal of Biogeography* 41 (6): 1036–54. https://doi.org/10.1111/jbi.12310.

Brown, Roy C., and H. L. Mogensen. 1972. "Late Ovule and Early Embryo Development in *Quercus gambelii*." *American Journal of Botany* 59 (3): 311–16.

Brown, Samantha, Diyendo Massilani, Maxim B. Kozlikin, Michael V. Shunkov, Anatoly P. Derevianko, Alexander Stoessel, Blair Jope-Street, et al. 2022. "The Earliest Denisovans and Their Cultural Adaptation." *Nature Ecology and Evolution* 6 (1): 28–35. https://doi.org/10.1038/s41559-021-01581-2.

Browne, Luke, Jessica W. Wright, Sorel Fitz-Gibbon, Paul F. Gugger, and Victoria L. Sork. 2019. "Adaptational Lag to Temperature in Valley Oak (*Quercus lobata*) Can Be Mitigated by Genome-Informed Assisted Gene Flow." *Proceedings of the National Academy of Sciences* 116 (50): 25179–85. https://doi.org/10.1073/pnas.1908771116.

Brundrett, Mark C., and Leho Tedersoo. 2018. "Evolutionary History of Mycorrhizal Symbioses and Global Host Plant Diversity." *New Phytologist* 220 (4): 1108–15. https://doi.org/10.1111/nph.14976.

Brusatte, Steve. 2018. *The Rise and Fall of the Dinosaurs: A New History of a Lost World*. New York: HarperCollins.

———. 2022. *The Rise and Reign of the Mammals: A New History, from the Shadow of the Dinosaurs to Us*. New York: Mariner Books.

Bruschi, Piero, Paolo Grossoni, and Filippo Bussotti. 2003. "Within- and among-Tree Variation in Leaf Morphology of *Quercus petraea* (Matt.) Liebl. Natural Populations." *Trees* 17 (2): 164–72. https://doi.org/10.1007/s00468-002-0218-y.

Bucher, Enrique H. 1992. "The Causes of Extinction of the Passenger Pigeon." In *Current Ornithology*, edited by Dennis M. Power, 1–36. Current Ornithology, vol. 9. Boston, MA: Springer. https://doi.org/10.1007/978-1-4757-9921-7_1.

Buck, Ryan, and Lluvia Flores-Rentería. 2022. "The Syngameon Enigma." *Plants* 11 (7): 895. https://doi.org/10.3390/plants11070895.

Buckley, Michael. 2015. "Ancient Collagen Reveals Evolutionary History of the Endemic South American 'Ungulates.'" *Proceedings of the Royal Society B: Biological Sciences* 282 (1806): 20142671. https://doi.org/10.1098/rspb.2014.2671.

Buckley, S. B. 1861. "Note on the Bartram Oak (*Quercus heterophylla*)." *Proceedings of the Academy of Natural Sciences of Philadelphia* 13:335–90.

Burckle, L., and H. D. Grissino-Mayer. 2003. "Stradivari, Violins, Tree Rings, and the Maunder Minimum: A Hypothesis." *Dendrochronologia* 21 (1): 41–45. https://doi.org/10.1078/1125-7865-00033.

Burge, Dylan O., V. Thomas Parker, Margaret Mulligan, and Victoria L. Sork. 2019. "Influence of a Climatic Gradient on Genetic Exchange between Two Oak Species." *American Journal of Botany* 106 (6): 864–78. https://doi.org/10.1002/ajb2.1315.

Burger, William C. 1975. "The Species Concept in *Quercus*." *Taxon* 24:45–50.

Burke, K. D., J. W. Williams, M. A. Chandler, A. M. Haywood, D. J. Lunt, and B. L. Otto-Bliesner. 2018. "Pliocene and Eocene Provide Best Analogs for Near-Future Climates." *Proceedings of the National Academy of Sciences* 115 (52): 13288–93. https://doi.org/10.1073/pnas.1809600115.

Buschbom, Jutta, Yulay Yanbaev, and Bernd Degen. 2011. "Efficient Long-Distance Gene Flow into an Isolated Relict Oak Stand." *Journal of Heredity* 102 (4): 464–72. https://doi.org/10.1093/jhered/esr023.

Caignard, Thomas, Antoine Kremer, Xavier P. Bouteiller, Julien Parmentier, Jean-Marc Louvet, Samuel Venner, and Sylvain Delzon. 2021. "Counter-Gradient Variation of Reproductive Effort in a Widely Distributed Temperate Oak." *Functional Ecology* 35 (8): 1745–55. https://doi.org/10.1111/1365-2435.13830.

Caignard, Thomas, Laura Truffaut, Sylvain Delzon, Benjamin Dencausse, Laura Lecacheux, José M. Torres-Ruiz, and Antoine Kremer. 2023. "Fluctuating Selection and Rapid Evolution of Oaks during Recent Climatic Transitions." *Plants, People, Planet* (blog). New Phytologist Foundation, August 28. https://doi.org/10.1002/ppp3.10422.

Callaham, R. Z. 1964. "Provenance Research: Investigation of Genetic Diversity Associated with Geography." *Unasylva* 18:2–12.

Cannon, Charles H., and Manuel Lerdau. 2015. "Variable Mating Behaviors and the Maintenance of Tropical Biodiversity." *Frontiers in Genetics* 6:183. https://doi.org/10.3389/fgene.2015.00183.

———. 2022. "Asking Half the Question in Explaining Tropical Diversity." *Trends in Ecology and Evolution* 37 (5): 392–93. https://doi.org/10.1016/j.tree.2022.01.006.

Cannon, Charles H., and Paul S. Manos. 2003. "Phylogeography of the Southeast Asian Stone Oaks (*Lithocarpus*)." *Journal of Biogeography* 30 (2): 211–26. https://doi.org/10.1046/j.1365-2699.2003.00829.x.

Cannon, Charles H., and Rémy J. Petit. 2020. "The Oak Syngameon: More Than the Sum of Its Parts." *New Phytologist* 226 (4): 978–83. https://doi.org/10.1111/nph.16091.

Cannon, Charles H., Gianluca Piovesan, and Sergi Munné-Bosch. 2022. "Old and Ancient Trees Are Life History Lottery Winners and Vital Evolutionary Resources for Long-Term Adaptive Capacity." *Nature Plants* (January): 1–10. https://doi.org/10.1038/s41477-021-01088-5.

Cao, Xianyong, Ulrike Herzschuh, Jian Ni, Yan Zhao, and Thomas Böhmer. 2015. "Spatial and Temporal Distributions of Major Tree Taxa in Eastern Continental Asia during the Last 22,000 Years." *Holocene* 25 (1): 79–91. https://doi.org/10.1177/0959683614556385.

Cardoni, Simone, Roberta Piredda, Thomas Denk, Guido W. Grimm, Aristotelis C. Papageorgiou, Ernst-Detlef Schulze, Anna Scoppola, et al. 2022. "5S-IGS rDNA in Wind-Pollinated Trees (*Fagus* L.) Encapsulates 55 Million Years of Reticulate Evolution and Hybrid Origins of Modern Species." *Plant Journal* 109 (4): 909–26. https://doi.org/10.1111/tpj.15601.

Carlisle, Emily, Christine M. Janis, Davide Pisani, Philip C. J. Donoghue, and Daniele Silvestro. 2023. "A Timescale for Placental Mammal Diversification Based on

Bayesian Modeling of the Fossil Record." *Current Biology* 33 (15): 3073–3082.E3. https://doi.org/10.1016/j.cub.2023.06.016.

Caron, Henri, Jean-François Molino, Daniel Sabatier, Patrick Léger, Philippe Chaumeil, Caroline Scotti-Saintagne, Jean-Marc Frigério, et al. 2019. "Chloroplast DNA Variation in a Hyperdiverse Tropical Tree Community." *Ecology and Evolution* 9 (8): 4897–4905. https://doi.org/10.1002/ece3.5096.

Carrero, Christina, Diana Jerome, Emily Beckman, Amy Byrne, Allen J. Coombes, Min Deng, Antonio González-Rodríguez, et al. 2020. *The Red List of Oaks 2020.* Lisle, IL: Morton Arboretum.

Cavender-Bares, Jeannine. 2016. "Diversity, Distribution, and Ecosystem Services of the North American Oaks." *International Oaks* 27:37–48.

———. 2019. "Diversification, Adaptation, and Community Assembly of the American Oaks (*Quercus*), a Model Clade for Integrating Ecology and Evolution." *New Phytologist* 221 (2): 669–92. https://doi.org/10.1111/nph.15450.

Cavender-Bares, Jeannine, David D. Ackerly, D. A. Baum, and F. A. Bazzaz. 2004. "Phylogenetic Overdispersion in Floridian Oak Communities." *American Naturalist* 163 (6): 823–43.

Cavender-Bares, Jeannine, David D. Ackerly, Sarah E. Hobbie, and Philip A. Townsend. 2016. "Evolutionary Legacy Effects on Ecosystems: Biogeographic Origins, Plant Traits, and Implications for Management in the Era of Global Change." *Annual Review of Ecology, Evolution, and Systematics* 47 (1): 433–62. https://doi.org /10.1146/annurev-ecolsys-121415-032229.

Cavender-Bares, Jeannine, Antonio González-Rodríguez, Deren A. R. Eaton, Andrew L. Hipp, Anne Beulke, and Paul S. Manos. 2015. "Phylogeny and Biogeography of the American Live Oaks (*Quercus* Subsection *Virentes*): A Genomic and Population Genetics Approach." *Molecular Ecology* 24:3668–87.

Cavender-Bares, Jeannine, and N. M. Holbrook. 2001. "Hydraulic Properties and Freezing-Induced Cavitation in Sympatric Evergreen and Deciduous Oaks with Contrasting Habitats." *Plant, Cell and Environment* 24 (12): 1243–56. https://doi .org/10.1046/j.1365-3040.2001.00797.x.

Cavender-Bares, Jeannine, A. Keen, and B. Miles. 2006. "Phylogenetic Structure of Floridian Plant Communities Depends on Taxonomic and Spatial Scale." Supplement. *Ecology* 87 (7): S109–S22.

Cavender-Bares, Jeannine, Kaoru Kitajima, and F. A. Bazzaz. 2004. "Multiple Trait Associations in Relation to Habitat Differentiation among 17 Floridian Oak Species." *Ecological Monographs* 74 (4): 635–62. https://doi.org/10.1890/03-4007.

Cavender-Bares, Jeannine, Shan Kothari, José Eduardo Meireles, Matthew A. Kaproth, Paul S. Manos, and Andrew L. Hipp. 2018. "The Role of Diversification in Community Assembly of the Oaks (*Quercus* L.) across the Continental U.S." *American Journal of Botany* 105 (3): 565–86. https://doi.org/10.1002/ajb2.1049.

Cavender-Bares, Jeannine, Erik Nelson, Jose Eduardo Meireles, Jesse R. Lasky, Daniela A. Miteva, David J. Nowak, William D. Pearse, et al. 2022. "The Hidden Value of Trees: Quantifying the Ecosystem Services of Tree Lineages and Their Major Threats across the Contiguous US." *PLOS Sustainability and Transformation* 1 (4): e0000010. https://doi.org/10.1371/journal.pstr.0000010.

Cavender-Bares, Jeannine, and Annette Pahlich. 2009. "Molecular, Morphological, and Ecological Niche Differentiation of Sympatric Sister Oak Species, *Quercus virginiana* and *Q. geminata* (Fagaceae)." *American Journal of Botany* 96 (9): 1690–1702. https://doi.org/10.3732/ajb.0800315.

Cavender-Bares, Jeannine, and José A. Ramírez-Valiente. 2017. "Physiological Evidence from Common Garden Experiments for Local Adaptation and Adaptive Plasticity to Climate in American Live Oaks (*Quercus* Section *Virentes*): Implications for Conservation under Global Change." In *Oaks Physiological Ecology. Exploring the Functional Diversity of Genus Quercus L.*, edited by Eustaquio Gil-Pelegrín, José Javier Peguero-Pina, and Domingo Sancho-Knapik, 107–35. Tree Physiology. Cham, Germany: Springer International. https://doi.org/10.1007/978-3-319-69099-5_4.

Cecich, Robert A. 1996. "The Reproductive Biology of *Quercus*, with an Emphasis on *Q. rubra*." *International Oaks: The Journal of the International Oak Society* 7:11.

———. 1997. "Notes: Pollen Tube Growth in *Quercus*." *Forest Science* 43 (1): 140–46. https://doi.org/10.1093/forestscience/43.1.140.

Cecich, Robert A., and William W. Haenchen. 1995. "Pollination Biology of Northern Red and Black Oak." In *Proceedings, 10th Central Hardwood Forest Conference; 1995 March 5–8; Morgantown, WV; Gen. Tech. Rep. NE-197*, edited by Kurt W. Gottschalk and Sandra L. C. Fosbroke, 238–46. Radnor, PA: U.S. Department of Agriculture, Forest Service, Northeastern Forest Experiment Station.

Chadburn, S. E., E. J. Burke, P. M. Cox, P. Friedlingstein, G. Hugelius, and S. Westermann. 2017. "An Observation-Based Constraint on Permafrost Loss as a Function of Global Warming." *Nature Climate Change* 7 (5): 340–44. https://doi.org/10.1038/nclimate3262.

Charra-Vaskou, Katline, Anna Lintunen, Thierry Améglio, Eric Badel, Hervé Cochard, Stefan Mayr, Yann Salmon, et al. 2023. "Xylem Embolism and Bubble Formation during Freezing Suggest Complex Dynamics of Pressure in *Betula pendula* Stems." *Journal of Experimental Botany* (July): erad275. https://doi.org/10.1093/jxb/erad275.

Chase, Mark W., Douglas E. Soltis, Richard G. Olmstead, David Morgan, Donald H. Les, Brent D. Mishler, Melvin R. Duvall, et al. 1993. "Phylogenetics of Seed Plants: An Analysis of Nucleotide Sequences from the Plastid Gene rbcL." *Annals of the Missouri Botanical Garden* 80 (3): 528–80. https://doi.org/10.2307/2399846.

Chassé, Béatrice. 2016. "Eating Acorns: What Story Do the Distant, Far, and Near Past Tell Us, and Why?" *International Oaks: The Journal of the International Oak Society* 27:107–35.

Chater, Arthur O., and Ray G. Woods. 2019. *The Powdery Mildews (Erysipha-*

les) of Wales: An Identification Guide and Census Catalogue. Aberystwyth, Wales: A. O. Chater. https://www.aber.ac.uk/waxcap/downloads/Chater19 -PowderyMildewsWalesCensus.pdf.

Chavez, Rodolfo Palacios, and Jerzy Rzedowski. 1993. "Estudio palinológico de las floras fósiles del Mioceno Inferior y principios del Mioceno Medio de la región de Pichucalco, Chiapas, México." Acta Botanica Mexicana, no. 24 (August): 1–96. https://doi.org/10.21829/abm24.1993.677.

Chávez-Vergara, Bruno M., Antonio González-Rodríguez, Jorge D. Etchevers, Ken Oyama, and Felipe García-Oliva. 2015. "Foliar Nutrient Resorption Constrains Soil Nutrient Transformations under Two Native Oak Species in a Temperate Deciduous Forest in Mexico." European Journal of Forest Research 134 (5): 803–17. https://doi.org/10.1007/s10342-015-0891-1.

Chen, Dongmei, Xianxian Zhang, Hongzhang Kang, Xiao Sun, Shan Yin, Hongmei Du, Norikazu Yamanaka, et al. 2012. "Phylogeography of Quercus variabilis Based on Chloroplast DNA Sequence in East Asia: Multiple Glacial Refugia and Mainland-Migrated Island Populations." PLOS ONE 7 (10): e47268. https://doi .org/10.1371/journal.pone.0047268.

Chen, Lu, Aaron B. Wolf, Wenqing Fu, Liming Li, and Joshua M. Akey. 2020. "Identifying and Interpreting Apparent Neanderthal Ancestry in African Individuals." Cell 180 (4): 677–687.e16. https://doi.org/10.1016/j.cell.2020.01.012.

Chen, Xi, Charles H. Cannon, and Nancy Lou Conklin-Brittan. 2012. "Evidence for a Trade-Off Strategy in Stone Oak (Lithocarpus) Seeds between Physical and Chemical Defense Highlights Fiber as an Important Antifeedant." PLOS ONE 7 (3): e32890. https://doi.org/10.1371/journal.pone.0032890.

Chen, Xi, and Takashi S. Kohyama. 2021. "Variation among 91 Stone Oak Species (Fagaceae, Lithocarpus) in Fruit and Vegetative Morphology in Relation to Climatic Factors." Flora (November): 151984. https://doi.org/10.1016/j.flora.2021 .151984.

Cheng-Yih, Wu, and K. Kubitzki. 1993. "Rhoipteleaceae." In Flowering Plants— Dicotyledons: Magnoliid, Hamamelid and Caryophyllid Families, edited by Klaus Kubitzki, Jens G. Rohwer, and Volker Bittrich, 584–85. The Families and Genera of Vascular Plants. Berlin: Springer. https://doi.org/10.1007/978-3-662-02899 -5_68.

Chiu, Wan-Ling, and Barbara B. Sears. 1985. "Recombination between Chloroplast DNAs Does Not Occur in Sexual Crosses of Oenothera." Molecular and General Genetics MGG 198 (3): 525–28. https://doi.org/10.1007/BF00332951.

Cho, S. E., S. H. Lee, S. K. Lee, S. T. Seo, and H. D. Shin. 2018. "Erysiphe alphitoides Causes Powdery Mildew on Eucalyptus gunnii." Forest Pathology 48 (1): e12377. https://doi.org/10.1111/efp.12377.

Chung, Kin Pan, Enrique Gonzalez-Duran, Stephanie Ruf, Pierre Endries, and Ralph Bock. 2023. "Control of Plastid Inheritance by Environmental and Genetic Factors." Nature Plants 9 (1): 68–80. https://doi.org/10.1038/s41477-022-01323-7.

Chung-MacCoubrey, Alice L., Ann E. Hagerman, and Roy L. Kirkpatrick. 1997. "Effects of Tannins on Digestion and Detoxification Activity in Gray Squirrels (*Sciurus carolinensis*)." *Physiological Zoology* 70 (3): 270–77. https://doi.org/10.1086 /639595.

Cinar-Yilmaz, H., and Ü. Akkemik. 2007. "Embryo Anatomy in *Quercus alnifolia* Poech." *Seed Science and Technology* 35 (2): 494–96. https://doi.org/10.15258/sst .2007.35.2.23.

Clark, James S., Chris Fastie, George Hurtt, Stephen T. Jackson, Carter Johnson, George A. King, Mark Lewis, et al. 1998. "Reid's Paradox of Rapid Plant Migration: Dispersal Theory and Interpretation of Paleoecological Records." *BioScience* 48 (1): 13–24. https://doi.org/10.2307/1313224.

Clark, Peter U., Arthur S. Dyke, Jeremy D. Shakun, Anders E. Carlson, Jorie Clark, Barbara Wohlfarth, Jerry X. Mitrovica, et al. 2009. "The Last Glacial Maximum." *Science* 325 (5941): 710–14. https://doi.org/10.1126/science.1172873.

Clausen, J., D. D. Keck, and W. M. Hiesey. 1940. *Experimental Studies on the Nature of Species. I. Effects of Varied Environments on Western North American Plants.* Publication number 520. Carnegie Institute of Washington, DC.

Clyde, William C., and Philip D. Gingerich. 1998. "Mammalian Community Response to the Latest Paleocene Thermal Maximum: An Isotaphonomic Study in the Northern Bighorn Basin, Wyoming." *Geology* 26 (11): 1011–14. https://doi.org /10.1130/0091-7613(1998)026<1011:MCRTTL>2.3.CO;2.

Cohen, K. M., S. C. Finney, P. L. Gibbard, and J.-X. Fan. 2013. "The ICS International Chronostratigraphic Chart [Updated June 2023]." *Episodes: Journal of International Geoscience* 36 (3): 199–204. https://doi.org/10.18814/epiiugs/2013/v36i3 /002. Latest version accessed January 29, 2024. https://stratigraphy.org/chart.

Coltharp, Erin, Chloe Knowd, Ella Abelli-Amen, Andrew Abounayan, Sophia Alcaraz, Rachael Auer, Sarah Beilman, et al. 2021. "Leaf Hair Tufts Function as Domatia for Mites in *Quercus agrifolia* (Fagaceae)." *Madroño* 67 (4): 165–69. https://doi.org/10.3120/0024-9637-67.4.165.

Condit, Carlton. 1944. "The Remington Hill Flora." In *Pliocene Floras of California and Oregon*, edited by R. W. Cheney, 21–55. Publication number 553. Carnegie Institution of Washington, DC.

Constabel, Peter C., Kazuko Yoshida, and Vincent Walker. 2014. "Diverse Ecological Roles of Plant Tannins: Plant Defense and Beyond." In *Recent Advances in Polyphenol Research*, edited by Annalisa Romani, Vincenzo Lattanzio, and Stéphane Quideau, 115–42. Hoboken, NJ: Wiley. https://doi.org/10.1002/9781118329634 .ch5.

Constance, Lincoln. 1955. "The Systematics of the Angiosperms." In *A Century of Progress in the Natural Sciences: 1853–1953*, edited by Ernest B. Babcock, J. Wyatt Durhan, and George S. Myers, 405–82. San Francisco: California Academy of Sciences.

Copolovici, Lucian, Fred Väärtnõu, Miguel Portillo Estrada, and Ülo Niinemets.

2014. "Oak Powdery Mildew (*Erysiphe alphitoides*)-Induced Volatile Emissions Scale with the Degree of Infection in *Quercus robur*." *Tree Physiology* 34 (12): 1399–1410. https://doi.org/10.1093/treephys/tpu091.

Corregidor, Victoria, Rita Viegas, Luís M. Ferreira, and Luís C. Alves. 2019. "Study of Iron Gall Inks, Ingredients and Paper Composition Using Non-Destructive Techniques." *Heritage* 2 (4): 2691–2703. https://doi.org/10.3390/heritage2040166.

Corriveau, Joseph L., and Annette W. Coleman. 1988. "Rapid Screening Method to Detect Potential Biparental Inheritance of Plastid DNA and Results for over 200 Angiosperm Species." *American Journal of Botany* 75 (10): 1443–58. https://doi.org/10.1002/j.1537-2197.1988.tb11219.x.

Cottam, Walter P., John M. Tucker, and Frank S. Santamour. 1982. *Oak Hybridization at the University of Utah*. Salt Lake City: State Arboretum of Utah.

Coyne, Jerry A. 2009. *Why Evolution Is True*. New York: Viking Penguin.

Coyne, Jerry A., and H. A. Orr. 2004. *Speciation*. Sunderland, MA: Sinauer Associates.

Craft, Kathleen J., and Mary V. Ashley. 2006. "Population Differentiation among Three Species of White Oak in Northeastern Illinois." *Canadian Journal of Forest Research* 26:206–15.

Craft, Kathleen J., Mary V. Ashley, and Walter D. Koenig. 2002. "Limited Hybridization between *Quercus lobata* and *Quercus douglasii* (Fagaceae) in a Mixed Stand in Central Coastal California." *American Journal of Botany* 89 (11): 1792–98.

Crepet, William L., and Kevin C. Nixon. 1989a. "Earliest Megafossil Evidence of Fagaceae: Phylogenetic and Biogeographic Implications." *American Journal of Botany* 76 (6): 842–55. https://doi.org/10.2307/2444540.

———. 1989b. "Extinct Transitional Fagaceae from the Oligocene and Their Phylogenetic Implications." *American Journal of Botany* 76 (10): 1493–1505. https://doi.org/10.2307/2444437.

Cronquist, Arthur. 1965. "The Status of the General System of Classification of Flowering Plants." *Annals of the Missouri Botanical Garden* 52 (3): 281–303. https://doi.org/10.2307/2394794.

Crowl, Andrew A., E. Bruno, Andrew L. Hipp, and Paul S. Manos. 2020. "Revisiting the Mystery of the Bartram Oak." *Arnoldia* 77 (4): 6–11.

Crowl, Andrew A., Paul S. Manos, John D. McVay, Alan R. Lemmon, Emily Moriarty Lemmon, and Andrew L. Hipp. 2020. "Uncovering the Genomic Signature of Ancient Introgression between White Oak Lineages (*Quercus*)." *New Phytologist* 226 (4): 1158–70. https://doi.org/10.1111/nph.15842.

Crowther, T. W., H. B. Glick, K. R. Covey, C. Bettigole, D. S. Maynard, S. M. Thomas, J. R. Smith, et al. 2015. "Mapping Tree Density at a Global Scale." *Nature* 525 (7568): 201–5. https://doi.org/10.1038/nature14967.

Cui, Ying, Brian A. Schubert, and A. Hope Jahren. 2020. "A 23 m.y. Record of Low Atmospheric CO_2." *Geology* 48 (9): 888–92. https://doi.org/10.1130/G47681.1.

Culley, Theresa M., Stephen G. Weller, and Ann K. Sakai. 2002. "The Evolution of

Wind Pollination in Angiosperms." *Trends in Ecology and Evolution* 17 (8): 361–69. https://doi.org/10.1016/S0169-5347(02)02540-5.

Currano, Ellen D., Peter Wilf, Scott L. Wing, Conrad C. Labandeira, Elizabeth C. Lovelock, and Dana L. Royer. 2008. "Sharply Increased Insect Herbivory during the Paleocene–Eocene Thermal Maximum." *Proceedings of the National Academy of Sciences* 105 (6): 1960–64. https://doi.org/10.1073/pnas.0708646105.

Daghlian, Charles P., and William L. Crepet. 1983. "Oak Catkins, Leaves and Fruits from the Oligocene Catahoula Formation and Their Evolutionary Significance." *American Journal of Botany* 70 (5): 639–49.

Dannemann, Michael, and Janet Kelso. 2017. "The Contribution of Neanderthals to Phenotypic Variation in Modern Humans." *American Journal of Human Genetics* 101 (4): 578–89. https://doi.org/10.1016/j.ajhg.2017.09.010.

Darley-Hill, Susan, and W. Carter Johnson. 1981. "Acorn Dispersal by the Blue Jay (*Cyanocitta cristata*)." *Oecologia* 50 (2): 231–32. https://doi.org/10.1007/BF00348043.

Darwin, Charles. 1872. *On the Origin of Species by Means of Natural Selection, or the Preservation of Favoured Races in the Struggle for Life.* Sixth Edition, with Additions and Corrections. London: John Murray.

———. 1874. *The Descent of Man, and Selection in Relation to Sex.* 2nd ed. London: John Murray.

———. 1958. *The Autobiography of Charles Darwin 1809–1882.* Edited by Nora Barlow. London: Collins.

———. Darwin, letter to Alphonse de Candolle, January 14, 1863. Darwin Correspondence Project. Accessed November 8, 2022. https://www.darwinproject.ac.uk/letter/?docId=letters/DCP-LETT-3917.xml.

Darwin, Leonard. 1913. "Heredity and Environment." *Eugenics Review* 5 (2): 153–54.

Davis, Frank W., Claudia M. Tyler, and Bruce E. Mahall. 2011. "Consumer Control of Oak Demography in a Mediterranean-Climate Savanna." *Ecosphere* 2 (10): 108. https://doi.org/10.1890/ES11-00187.1.

Davis, Mark A., and Richard Condit. 2022. "Neighbours Consistently Influence Tree Growth and Survival in a Frequently Burned Open Oak Landscape." *Journal of Ecology* 110 (8): 1802–12. https://doi.org/10.1111/1365-2745.13906.

Dawson, M. R., R. M. West, W. Langston, and J. H. Hutchison. 1976. "Paleogene Terrestrial Vertebrates: Northernmost Occurrence, Ellesmere Island, Canada." *Science* 192 (4241): 781–82. https://doi.org/10.1126/science.192.4241.781.

Deacon, Nicholas John, and Jeannine Cavender-Bares. 2015. "Limited Pollen Dispersal Contributes to Population Genetic Structure but Not Local Adaptation in *Quercus oleoides* Forests of Costa Rica." *PLOS ONE* 10 (9): e0138783. https://doi.org/10.1371/journal.pone.0138783.

Degroot, Dagomar. 2018. "Some Places Flourished in the Little Ice Age. There Are Lessons for Us Now." *Washington Post*, February 19.

———. 2019. "The Little Ice Age Is a History of Resilience and Surprises." *Aeon*, November.

de Heredia, Unai López, Pilar Jiménez, Carmen Collada, Marco C. Simeone, Rosanna Bellarosa, Bartolomeo Schirone, María T. Cervera, et al. 2007. "Multi-Marker Phylogeny of Three Evergreen Oaks Reveals Vicariant Patterns in the Western Mediterranean." *Taxon* 56 (November): 1209–9E.

De Kort, H., J. G. Prunier, S. Ducatez, O. Honnay, M. Baguette, V. M. Stevens, and S. Blanchet. 2021. "Life History, Climate and Biogeography Interactively Affect Worldwide Genetic Diversity of Plant and Animal Populations." *Nature Communications* 12 (1): 516. https://doi.org/10.1038/s41467-021-20958-2.

De Kort, Selvino R., and Nicola S. Clayton. 2005. "An Evolutionary Perspective on Caching by Corvids." *Proceedings of the Royal Society B: Biological Sciences* 273 (1585): 417–23. https://doi.org/10.1098/rspb.2005.3350.

Delcourt, H. R., and P. A. Delcourt. 1984. "Ice Age Haven for Hardwoods." *Natural History* 93 (9): 22–28.

Delcourt, Paul A., and Hazel R. Delcourt. 1987. *Long-Term Forest Dynamics of the Temperate Zone: A Case Study of Late-Quaternary Forests in Eastern North America.* New York: Springer-Verlag.

Delgado, Mikel M., and Lucia F. Jacobs. 2017. "Caching for Where and What: Evidence for a Mnemonic Strategy in a Scatter-Hoarder." *Royal Society Open Science* 4 (9): 170958. https://doi.org/10.1098/rsos.170958.

de Lima Navarro, Pedro, and Cristina de Amorim Machado. 2020. "An Origin of Citations: Darwin's Collaborators and Their Contributions to the Origin of Species." *Journal of the History of Biology* 53 (1): 45–79. https://doi.org/10.1007/s10739-020 -09592-8.

Del Tredici, Peter. 2006. "The Other Kinsey Report." *Natural History*, August.

Delzon, Sylvain, Morgane Urli, Jean-Charles Samalens, Jean-Baptiste Lamy, Heike Lischke, Fabrice Sin, Niklaus E. Zimmermann, et al. 2013. "Field Evidence of Colonisation by Holm Oak, at the Northern Margin of Its Distribution Range, during the Anthropocene Period." *PLOS ONE* 8 (11): e80443. https://doi.org/10 .1371/journal.pone.0080443.

Deng, Min, Xiao-Long Jiang, Andrew L. Hipp, Paul S. Manos, and Marlene Hahn. 2018. "Phylogeny and Biogeography of East Asian Evergreen Oaks (*Quercus* Section *Cyclobalanopsis*; Fagaceae): Insights into the Cenozoic History of Evergreen Broad-Leaved Forests in Subtropical Asia." *Molecular Phylogenetics and Evolution* 119: 170–81. https://doi.org/10.1016/j.ympev.2017.11.003.

Deng, Min, Kaiping Yao, Chengcheng Shi, Wen Shao, and Qiansheng Li. 2022. "Development of *Quercus acutissima* (Fagaceae) Pollen Tubes inside Pistils during the Sexual Reproduction Process." *Planta* 256 (1): 16. https://doi.org/10.1007/s00425 -022-03937-9.

Deng, Min, Zhe-Kun Zhou, Ya-Qiong Chen, and Wei-Bang Sun. 2008. "Systematic Significance of the Development and Anatomy of Flowers and Fruit of *Quercus schottkyana* (Subgenus *Cyclobalanopsis*: Fagaceae)." *International Journal of Plant Sciences* 169 (9): 1261–77. https://doi.org/10.1086/591976.

Deng, Min, Zhe-Kun Zhou, and Qiansheng Li. 2013. "Taxonomy and Systematics of *Quercus* Subgenus *Cyclobalanopsis*." *International Oaks* 24:49–60.

Denk, Thomas, and Guido W. Grimm. 2009. "The Biogeographic History of Beech Trees." *Review of Palaeobotany and Palynology* 158 (1): 83–100. https://doi.org/10.1016/j.revpalbo.2009.08.007.

———. 2010. "The Oaks of Western Eurasia: Traditional Classifications and Evidence from Two Nuclear Markers." *Taxon* 59:351–66.

Denk, Thomas, Guido W. Grimm, Andrew L. Hipp, Johannes M. Bouchal, Ernst-Detlef Schulze, and Marco C. Simeone. 2023. "Niche Evolution in a Northern Temperate Tree Lineage: Biogeographical Legacies in Cork Oaks (*Quercus* Section *Cerris*)." *Annals of Botany* 131 (5): 769–87. https://doi.org/10.1093/aob/mcad032.

Denk, Thomas, Guido W. Grimm, Paul S. Manos, Min Deng, and Andrew L. Hipp. 2017. "An Updated Infrageneric Classification of the Oaks: Review of Previous Taxonomic Schemes and Synthesis of Evolutionary Patterns." In *Oaks Physiological Ecology. Exploring the Functional Diversity of Genus* Quercus L., edited by Eustaquio Gil-Pelegrín, José Javier Peguero-Pina, and Domingo Sancho-Knapik, 13–38. Tree Physiology. Cham, Germany: Springer International. https://doi.org/10.1007/978-3-319-69099-5_2.

Denk, Thomas, Friðgeir Grímsson, and Reinhard Zetter. 2010. "Episodic Migration of Oaks to Iceland: Evidence for a North Atlantic 'Land Bridge' in the Latest Miocene." *American Journal of Botany* 97 (2): 276–87.

Denk, Thomas, Friðgeir Grímsson, Reinhard Zetter, and Leifur A. Símonarson. 2011a. "The Biogeographic History of Iceland—The North Atlantic Land Bridge Revisited." In *Late Cainozoic Floras of Iceland: 15 Million Years of Vegetation and Climate History in the Northern North Atlantic*, edited by Thomas Denk, Friðgeir Grímsson, Reinhard Zetter, and Leifur A. Símonarson, 647–68. Topics in Geobiology. Dordrecht: Springer Netherlands. https://doi.org/10.1007/978-94-007-0372-8_12.

Denk, Thomas, Friðgeir Grímsson, Reinhard Zetter, and Leifur A. Símonarson. 2011b. "Climate Evolution in the Northern North Atlantic—15 Ma to Present." In *Late Cainozoic Floras of Iceland: 15 Million Years of Vegetation and Climate History in the Northern North Atlantic*, edited by Thomas Denk, Friðgeir Grímsson, Reinhard Zetter, and Leifur A. Símonarson, 669–721. Topics in Geobiology. Dordrecht: Springer Netherlands. https://doi.org/10.1007/978-94-007-0372-8_13.

Denk, Thomas, Robert S. Hill, Marco C. Simeone, Chuck Cannon, Mary E. Dettmann, and Paul S. Manos. 2019. "Comment on 'Eocene Fagaceae from Patagonia and Gondwanan Legacy in Asian Rainforests.'" *Science* 366 (6467): eaaz2189. https://doi.org/10.1126/science.aaz2189.

DePalma, Robert A., Jan Smit, David A. Burnham, Klaudia Kuiper, Phillip L. Manning, Anton Oleinik, Peter Larson, et al. 2019. "A Seismically Induced Onshore Surge Deposit at the KPg Boundary, North Dakota." *Proceedings of the National Academy of Sciences* 116 (17): 8190–99. https://doi.org/10.1073/pnas.1817407116.

Desprez-Loustau, Marie-Laure, Marie Massot, Nicolas Feau, Tania Fort, Antonio de Vicente, Juan Antonio Torés, and Dolores Fernández Ortuño. 2017. "Further Support of Conspecificity of Oak and Mango Powdery Mildew and First Report of *Erysiphe quercicola* and *Erysiphe alphitoides* on Mango in Mainland Europe." *Plant Disease* 101 (7): 1086–93. https://doi.org/10.1094/PDIS-01-17-0116-RE.

Desprez-Loustau, Marie-Laure, Gilles Saint-Jean, Benoît Barrès, Cécile Françoise Dantec, and Cyril Dutech. 2014. "Oak Powdery Mildew Changes Growth Patterns in Its Host Tree: Host Tolerance Response and Potential Manipulation of Host Physiology by the Parasite." *Annals of Forest Science* 71 (5): 563–73. https://doi.org/10.1007/s13595-014-0364-6.

De Vos, Jurriaan M., Lucas N. Joppa, John L. Gittleman, Patrick R. Stephens, and Stuart L. Pimm. 2015. "Estimating the Normal Background Rate of Species Extinction." *Conservation Biology* 29 (2): 452–62. https://doi.org/10.1111/cobi.12380.

Dhawan, Siddhant Suri, Vivek Yedavalli, and Tarik F. Massoud. 2023. "Atavistic and Vestigial Anatomical Structures in the Head, Neck, and Spine: An Overview." *Anatomical Science International* 98 (3): 370–90. https://doi.org/10.1007/s12565-022-00701-7.

Díaz-Fernández, Pedro M., José Climent, and Luis Gil. 2004. "Biennial Acorn Maturation and Its Relationship with Flowering Phenology in Iberian Populations of *Quercus suber*." *Trees* 18 (6): 615–21. https://doi.org/10.1007/s00468-004-0325-z.

Dillard, Annie. 1974. *Pilgrim at Tinker Creek*. New York: Harper's Magazine Press.

Ding, Jingyi, Manuel Delgado-Baquerizo, Jun-Tao Wang, and David J. Eldridge. 2021. "Ecosystem Functions Are Related to Tree Diversity in Forests but Soil Biodiversity in Open Woodlands and Shrublands." *Journal of Ecology* 109 (12): 4158–70. https://doi.org/10.1111/1365-2745.13788.

Ding, Ya-Mei, Xiao-Xu Pang, Yu Cao, Wei-Ping Zhang, Susanne S. Renner, Da-Yong Zhang, and Wei-Ning Bai. 2023. "Genome Structure-Based Juglandaceae Phylogenies Contradict Alignment-Based Phylogenies and Substitution Rates Vary with DNA Repair Genes." *Nature Communications* 14 (1): 617. https://doi.org/10.1038/s41467-023-36247-z.

Dobzhansky, Theodosius. 1935. "A Critique of the Species Concept in Biology." *Philosophy of Science* 2 (3): 344–55. https://doi.org/10.1086/286379.

———. 1982. *Genetics and the Origin of Species*. New York: Columbia University Press.

———. 1965. "Mendelism, Darwinism, and Evolutionism." *Proceedings of the American Philosophical Society* 109 (4): 205–15.

Dodd, Richard S., and Zara Afzal-Rafii. 2004. "Selection and Dispersal in a Multispecies Oak Hybrid Zone." *Evolution* 58 (2): 261–69.

Dodd, Richard S., and N. Kashani. 2003. "Molecular Differentiation and Diversity among the California Red Oaks (Fagaceae; *Quercus* Section *Lobatae*)." *Theoretical and Applied Genetics* 107 (5): 884–92.

Dodd, Richard S., Nasser Kashani, and Zara Afzal-Rafii. 2002. "Population Diversity and Evidence of Introgression among the Black Oaks of California." In *Proceed-*

ings of the Fifth Symposium on Oak Woodlands: Oaks in California's Challenging Landscape.; Gen. Tech. Rep. PSW-GTR-184. edited by Richard B. Standiford, D. McCreary, and Kathryn L. Purcell, 775–85. Albany, CA: U.S. Department of Agriculture, Forest Service, Pacific Southwest Research Station.

Donoghue, Michael J. 2008. "A Phylogenetic Perspective on the Distribution of Plant Diversity." Supplement. *Proceedings of the National Academy of Sciences* 105 (S1): 11549–55.

Donoghue, Michael J., Deren A. R. Eaton, Carlos A. Maya-Lastra, Michael J. Landis, Patrick W. Sweeney, Mark E. Olson, N. Ivalú Cacho, et al. 2022. "Replicated Radiation of a Plant Clade along a Cloud Forest Archipelago." *Nature Ecology and Evolution* (July): 1–12. https://doi.org/10.1038/s41559-022-01823-x.

Dow, Beverly D., and Mary V. Ashley. 1998. "High Levels of Gene Flow in Bur Oak Revealed by Paternity Analysis Using Microsatellites." *Journal of Heredity* 89:62–70.

Driver, Harold E. 1952. "The Acorn in North American Indian Diet." *Proceedings of the Indiana Academy of Science* 62:56–62.

Drori, Jonathan. 2018. *Around the World in 80 Trees*. London: Laurence King.

Du, Fang K., Tianrui Wang, Yuyao Wang, Saneyoshi Ueno, and Guillaume de Lafontaine. 2020. "Contrasted Patterns of Local Adaptation to Climate Change across the Range of an Evergreen Oak, *Quercus aquifolioides*." *Evolutionary Applications* 13 (9): 2377–91. https://doi.org/10.1111/eva.13030.

Ducousso, A., H. Michaud, and R. Lumaret. 1993. "Reproduction and Gene Flow in the Genus *Quercus* L." Supplement. *Annales Des Sciences Forestières* 50:91s–106s. https://doi.org/10.1051/forest:19930708.

Dumolin-Lapegue, S., B. Demesure, S. Fineschi, V. Le Come, and R. J. Petit. 1997. "Phylogeographic Structure of White Oaks throughout the European Continent." *Genetics* 146:1475–87.

Dumolin-Lapegue, S., B. Demesure, and R. J. Petit. 1995. "Inheritance of Chloroplast and Mitochondrial Genomes in Pedunculate Oak Investigated with an Efficient PCR Method." *Theoretical and Applied Genetics* 91 (8): 1253–56. https://doi.org/10.1007/BF00220937.

Dumolin-Lapegue, S., Antoine Kremer, and Rémy J. Petit. 1999. "Are Chloroplast and Mitochondrial DNA Variation Species Independent in Oaks?" *Evolution* 53 (5): 1406–13.

Dutech, Cyril, Victoria L. Sork, Andrew J. Irwin, Peter E. Smouse, and Frank W. Davis. 2005. "Gene Flow and Fine-Scale Genetic Structure in a Wind-Pollinated Tree Species *Quercus lobata* (Fagaceaee)." *American Journal of Botany* 92 (2): 252–61.

Eaton, Deren A. R., Andrew L. Hipp, Antonio González-Rodríguez, and Jeannine Cavender-Bares. 2015. "Historical Introgression among the American Live Oaks and the Comparative Nature of Tests for Introgression." *Evolution* 69 (10): 2587–2601. https://doi.org/10.1111/evo.12758.

Edwards, Erika J., and Stephen A. Smith. 2010. "Phylogenetic Analyses Reveal the Shady History of C$_4$ Grasses." *Proceedings of the National Academy of Sciences*

of the United States of America 107 (6): 2532–37. https://doi.org/10.1073/pnas
.0909672107.

Edwards, Erika J., Elizabeth L. Spriggs, David S. Chatelet, and Michael J. Donoghue.
2016. "Unpacking a Century-Old Mystery: Winter Buds and the Latitudinal Gradient in Leaf Form." *American Journal of Botany* 103 (6): 975–78. https://doi.org
/10.3732/ajb.1600129.

Egan, Scott P., Glen R. Hood, Ellen O. Martinson, and James R. Ott. 2018. "Cynipid
Gall Wasps." *Current Biology* 28 (24): R1370–74. https://doi.org/10.1016/j.cub
.2018.10.028.

Egerton-Warburton, Louise M., José Ignacio Querejeta, and Michael F. Allen. 2007.
"Common Mycorrhizal Networks Provide a Potential Pathway for the Transfer of
Hydraulically Lifted Water between Plants." *Journal of Experimental Botany* 58 (6):
1473–83. https://doi.org/10.1093/jxb/erm009.

Egger, H., C. Heilmann-Clausen, and B. Schmitz. 2009. "From Shelf to Abyss: Record of the Paleocene/Eocene-Boundary in the Eastern Alps (Austria)." *Geologica
Acta* 7 (1–2) (March–June): 215–27.

Ekholm, Adam, Tomas Roslin, Pertti Pulkkinen, and Ayco J. M. Tack. 2017. "Dispersal, Host Genotype and Environment Shape the Spatial Dynamics of a Parasite in
the Wild." *Ecology* 98 (10): 2574–84. https://doi.org/10.1002/ecy.1949.

Encinas-Valero, Manuel, Raquel Esteban, Ana-Maria Hereş, María Vivas, Dorra
Fakhet, Iker Aranjuelo, Alejandro Solla, et al. 2022. "Holm Oak Decline Is Deter-
mined by Shifts in Fine Root Phenotypic Plasticity in Response to Belowground
Stress." *New Phytologist* 235 (6): 2237–51. https://doi.org/10.1111/nph.18182.

Endersby, Jim. 2018. "Descriptive and Prescriptive Taxonomies." In *Worlds of Natural History*, edited by Emma C. Spary, Helen Anne Curry, James Andrew Secord,
and Nicholas Jardine, 447–59. Cambridge: Cambridge University Press. https://
doi.org/10.1017/9781108225229.028.

Engelmann, George. 1878. "About the Oaks of the United States." *Transactions of the
Academy of Sciences of St. Louis* 3:372–400, 539–43.

Escudero, Marcial, Marlene Hahn, Bethany H. Brown, Kate Lueders, and Andrew L.
Hipp. 2016. "Chromosomal Rearrangements in Holocentric Organisms Lead to
Reproductive Isolation by Hybrid Dysfunction: The Correlation between Karyo-
type Rearrangements and Germination Rates in Sedges." *American Journal of Botany* 103 (8): 1529–36. https://doi.org/10.3732/ajb.1600051.

Escudero, Marcial, André Marques, Kay Lucek, and Andrew L. Hipp. 2023. "Genomic
Hotspots of Chromosome Rearrangements Explain Conserved Synteny despite
High Rates of Chromosome Evolution in a Holocentric Lineage." *Molecular Ecol-
ogy*, 00:1–12. 10.1111/mec.17086. https://doi.org/10.1111/mec.17086.

Faison, Edward K., and David R. Foster. 2014. "Did American Chestnut Really Dominate the Eastern Forest?" *Arnoldia* 72 (2): 18–32.

Fallon, Beth, and Jeannine Cavender-Bares. 2018. "Leaf-Level Trade-Offs between

Drought Avoidance and Desiccation Recovery Drive Elevation Stratification in Arid Oaks." *Ecosphere* 9 (3): e02149. https://doi.org/10.1002/ecs2.2149.

Fallon, Beth, Anna Yang, Cathleen Lapadat, Isabella Armour, Jennifer Juzwik, Rebecca A. Montgomery, and Jeannine Cavender-Bares. 2020. "Spectral Differentiation of Oak Wilt from Foliar Fungal Disease and Drought Is Correlated with Physiological Changes." *Tree Physiology* 40 (3): 377–90. https://doi.org/10.1093/treephys/tpaa005.

Feeley, Kenneth, and James Stroud. 2018. "Where on Earth Are the 'Tropics'?" *Frontiers of Biogeography* 10 (1–2): e38649. https://doi.org/10.21425/F5FBG38649.

Ferris, C., R. P. Oliver, A. J. Davy, and G. M. Hewitt. 1993. "Native Oak Chloroplasts Reveal an Ancient Divide across Europe." *Molecular Ecology* 2 (6): 337–43. https://doi.org/10.1111/j.1365-294X.1993.tb00026.x.

Fey, Beat S., and Peter K. Endress. 1983. "Development and Morphological Interpretation of the Cupule in Fagaceae." *Flora* 173 (5–6): 451–68. https://doi.org/10.1016/S0367-2530(17)32023-6.

Figueroa-Rangel, Blanca Lorena, and Miguel Olvera-Vargas. 2022. "Environmental and Spatial Processes Shaping *Quercus*-Dominated Forest Communities in the Neotropics." *Ecosphere* 13 (6): e4103. https://doi.org/10.1002/ecs2.4103.

Firmat, C., S. Delzon, J.-M. Louvet, J. Parmentier, and A. Kremer. 2017. "Evolutionary Dynamics of the Leaf Phenological Cycle in an Oak Metapopulation along an Elevation Gradient." *Journal of Evolutionary Biology* 30 (12): 2116–31. https://doi.org/10.1111/jeb.13185.

Fitz-Gibbon, Sorel, Andrew L. Hipp, Kasey Khanh Pham, Paul S. Manos, and Victoria Sork. 2017. "Phylogenomic Inferences from Reference-Mapped and de Novo Assembled Short-Read Sequence Data Using RADseq Sequencing of California White Oaks (*Quercus* Subgenus *Quercus*)." *Genome* 60 (March): 743–55. https://doi.org/10.1139/gen-2016-0202.

Flanagan, Erin. 2021. "The Global Carbon Budget and Permafrost Feedback Loops in the Arctic." *Arctic Institute* (blog). February 25. https://www.thearcticinstitute.org/global-carbon-budget-permafrost-feedback-loops-arctic/.

Fleurot, Emilie, Jean R. Lobry, Vincent Boulanger, François Debias, Camille Mermet-Bouvier, Thomas Caignard, Sylvain Delzon, et al. 2023. "Oak Masting Drivers Vary between Populations Depending on Their Climatic Environments." *Current Biology* 33 (6): 1117–24.e4. https://doi.org/10.1016/j.cub.2023.01.034.

Folk, R. A., C. M. Siniscalchi, J. Doby, H. R. Kates, S. R. Manchester, P. S. Soltis, D. E. Soltis, et al. 2023. "Spatial Phylogenetics of Fagales: Investigating the History of Temperate Forests." bioRxiv. https://doi.org/10.1101/2023.04.17.537249.

Forman, L. L. 1964. "*Trigonobalanus*, a New Genus of Fagaceae, with Notes on the Classification of the Family." *Kew Bulletin* 17 (3): 381–96. https://doi.org/10.2307/4113784.

Fort, Tania, Charlie Pauvert, Amy E. Zanne, Otso Ovaskainen, Thomas Caignard,

Matthieu Barret, Stéphane Compant, et al. 2021. "Maternal Effects Shape the Seed Mycobiome in *Quercus petraea.*" *New Phytologist* 230 (4): 1594–1608. https://doi.org/10.1111/nph.17153.

Frantz, Laurent A. F., Victoria E. Mullin, Maud Pionnier-Capitan, Ophélie Lebrasseur, Morgane Ollivier, Angela Perri, Anna Linderholm, et al. 2016. "Genomic and Archaeological Evidence Suggest a Dual Origin of Domestic Dogs." *Science* 352 (6290): 1228–31. https://doi.org/10.1126/science.aaf3161.

Freeman, Benjamin G., and Nicholas A. Mason. 2015. "The Geographic Distribution of a Tropical Montane Bird Is Limited by a Tree: Acorn Woodpeckers (*Melanerpes formicivorus*) and Colombian Oaks (*Quercus humboldtii*) in the Northern Andes." *PLOS ONE* 10 (6): e0128675. https://doi.org/10.1371/journal.pone.0128675.

Freking, Brad A., Susan K. Murphy, Andrew A. Wylie, Simon J. Rhodes, John W. Keele, Kreg A. Leymaster, Randy L. Jirtle, et al. 2002. "Identification of the Single Base Change Causing the Callipyge Muscle Hypertrophy Phenotype, the Only Known Example of Polar Overdominance in Mammals." *Genome Research* 12 (10): 1496–1506. https://doi.org/10.1101/gr.571002.

Friedlingstein, Pierre, Matthew W. Jones, Michael O'Sullivan, Robbie M. Andrew, Dorothee C. E. Bakker, Judith Hauck, Corinne Le Quéré, et al. 2022. "Global Carbon Budget 2021." *Earth System Science Data* 14 (4): 1917–2005. https://doi.org/10.5194/essd-14-1917-2022.

Friis, Else Marie, Peter R. Crane, and Kaj Raunsgaard Pedersen. 2011a. "Fossils of Core Eudicots: Rosids." In *Early Flowers and Angiosperm Evolution*, 1st ed., 327–60. Cambridge: Cambridge University Press. https://doi.org/10.1017/CBO9780511980206.

———. 2011b. "Vegetational Context of Early Angiosperm Diversification." In *Early Flowers and Angiosperm Evolution*, 1st ed., 461–73. Cambridge: Cambridge University Press. https://doi.org/10.1017/CBO9780511980206.

Friis, Else Marie, K. Raunsgaard Pedersen, and J. Schönenberger. 2006. "Normapolles Plants: A Prominent Component of the Cretaceous Rosid Diversification." *Plant Systematics and Evolution* 260 (2): 107–40. https://doi.org/10.1007/s00606-006-0440-y.

Fu, Qiaomei, Heng Li, Priya Moorjani, Flora Jay, Sergey M. Slepchenko, Aleksei A. Bondarev, Philip L. F. Johnson, et al. 2014. "Genome Sequence of a 45,000-Year-Old Modern Human from Western Siberia." *Nature* 514 (7523): 445–49. https://doi.org/10.1038/nature13810.

Fu, Ruirui, Yuxiang Zhu, Ying Liu, Yu Feng, Rui-Sen Lu, Yao Li, Pan Li, et al. 2022. "Genome-Wide Analyses of Introgression between Two Sympatric Asian Oak Species." *Nature Ecology and Evolution* 6 (May): 924–35. https://doi.org/10.1038/s41559-022-01754-7.

Gailing, Oliver, Andrew L. Hipp, Christophe Plomion, and John E. Carlson. 2022. "Oak Population Genomics." In *Population Genomics: Forest Trees*, edited by Om

P. Rajora. Cham, Germany: Springer International. https://doi.org/10.1007/13836_2021_100.

Gaiman, N. 2017. *Norse Mythology*. New York and London: W. W. Norton.

Gale, Andrew S. 2000. "The Cretaceous World." In *Biotic Response to Global Change: The Last 145 Million Years*, edited by S. J. Culver and P. F. Rawson, 145:4–19. Cambridge: Cambridge University Press.

Gale, L. D. 1856. "On the Oaks of the District of Columbia." *Proceedings of the National Institute, Washington, D.C.* 1: 67–78.

Gandolfo, Maria A., Kevin C. Nixon, William L. Crepet, and David A. Grimaldi. 2018. "A Late Cretaceous Fagalean Inflorescence Preserved in Amber from New Jersey." *American Journal of Botany* 105 (8): 1424–35. https://doi.org/10.1002/ajb2.1103.

Gao, Jie, Zhi-Long Liu, Wei Zhao, Kyle W. Tomlinson, Shang-Wen Xia, Qing-Yin Zeng, Xiao-Ru Wang, et al. 2021. "Combined Genotype and Phenotype Analyses Reveal Patterns of Genomic Adaptation to Local Environments in the Subtropical Oak *Quercus acutissima*." *Journal of Systematics and Evolution* 59 (3): 541–56. https://doi.org/10.1111/jse.12568.

García-de la Cruz, Yureli, Joaquín Becerra-Zavaleta, Alejandra Quintanar-Isaías, José María Ramos-Prado, and Angélica María Hernández-Ramírez. 2014. "La bellota de *Quercus insignis* Martens & Galeotti, 1843, la más grande del mundo." Universidad de Alicante, Cuadernos de Biodiversidad. 46:1–8. https://doi.org/10.14198/cdbio.2014.46.01.

García-Hernández, María de los Ángeles, Fabiola López-Barrera, and Ramón Perea. 2023. "Simulated Partial Predation on the Largest-Seeded Oak: Effects of Seed Morphology and Size on Early Establishment." *Forest Ecology and Management* 534 (April): 120863. https://doi.org/10.1016/j.foreco.2023.120863.

Garrison, W. J., and C. K. Augspurger. 1983. "Double- and Single-Seeded Acorns of Bur Oak (*Quercus macrocarpa*): Frequency and Some Ecological Consequences." *Bulletin of the Torrey Botanical Club* 110 (2): 154–60. https://doi.org/10.2307/2996335.

Gelbart, Galatea, and Patrick von Aderkas. 2002. "Ovular Secretions as Part of Pollination Mechanisms in Conifers." *Annals of Forest Science* 59 (4): 345–57. https://doi.org/10.1051/forest:2002011.

Gellis, Jason J., and Robert A. Foley. 2023. "Hominin Evolution." In *Handbook of Archaeological Sciences*, edited by A. Mark Pollard, Ruth Ann Armitage, and Cheryl A. Makarewicz, 359–86. Hoboken, NJ: Wiley. https://doi.org/10.1002/9781119592112.ch17.

George, Jan-Peter, Guillaume Theroux-Rancourt, Kanin Rungwattana, Susanne Scheffknecht, Nevena Momirovic, Lea Neuhauser, Lambert Weißenbacher, et al. 2020. "Assessing Adaptive and Plastic Responses in Growth and Functional Traits in a 10-Year-Old Common Garden Experiment with Pedunculate Oak (*Quercus robur* L.) Suggests That Directional Selection Can Drive Climatic Adaptation." *Evolutionary Applications* 13 (9): 2422–38. https://doi.org/10.1111/eva.13034.

Giannetti, Daniele, Enrico Schifani, Cristina Castracani, Fiorenza Augusta Spotti, Alessandra Mori, and Donato Antonio Grasso. 2022. "The Introduced Oak *Quercus rubra* and Acorn-Associated Arthropods in Europe: An Opportunity for Both Carpophagous Insects and Their Ant Predators." *Ecological Entomology* 47 (4): 515–26. https://doi.org/10.1111/een.13136.

Giesecke, Thomas. 2016. "Did Thermophilous Trees Spread into Central Europe during the Late Glacial?" *New Phytologist* 212 (1): 15–18. https://doi.org/10.1111/nph.14149.

Gil-Pelegrín, Eustaquio, Miguel Ángel Saz, Jose María Cuadrat, José Javier Peguero-Pina, and Domingo Sancho-Knapik. 2017. "Oaks under Mediterranean-Type Climates: Functional Response to Summer Aridity." In *Oaks Physiological Ecology. Exploring the Functional Diversity of Genus Quercus L.*, edited by Eustaquio Gil-Pelegrín, José Javier Peguero-Pina, and Domingo Sancho-Knapik, 137–93. Tree Physiology. Cham, Germany: Springer International. https://doi.org/10.1007/978-3-319-69099-5_5.

Givnish, Thomas J., J. Chris Pires, Sean W. Graham, Marc A. McPherson, Linda M. Prince, Thomas B. Patterson, Hardeep S. Rai, et al. 2005. "Repeated Evolution of Net Venation and Fleshy Fruits among Monocots in Shaded Habitats Confirms a Priori Predictions: Evidence from an ndhF Phylogeny." *Proceedings of the Royal Society B: Biological Sciences* 272 (1571): 1481–90. https://doi.org/10.1098/rspb.2005.3067.

Gómez, José María. 2003. "Spatial Patterns in Long-Distance Dispersal of *Quercus ilex* Acorns by Jays in a Heterogeneous Landscape." *Ecography* 26 (5): 573–84. https://doi.org/10.1034/j.1600-0587.2003.03586.x.

Gómez-Casero, María Teresa, Pablo J. Hidalgo, Herminia García-Mozo, Eugenio Domínguez, and Carmen Galán. 2004. "Pollen Biology in Four Mediterranean *Quercus* Species." *Grana* 43 (1): 22–30. https://doi.org/10.1080/00173130410018957.

Gómez-Laurito, Jorge, and Luis D. Gómez P. 1989. "*Ticodendron*: A New Tree from Central America." *Annals of the Missouri Botanical Garden* 76 (4): 1148–51. https://doi.org/10.2307/2399700.

González-Rodríguez, Antonio, Dulce M. Arias, Susana Valencia-Avalos, and Ken Oyama. 2004. "Morphological and RAPD Analysis of Hybridization between *Quercus affinis* and *Q. laurina* (Fagaceae), Two Mexican Red Oaks." *American Journal of Botany* 91 (3): 401–9.

González-Rodríguez, Antonio, Felipe García-Oliva, Yunuen Tapie-Torres, Alberto Morón-Cruz, Bruno Chávez-Vergara, Brenda Baca-Patiño, and Pablo Cuevas-Reyes. 2019. "Oak Community Diversity Affects Nitrogen in Litter and Soil." *International Oaks: The Journal of the International Oak Society* 30:125–30.

Gott, J. Richard. 1993. "Implications of the Copernican Principle for Our Future Prospects." *Nature* 363 (6427): 315–19. https://doi.org/10.1038/363315a0.

Gottschalk, Kurt W., and Philip M. Wargo. 1997. "Oak Decline around the World." In *Proceedings, U.S. Department of Agriculture Interagency Gypsy Moth Research Forum 1996, 1996 January 16–19; Annapolis, MD; Gen. Tech. Rep. NE-230*, edited by S. L. C. Fosbroke and K. W. Gottschalk, 3–13. Radnor, PA: U.S. Department of Agriculture, Forest Service, Northeastern Forest Experiment Station.

Grabowski, Mark, and William L. Jungers. 2017. "Evidence of a Chimpanzee-Sized Ancestor of Humans but a Gibbon-Sized Ancestor of Apes." *Nature Communications* 8 (1): 880. https://doi.org/10.1038/s41467-017-00997-4.

Graham, Alan. 1999a. *Late Cretaceous and Cenozoic History of North American Vegetation, North of Mexico.* New York: Oxford University Press.

———. 1999b. "The Tertiary History of the Northern Temperate Element in the Northern Latin American Biota." *American Journal of Botany* 86 (1): 32–38. https://doi.org/10.2307/2656952.

———. 2011. *A Natural History of the New World: The Ecology and Evolution of Plants in the Americas.* Chicago: University of Chicago Press.

Grant, Verne E. 1981. *Plant Speciation.* 2nd ed. New York: Columbia University Press.

Grimm, Guido W., and Susanne S. Renner. 2013. "Harvesting Betulaceae Sequences from GenBank to Generate a New Chronogram for the Family." *Botanical Journal of the Linnean Society* 172 (4): 465–77. https://doi.org/10.1111/boj.12065.

Grímsson, Friðgeir, Guido W. Grimm, Reinhard Zetter, and Thomas Denk. 2016. "Cretaceous and Paleogene Fagaceae from North America and Greenland: Evidence for a Late Cretaceous Split between *Fagus* and the Remaining Fagaceae." *Acta Palaeobotanica* 56 (2): 247–305. https://doi.org/10.1515/acpa-2016-0016.

Grímsson, Friðgeir, Reinhard Zetter, Guido W. Grimm, Gunver Krarup Pedersen, Asger Ken Pedersen, and Thomas Denk. 2015. "Fagaceae Pollen from the Early Cenozoic of West Greenland: Revisiting Engler's and Chaney's Arcto-Tertiary Hypotheses." *Plant Systematics and Evolution* 301 (2): 809–32. https://doi.org/10.1007/s00606-014-1118-5.

Grivet, Delphine, Marie-France Deguilloux, Rémy J. Petit, and Victoria L. Sork. 2006. "Contrasting Patterns of Historical Colonization in White Oaks (*Quercus* Spp.) in California and Europe." *Molecular Ecology* 15 (13): 4085–93. https://doi.org/10.1111/j.1365-294X.2006.03083.x.

Grover, Corrinne E., and Jonathan F. Wendel. 2010. "Recent Insights into Mechanisms of Genome Size Change in Plants." *Journal of Botany* 2010 (May): e382732. https://doi.org/10.1155/2010/382732.

Gugger, Paul F., Sorel T. Fitz-Gibbon, Ana Albarrán-Lara, Jessica W. Wright, and Victoria L. Sork. 2021. "Landscape Genomics of *Quercus lobata* Reveals Genes Involved in Local Climate Adaptation at Multiple Spatial Scales." *Molecular Ecology* 30 (2): 406–23. https://doi.org/10.1111/mec.15731.

Gutjahr, Marcus, Andy Ridgwell, Philip F. Sexton, Eleni Anagnostou, Paul N. Pearson, Heiko Pälike, Richard D. Norris, et al. 2017. "Very Large Release of Mostly Volcanic Carbon during the Palaeocene–Eocene Thermal Maximum." *Nature* 548 (7669): 573–77. https://doi.org/10.1038/nature23646.

Guttman, Sheldon I., and Lee A. Weigt. 1989. "Electrophoretic Evidence of Relationships among *Quercus* (Oaks) of Eastern North America." *Canadian Journal of Botany* 67.

Hackett, Shannon J., Rebecca T. Kimball, Sushma Reddy, Rauri C. K. Bowie, Edward L. Braun, Michael J. Braun, Jena L. Chojnowski, et al. 2008. "A Phylogenomic Study of Birds Reveals Their Evolutionary History." *Science* 320 (5884): 1763–68. https://doi.org/10.1126/science.1157704.

Hagen, Joel B. 1983. "The Development of Experimental Methods in Plant Taxonomy, 1920–1950." *Taxon* 32 (3): 406–16. https://doi.org/10.2307/1221497.

Hajji, Mostafa, Erwin Dreyer, and Benoit Marçais. 2009. "Impact of *Erysiphe alphitoides* on Transpiration and Photosynthesis in *Quercus robur* Leaves." *European Journal of Plant Pathology* 125 (1): 63–72. https://doi.org/10.1007/s10658-009 -9458-7.

Hampe, Arndt, Marie-Hélène Pemonge, and Rémy J. Petit. 2013. "Efficient Mitigation of Founder Effects during the Establishment of a Leading-Edge Oak Population." *Proceedings of the Royal Society of London B: Biological Sciences* 280 (1764): 20131070. https://doi.org/10.1098/rspb.2013.1070.

Hamrick, J. L., and M. J. W. Godt. 1996. "Effects of Life History Traits on Genetic Diversity in Plant Species." *Philosophical Transactions: Biological Sciences* 351 (1345): 1291–98.

Hamrick, J. L., Mary Jo W. Godt, and Susan L. Sherman-Broyles. 1992. "Factors Influencing Levels of Genetic Diversity in Woody Plant Species." *New Forests* 6 (1): 95–124. https://doi.org/10.1007/BF00120641.

Han, Biao, Longxin Wang, Yang Xian, Xiao-Man Xie, Wen-Qing Li, Ye Zhao, Ren-Gang Zhang, et al. 2022. "A Chromosome-Level Genome Assembly of the Chinese Cork Oak (*Quercus variabilis*)." *Frontiers in Plant Science* 13 (September). https://doi.org/10.3389/fpls.2022.1001583.

Hansen, Thomas F., Christophe Pélabon, and David Houle. 2011. "Heritability Is Not Evolvability." *Evolutionary Biology* 38 (3): 258–77. https://doi.org/10.1007/s11692 -011-9127-6.

Hao, Qian, Hongyan Liu, Ying Cheng, and Zhaoliang Song. 2022. "The LGM Refugia of Deciduous Oak and Distribution Development since the LGM in China." *Science China Earth Sciences* (October). https://doi.org/10.1007/s11430-021-9981-9.

Hardin, J. W. 1975. "Hybridization and Introgression in *Quercus alba*." *Journal of the Arnold Arboretum* 56:336–63.

Harrington, Guy J., Jaelyn Eberle, Ben A. Le-Page, Mary Dawson, and J. Howard Hutchison. 2012. "Arctic Plant Diversity in the Early Eocene Greenhouse." *Pro-*

ceedings of the Royal Society B: Biological Sciences 279 (1733): 1515–21. https://doi.org/10.1098/rspb.2011.1704.

Harrington, Thomas C., Doug McNew, and Hye Young Yun. 2012. "Bur Oak Blight, a New Disease on *Quercus macrocarpa* Caused by *Tubakia iowensis* Sp. Nov." *Mycologia* 104 (1): 79–92. https://doi.org/10.3852/11-112.

Hauser, Duncan A., Al Keuter, John D. McVay, Andrew L. Hipp, and Paul S. Manos. 2017. "The Evolution and Diversification of the Red Oaks of the California Floristic Province (*Quercus* Section *Lobatae*, Series *Agrifoliae*)." *American Journal of Botany* 104 (10): 1581–95. https://doi.org/10.3732/ajb.1700291.

Head, Martin J., Marie-Pierre Aubry, Mike Walker, Kenneth G. Miller, and Brian R. Pratt. 2017. "A Case for Formalizing Subseries (Subepochs) of the Cenozoic Era(a)." *Episodes: Journal of International Geoscience* 40 (1): 22–27. https://doi.org/10.18814/epiiugs/2017/v40i1/017004.

Hearn, Jack, Mark Blaxter, Karsten Schönrogge, José-Luis Nieves-Aldrey, Juli Pujade-Villar, Elisabeth Huguet, Jean-Michel Drezen, et al. 2019. "Genomic Dissection of an Extended Phenotype: Oak Galling by a Cynipid Gall Wasp." *PLOS Genetics* 15 (11): e1008398. https://doi.org/10.1371/journal.pgen.1008398.

Herbert, Jane. 2005. "Systematics and Biogeography of Myricaceae." Thesis, University of St. Andrews, Scotland.

Herendeen, Patrick S., Peter R. Crane, and Andrew N. Drinnan. 1995. "Fagaceous Flowers, Fruits, and Cupules from the Campanian (Late Cretaceous) of Central Georgia, USA." *International Journal of Plant Sciences* 156 (1): 93–116. https://doi.org/10.1086/297231.

Herendeen, Patrick S., Else Marie Friis, Kaj Raunsgaard Pedersen, and Peter R. Crane. 2017. "Palaeobotanical Redux: Revisiting the Age of the Angiosperms." *Nature Plants* 3 (3): 17015. https://doi.org/10.1038/nplants.2017.15.

Heřmanová, Zuzana, Jiří Kvaček, and Else Marie Friis. 2011. "*Budvaricarpus serialis* Knobloch & Mai, An Unusual New Member of the Normapolles Complex from the Late Cretaceous of the Czech Republic." *International Journal of Plant Sciences* 172 (2): 285–93. https://doi.org/10.1086/657278.

Hill, Robert S. 1992. "*Nothofagus*: Evolution from a Southern Perspective." *Trends in Ecology and Evolution* 7 (6): 190–94. https://doi.org/10.1016/0169-5347(92)90071-I.

Hipp, Andrew L. 2010. "Hill's Oak: The Taxonomy and Dynamics of a Western Great Lakes Endemic." *Arnoldia* 67 (4): 2–14.

Hipp, Andrew L. 2021a. "Fields of View." *The Learned Pig* (blog), June. https://thelearnedpig.org/fields-of-view/10187.

———. 2021b. "Taking the Measure of a Forest." *Places Journal*, February. https://placesjournal.org/article/taking-the-measure-of-a-suburban-forest-preserve/?cn-reloaded=1.

Hipp, Andrew L., Kieran Althaus, Allen J. Coombes, González-Elizondo M. Socorro,

Antonio González-Rodríguez, Marlene Hahn, Paul S. Manos, et al. 2023. "Time, Space, Function: Biogeography of the Mexican Oaks." *International Oaks: The Journal of the International Oak Society* 34:125–39.

Hipp, Andrew L., Deren A. R. Eaton, Jeannine Cavender-Bares, Elisabeth Fitzek, Rick Nipper, and Paul S. Manos. 2014. "A Framework Phylogeny of the American Oak Clade Based on Sequenced RAD Data." *PLOS ONE* 9:e93975.

Hipp, Andrew L., Paul S. Manos, Antonio González-Rodríguez, Marlene Hahn, Matthew Kaproth, John D. McVay, Susana Valencia-Avalos, et al. 2018. "Sympatric Parallel Diversification of Major Oak Clades in the Americas and the Origins of Mexican Species Diversity." *New Phytologist* 217 (1): 439–52. https://doi.org/10.1111/nph.14773.

Hipp, Andrew L., Paul S. Manos, Marlene Hahn, Michael Avishai, Cathérine Bodénès, Jeannine Cavender-Bares, Andrew A. Crowl, et al. 2020. "Genomic Landscape of the Global Oak Phylogeny." *New Phytologist* 226 (4): 1198–1212. https://doi.org/10.1111/nph.16162.

Hipp, Andrew L., and Jaime A. Weber. 2008. "Taxonomy of Hill's Oak (*Quercus ellipsoidalis*: Fagaceae): Evidence from AFLP Data." *Systematic Botany* 33 (March): 148–58. https://doi.org/10.1600/036364408783887320.

Hipp, Andrew L., Alan T. Whittemore, Mira Garner, Marlene Hahn, Elizabeth Fitzek, Erwan Guichoux, Jeannine Cavender-Bares, et al. 2019. "Genomic Identity of White Oak Species in an Eastern North American Syngameon." *Annals of the Missouri Botanical Garden* 104 (3): 455–77. https://doi.org/10.3417/2019434.

Hoekstra, Hopi E., Rachel J. Hirschmann, Richard A. Bundey, Paul A. Insel, and Janet P. Crossland. 2006. "A Single Amino Acid Mutation Contributes to Adaptive Beach Mouse Color Pattern." *Science* 313 (5783): 101–4. https://doi.org/10.1126/science.1126121.

Hofmann, Christa-Ch., Omar Mohamed, and Hans Egger. 2011. "A New Terrestrial Palynoflora from the Palaeocene/Eocene Boundary in the Northwestern Tethyan Realm (St. Pankraz, Austria)." *Review of Palaeobotany and Palynology* 166 (3): 295–310. https://doi.org/10.1016/j.revpalbo.2011.06.003.

Hokanson, Stan C., J. G. Isebrands, Richard J. Jensen, and James F. Hancock. 1993. "Isozyme Variation in Oaks of the Apostle Islands in Wisconsin: Genetic Structure and Levels of Inbreeding in *Quercus rubra* and *Q. ellipsoidalis* (Fagaceae)." *American Journal of Botany* 80 (11): 1349–57.

Hollick, Arthur. 1919. "The Story of the Bartram Oak." *Scientific American* 121 (17): 422–32.

Holliday, T. W. 2006. "Neanderthals and Modern Humans: An Example of a Mammalian Syngameon?" In *Neanderthals Revisited: New Approaches and Perspectives*, edited by Jean-Jacques Hublin, Katerina Harvati, and Terry Harrison, 281–97. Vertebrate Paleobiology and Paleoanthropology. Dordrecht: Springer Netherlands. https://doi.org/10.1007/978-1-4020-5121-0_16.

Holt, Jim. 2018. *When Einstein Walked with Gödel: Excursions to the Edge of Thought*. New York: Farrar, Straus and Giroux.

Homer. 1919. *The Odyssey*. Translated by A. T. Murray. Loeb Classical Library. Cambridge, MA: Harvard University Press.

Hooghiemstra, Henry, Antoine M. Cleef, and Suzette G. A. Flantua. 2022. "A Paleoecological Context to Assess the Development of Oak Forest in Colombia: A Comment on Zorilla-Azcué, S., González-Rodríguez, A., Oyama, K., González, M. A., & Rodríguez-Correa, H., The DNA History of a Lonely Oak: *Quercus humboldtii* Phylogeography in the Colombian Andes. *Ecology and Evolution* 2021, Doi: 10.100−2/Ece3.7529." *Ecology and Evolution* 12 (3): e8702. https://doi.org/10.1002/ece3.8702.

Howard, D. J., R. W. Preszler, J. Williams, S. Fenchel, and W. J. Boecklen. 1997. "How Discrete Are Oak Species? Insights from a Hybrid Zone between *Quercus grisea* and *Quercus gambelii*." *Evolution* 51:747−55.

Hren, Michael T., Nathan D. Sheldon, Stephen T. Grimes, Margaret E. Collinson, Jerry J. Hooker, Melanie Bugler, and Kyger C. Lohmann. 2013. "Terrestrial Cooling in Northern Europe during the Eocene−Oligocene Transition." *Proceedings of the National Academy of Sciences* 110 (19): 7562−67. https://doi.org/10.1073/pnas.1210930110.

Huang, Yen-Ning, Hao Zhang, Scott Rogers, Mark Coggeshall, and Keith Woeste. 2016. "White Oak Growth after 23 Years in a Three-Site Provenance/Progeny Trial on a Latitudinal Gradient in Indiana." *Forest Science* 62 (1): 99−106. https://doi.org/10.5849/forsci.15-013.

Hubbard, J. Andrew, and Guy R. McPherson. 1997. "Acorn Selection by Mexican Jays: A Test of a Tri-Trophic Symbiotic Relationship Hypothesis." *Oecologia* 110 (1): 143−46. https://doi.org/10.1007/s004420050142.

Hubert, François, Guido W. Grimm, Emmanuelle Jousselin, Vincent Berry, Alain Franc, and Antoine Kremer. 2014. "Multiple Nuclear Genes Stabilize the Phylogenetic Backbone of the Genus *Quercus*." *Systematics and Biodiversity* 12 (4): 405−23. https://doi.org/10.1080/14772000.2014.941037.

Huerta-Sánchez, Emilia, Xin Jin, Asan, Zhuoma Bianba, Benjamin M. Peter, Nicolas Vinckenbosch, Yu Liang, et al. 2014. "Altitude Adaptation in Tibetans Caused by Introgression of Denisovan-like DNA." *Nature* 512 (7513): 194−97. https://doi.org/10.1038/nature13408.

Hughes, Joseph, and Alfried P. Vogler. 2004. "Ecomorphological Adaptation of Acorn Weevils to Their Oviposition Site." *Evolution* 58 (9): 1971−83.

Hughes, Philip D., and Philip L. Gibbard. 2018. "Global Glacier Dynamics during 100 Ka Pleistocene Glacial Cycles." *Quaternary Research* 90 (1): 222−43. https://doi.org/10.1017/qua.2018.37.

Humphreys, Aelys M., Rafaël Govaerts, Sarah Z. Ficinski, Eimear Nic Lughadha, and Maria S. Vorontsova. 2019. "Global Dataset Shows Geography and Life Form

Predict Modern Plant Extinction and Rediscovery." *Nature Ecology and Evolution* 3 (7): 1043–47. https://doi.org/10.1038/s41559-019-0906-2.

Hung, Chih-Ming, Pei-Jen L. Shaner, Robert M. Zink, Wei-Chung Liu, Te-Chin Chu, Wen-San Huang, and Shou-Hsien Li. 2014. "Drastic Population Fluctuations Explain the Rapid Extinction of the Passenger Pigeon." *Proceedings of the National Academy of Sciences* (June). https://doi.org/10.1073/pnas.1401526111.

Hutchinson, David K., Helen K. Coxall, Daniel J. Lunt, Margret Steinthorsdottir, Agatha M. de Boer, Michiel Baatsen, Anna von der Heydt, et al. 2021. "The Eocene-Oligocene Transition: A Review of Marine and Terrestrial Proxy Data, Models and Model–Data Comparisons." *Climate of the Past* 17 (1): 269–315. https://doi.org/10.5194/cp-17-269-2021.

Irgens-Moller, H. 1955. "Forest-Tree Genetics Research: *Quercus* L." *Economic Botany* 9 (1): 53. https://doi.org/10.1007/BF02984960.

Ivany, Linda C., William P. Patterson, and Kyger C. Lohmann. 2000. "Cooler Winters as a Possible Cause of Mass Extinctions at the Eocene/Oligocene Boundary." *Nature* 407 (6806): 887–90. https://doi.org/10.1038/35038044.

Iverson, Louis R., Matthew P. Peters, Anantha M. Prasad, and Stephen N. Matthews. 2019. "Analysis of Climate Change Impacts on Tree Species of the Eastern US: Results of DISTRIB-II Modeling." *Forests* 10 (4): 302. https://doi.org/10.3390/f10040302.

Iverson, Louis R., and Anantha M. Prasad. 1998. "Predicting Abundance of 80 Tree Species Following Climate Change in the Eastern United States." *Ecological Monographs* 68 (4): 21.

Jackson, Stephen T., Robert S. Webb, Katharine H. Anderson, Jonathan T. Overpeck, Thompson Webb III, John W. Williams, and Barbara C. S. Hansen. 2000. "Vegetation and Environment in Eastern North America during the Last Glacial Maximum." *Quaternary Science Reviews* 19 (6): 489–508. https://doi.org/10.1016/S0277-3791(99)00093-1.

Jacobs, Lucia F, and Emily R Liman. 1991. "Grey Squirrels Remember the Locations of Buried Nuts." *Animal Behaviour* 41:103–10.

Jaramillo, Carlos, Diana Ochoa, Lineth Contreras, Mark Pagani, Humberto Carvajal-Ortiz, Lisa M. Pratt, Srinath Krishnan, et al. 2010. "Effects of Rapid Global Warming at the Paleocene-Eocene Boundary on Neotropical Vegetation." *Science* 330 (6006): 957–61. https://doi.org/10.1126/science.1193833.

Jerome, Diana. 2018. "*Quercus insignis*." *IUCN Red List of Threatened Species* 2018: e.T194177A2302931. https://doi.org/10.2305/IUCN.UK.2018-1.RLTS.T194177A2302931.en.

Jia, Hui, Peihong Jin, Jingyu Wu, Zixi Wang, and Bainian Sun. 2015. "*Quercus* (Subg. *Cyclobalanopsis*) Leaf and Cupule Species in the Late Miocene of Eastern China and Their Paleoclimatic Significance." *Review of Palaeobotany and Palynology* 219 (August): 132–46. https://doi.org/10.1016/j.revpalbo.2015.01.011.

Jiang, Lu, Qin Bao, Wei He, Deng-Mei Fan, Shan-Mei Cheng, Jordi López-Pujol,

Myong Gi Chung, et al. 2022. "Phylogeny and Biogeography of *Fagus* (Fagaceae) Based on 28 Nuclear Single/Low-Copy Loci." *Journal of Systematics and Evolution* 60 (4): 759–72. https://doi.org/10.1111/jse.12695.

Jiang, Xiao-Long, Andrew L. Hipp, Min Deng, Tao Su, Zhe-Kun Zhou, and Meng-Xiao Yan. 2019. "East Asian Origins of European Holly Oaks via the Tibet-Himalayas." *Journal of Biogeography* 46: 2203–14. https://doi.org/10.1111/jbi.13654.

Jin, Wei-Tao, David S. Gernandt, Christian Wehenkel, Xiao-Mei Xia, Xiao-Xin Wei, and Xiao-Quan Wang. 2021. "Phylogenomic and Ecological Analyses Reveal the Spatiotemporal Evolution of Global Pines." *Proceedings of the National Academy of Sciences* 118 (20): e2022302118. https://doi.org/10.1073/pnas.2022302118.

Johnson, P. S., S. R. Shifley, and R. Rogers. 2009. *The Ecology and Silviculture of Oaks*. 2nd ed. Wallingford, UK: CABI.

Johnson, W. Carter, and Thompson Webb. 1989. "The Role of Blue Jays (*Cyanocitta cristata* L.) in the Postglacial Dispersal of Fagaceous Trees in Eastern North America." *Journal of Biogeography* 16 (6): 561–71. https://doi.org/10.2307/2845211.

Jones, Stephen M., Murray Hoggett, Sarah E. Greene, and Tom Dunkley Jones. 2019. "Large Igneous Province Thermogenic Greenhouse Gas Flux Could Have Initiated Paleocene-Eocene Thermal Maximum Climate Change." *Nature Communications* 10 (1): 5547. https://doi.org/10.1038/s41467-019-12957-1.

Jónsson, Hákon, Mikkel Schubert, Andaine Seguin-Orlando, Aurélien Ginolhac, Lillian Petersen, Matteo Fumagalli, Anders Albrechtsen, et al. 2014. "Speciation with Gene Flow in Equids despite Extensive Chromosomal Plasticity." *Proceedings of the National Academy of Sciences* 111 (52): 18655–60. https://doi.org/10.1073/pnas.1412627111.

Juzwik, Jennifer, Thomas C. Harrington, William L. MacDonald, and David N. Appel. 2008. "The Origin of *Ceratocystis fagacearum*, the Oak Wilt Fungus." *Annual Review of Phytopathology* 46:13–26.

Källander, Hans. 2007. "Food Hoarding and Use of Stored Food by Rooks *Corvus frugilegus*." *Bird Study* 54 (2): 192–98. https://doi.org/10.1080/00063650709461475.

Kapoor, Beant, Jerry Jenkins, Jeremy Schmutz, Tatyana Zhebentyayeva, Carsten Kuelheim, Mark Coggeshall, Chris Heim, et al. 2023. "A Haplotype-Resolved Chromosome-Scale Genome for *Quercus rubra* L. Provides Insights into the Genetics of Adaptive Traits for Red Oak Species." *G3 Genes|Genomes|Genetics* (September) 13 (11): jkad209. https://doi.org/10.1093/g3journal/jkad209.

Kaproth, Matthew A., Brett W. Fredericksen, Antonio González-Rodríguez, Andrew L. Hipp, and Jeannine Cavender-Bares. 2023. "Drought Response Strategies Are Coupled with Leaf Habit in 35 Evergreen and Deciduous Oak (*Quercus*) Species across a Climatic Gradient in the Americas." *New Phytologist* 239 (3): 888–904. https://doi.org/10.1111/nph.19019.

Karban, Richard. 1980. "Periodical Cicada Nymphs Impose Periodical Oak Tree

Wood Accumulation." *Nature* 287 (5780): 326–27. https://doi.org/10.1038/287326a0.

Karlik, Joseph, Mary Jane Epps, Robert R. Dunn, and Clint A. Penick. 2016. "Life inside an Acorn: How Microclimate and Microbes Influence Nest Organization in *Temnothorax* Ants." *Ethology* 122 (10): 790–97. https://doi.org/10.1111/eth.12525.

Karst, Justine, Melanie D. Jones, and Jason D. Hoeksema. 2023. "Positive Citation Bias and Overinterpreted Results Lead to Misinformation on Common Mycorrhizal Networks in Forests." *Nature Ecology and Evolution* (February): 1–11. https://doi.org/10.1038/s41559-023-01986-1.

Keator, Glenn, and Susan Bazell. 1998. *The Life of an Oak: An Intimate Portrait*. Berkeley: Heyday Books.

Keeley, Jon E., and Philip W. Rundel. 2005. "Fire and the Miocene Expansion of C_4 Grasslands." *Ecology Letters* 8 (7): 683–90. https://doi.org/10.1111/j.1461-0248.2005.00767.x.

Kender, Sev, Kara Bogus, Gunver K. Pedersen, Karen Dybkjær, Tamsin A. Mather, Erica Mariani, Andy Ridgwell, et al. 2021. "Paleocene/Eocene Carbon Feedbacks Triggered by Volcanic Activity." *Nature Communications* 12 (1): 5186. https://doi.org/10.1038/s41467-021-25536-0.

Khodwekar, Sudhir, and Oliver Gailing. 2017. "Evidence for Environment-Dependent Introgression of Adaptive Genes between Two Red Oak Species with Different Drought Adaptations." *American Journal of Botany* 104 (7): 1088–98. https://doi.org/10.3732/ajb.1700060.

Kim, Bernard Y., Xinzeng Wei, Sorel Fitz-Gibbon, Kirk E. Lohmueller, Joaquín Ortego, Paul F. Gugger, and Victoria L. Sork. 2018. "RADseq Data Reveal Ancient, but Not Pervasive, Introgression between Californian Tree and Scrub Oak Species (*Quercus* Sect. *Quercus*: Fagaceae)." *Molecular Ecology* 27 (22): 4556–71. https://doi.org/10.1111/mec.14869.

Kleinman, Kim. 2013. "Systematics and the Origin of Species from the Viewpoint of a Botanist: Edgar Anderson Prepares the 1941 Jesup Lectures with Ernst Mayr." *Journal of the History of Biology* 46 (1): 73–101. https://doi.org/10.1007/s10739-012-9325-9.

Kleinschmit, J. 1993. "Intraspecific Variation of Growth and Adaptive Traits in European Oak Species." Supplement. *Annales Des Sciences Forestières* 50:166s–85s. https://doi.org/10.1051/forest:19930716.

Knapp, Michael, Karen Stöckler, David Havell, Frédéric Delsuc, Federico Sebastiani, and Peter J. Lockhart. 2005. "Relaxed Molecular Clock Provides Evidence for Long-Distance Dispersal of *Nothofagus* (Southern Beech)." *PLOS Biology* 3 (1): e14. https://doi.org/10.1371/journal.pbio.0030014.

Koehler, Kari, Alyson Center, and Jeannine Cavender-Bares. 2012. "Evidence for a Freezing Tolerance–Growth Rate Trade-Off in the Live Oaks (*Quercus* Series *Virentes*) across the Tropical–Temperate Divide." *New Phytologist* 193 (3): 730–44. https://doi.org/10.1111/j.1469-8137.2011.03992.x.

Koenig, Walter D. 2021. "A Brief History of Masting Research." *Philosophical Transactions of the Royal Society B: Biological Sciences* 376 (1839): 20200423. https://doi.org/10.1098/rstb.2020.0423.

Koenig, Walter D., and Joseph Haydock. 1999. "Oaks, Acorns, and the Geographical Ecology of Acorn Woodpeckers." *Journal of Biogeography* 26 (1): 159–65. https://doi.org/10.1046/j.1365-2699.1999.00256.x.

Koenig, Walter D., and M. Katy Heck. 1988. "Ability of Two Species of Oak Woodland Birds to Subsist on Acorns." *Condor* 90 (3): 705. https://doi.org/10.2307/1368361.

Koenig, Walter D., Johannes M. H. Knops, William J. Carmen, and Ian S. Pearse. 2015. "What Drives Masting? The Phenological Synchrony Hypothesis." *Ecology* 96 (1): 184–92. https://doi.org/10.1890/14-0819.1.

Koenig, Walter D., and Andrew M. Liebhold. 2003. "Regional Impacts of Periodical Cicadas on Oak Radial Increment." *Canadian Journal of Forest Research* 33 (6): 1084–89. https://doi.org/10.1139/x03-037.

Koenig, Walter D., Andrew M. Liebhold, Jalene M. LaMontagne, and Ian S. Pearse. 2022. "Periodical Cicada Emergences Affect Masting Behavior of Oaks." *American Naturalist* (December). https://doi.org/10.1086/723735.

Koenig, Walter D., and Ronald L. Mumme. 1987. *Population Ecology of the Cooperatively Breeding Acorn Woodpecker*. Monographs in Population Biology 24. Princeton: Princeton University Press.

Koenig, Walter D., Mario B. Pesendorfer, Ian S. Pearse, William J. Carmen, and Johnannes M. H. Knops. 2021. "Budburst Timing of Valley Oaks at Hastings Reservation, Central Coastal California." *Madroño* 68 (4): 434–42. https://doi.org/10.3120/0024-9637-68.4.434.

Konar, Arpita, Olivia Choudhury, Rebecca Bullis, Lauren Fiedler, Jacqueline M. Kruser, Melissa T. Stephens, Oliver Gailing, et al. 2017. "High-Quality Genetic Mapping with ddRADseq in the Non-Model Tree *Quercus rubra*." *BMC Genomics* 18:417. https://doi.org/10.1186/s12864-017-3765-8.

Korstian, Clarence F. 1927. *Factors Controlling Germination and Early Survival in Oaks*. New Haven: Yale University Press.

Kremer, Antoine. 2021. "In What Direction Is Evolution Shaping Populations?" IUFRO Fagaceae and Nothofagaceae Working Group. Online meeting talk. Accessed October 13, 2022. https://www.youtube.com/watch?v=ZuLEcOwBJT0.

Kremer, Antoine, and Andrew L. Hipp. 2020. "Oaks: An Evolutionary Success Story." *New Phytologist* 226 (4): 987–1011. https://doi.org/10.1111/nph.16274.

Kremer, Antoine, Valérie Le Corre, Rémy J. Petit, and Alexis Ducousso. 2010. "Historical and Contemporary Dynamics of Adaptive Differentiation in European Oaks." In *Molecular Approaches in Natural Resource Conservation and Management*, edited by Charles H. Michler, Gene E. Rhodes, J. Andrew DeWoody, John W. Bickham, Keith E. Woeste, and Krista M. Nichols, 101–22. Cambridge: Cambridge University Press. https://doi.org/10.1017/CBO9780511777592.006.

Kring, David A. 1995. "The Dimensions of the Chicxulub Impact Crater and Impact

Melt Sheet." *Journal of Geophysical Research: Planets* 100 (E8): 16979–86. https://doi.org/10.1029/95JE01768.

Krishnan, Jaya, and Nicolas Rohner. 2017. "Cavefish and the Basis for Eye Loss." *Philosophical Transactions of the Royal Society B: Biological Sciences* 372 (1713): 20150487. https://doi.org/10.1098/rstb.2015.0487.

Kusi, Joseph, and Istvan Karsai. 2020. "Plastic Leaf Morphology in Three Species of *Quercus*: The More Exposed Leaves Are Smaller, More Lobated and Denser." *Plant Species Biology* 35 (1): 24–37. https://doi.org/10.1111/1442-1984.12253.

Labandeira, Conrad C. 2021. "Ecology and Evolution of Gall-Inducing Arthropods: The Pattern from the Terrestrial Fossil Record." *Frontiers in Ecology and Evolution* 9.

Ladant, Jean-Baptiste, and Yannick Donnadieu. 2016. "Palaeogeographic Regulation of Glacial Events during the Cretaceous Supergreenhouse." *Nature Communications* 7 (1): 12771. https://doi.org/10.1038/ncomms12771.

LaMontagne, Jalene M., Ian S. Pearse, David F. Greene, and Walter D. Koenig. 2020. "Mast Seeding Patterns Are Asynchronous at a Continental Scale." *Nature Plants* 6 (5): 460–65. https://doi.org/10.1038/s41477-020-0647-x.

Lanfear, Robert, Simon Y. W. Ho, T. Jonathan Davies, Angela T. Moles, Lonnie Aarssen, Nathan G. Swenson, Laura Warman, Amy E. Zanne, and Andrew P. Allen. 2013. "Taller Plants Have Lower Rates of Molecular Evolution." *Nature Communications* 4 (1): 1879. https://doi.org/10.1038/ncomms2836.

Lang, Ping, Fenny Dane, Thomas L. Kubisiak, and Hongwen Huang. 2007. "Molecular Evidence for an Asian Origin and a Unique Westward Migration of Species in the Genus *Castanea* via Europe to North America." *Molecular Phylogenetics and Evolution* 43 (1): 49–59. https://doi.org/10.1016/j.ympev.2006.07.022.

Larsen, Brendan B., Elizabeth C. Miller, Matthew K. Rhodes, and John J. Wiens. 2017. "Inordinate Fondness Multiplied and Redistributed: The Number of Species on Earth and the New Pie of Life." *Quarterly Review of Biology* 92 (3): 229–65. https://doi.org/10.1086/693564.

Larson, Drew A., Oscar M. Vargas, Alberto Vicentini, and Christopher W. Dick. 2021. "Admixture May Be Extensive among Hyperdominant Amazon Rainforest Tree Species." *New Phytologist* 232 (6): 2520–34. https://doi.org/10.1111/nph.17675.

Larson, James L. 1967. "Linnaeus and the Natural Method." *Isis* 58 (3): 304–20.

Larson-Johnson, Kathryn. 2016. "Phylogenetic Investigation of the Complex Evolutionary History of Dispersal Mode and Diversification Rates across Living and Fossil Fagales." *New Phytologist* 209 (1): 418–35. https://doi.org/10.1111/nph.13570.

Lauretano, Vittoria, Alan T. Kennedy-Asser, Vera A. Korasidis, Malcolm W. Wallace, Paul J. Valdes, Daniel J. Lunt, Richard D. Pancost, et al. 2021. "Eocene to Oligocene Terrestrial Southern Hemisphere Cooling Caused by Declining pCO$_2$." *Nature Geoscience* 14 (9): 659–64. https://doi.org/10.1038/s41561-021-00788-z.

Le Corre, Valérie, Nathalie Machon, Rémy J. Petit, and Antoine Kremer. 1997. "Colonization with Long-Distance Seed Dispersal and Genetic Structure of Maternally Inherited Genes in Forest Trees: A Simulation Study." *Genetics Research* 69 (2): 117–25.

Leroy, Thibault, Jean-Marc Louvet, Céline Lalanne, Grégoire Le Provost, Karine Labadie, Jean-Marc Aury, Sylvain Delzon, et al. 2020. "Adaptive Introgression as a Driver of Local Adaptation to Climate in European White Oaks." *New Phytologist* 226 (4): 1171–82. https://doi.org/10.1111/nph.16095.

Leroy, Thibault, Quentin Rougemont, Jean-Luc Dupouey, Catherine Bodénès, Céline Lalanne, Caroline Belser, Karine Labadie, et al. 2020. "Massive Postglacial Gene Flow between European White Oaks Uncovered Genes Underlying Species Barriers." *New Phytologist* 226 (4): 1183–97. https://doi.org/10.1111/nph.16039.

Leroy, Thibault, Camille Roux, Laure Villate, Catherine Bodénès, Jonathan Romiguier, Jorge A. P. Paiva, Carole Dossat, et al. 2017. "Extensive Recent Secondary Contacts between Four European White Oak Species." *New Phytologist* 214 (2): 865–78. https://doi.org/10.1111/nph.14413.

Lewis, Isaac M. 1911. "The Seedling of '*Quercus virginiana.*'" *Plant World* 14 (5): 119–23.

Li, R.-Q., Z.-D. Chen, A.-M. Lu, D. E. Soltis, P. S. Soltis, and P. S. Manos. 2004. "Phylogenetic Relationships in Fagales Based on DNA Sequences from Three Genomes." *International Journal of Plant Sciences* 165:311–24.

Li, Xuan, Yongfu Li, Mingyue Zang, Mingzhi Li, and Yanming Fang. 2018. "Complete Chloroplast Genome Sequence and Phylogenetic Analysis of *Quercus acutissima.*" *International Journal of Molecular Sciences* 19 (8): 2443. https://doi.org/10.3390/ijms19082443.

Li, Xuan, Gaoming Wei, Yousry A. El-Kassaby, and Yanming Fang. 2021. "Hybridization and Introgression in Sympatric and Allopatric Populations of Four Oak Species." *BMC Plant Biology* 21 (1): 266. https://doi.org/10.1186/s12870-021-03007-4.

Li, Yao, Xingwang Zhang, and Yanming Fang. 2019. "Landscape Features and Climatic Forces Shape the Genetic Structure and Evolutionary History of an Oak Species (*Quercus chenii*) in East China." *Frontiers in Plant Science* 10.

Li, Yao, Xingwang Zhang, Lu Wang, Victoria L Sork, Lingfeng Mao, and Yanming Fang. 2021. "Influence of Pliocene and Pleistocene Climates on Hybridization Patterns between Two Closely Related Oak Species in China." *Annals of Botany* (December): mcab140. https://doi.org/10.1093/aob/mcab140.

Lind-Riehl, Jennifer F., Alexis R. Sullivan, and Oliver Gailing. 2014. "Evidence for Selection on a CONSTANS-like Gene between Two Red Oak Species." *Annals of Botany* 113 (6): 967–75. https://doi.org/10.1093/aob/mcu019.

Linné, Carl von. 2003. *Linnaeus' Philosophia Botanica.* Translated by Stephen Freer. 1st English ed. Oxford: Oxford University Press.

Liu, Xiaoyan, Hanzhang Song, and Jianhua Jin. 2020. "Diversity of Fagaceae on Hainan Island of South China during the Middle Eocene: Implications for Phytogeography and Paleoecology." *Frontiers in Ecology and Evolution* 8.

Liu, Xiao-Yan, Sheng-Lan Xu, Meng Han, and Jian-Hua Jin. 2019. "An Early Oligocene Fossil Acorn, Associated Leaves and Pollen of the Ring-Cupped Oaks (*Quercus* Subg. *Cyclobalanopsis*) from Maoming Basin, South China." *Journal of Systematics and Evolution* 57 (2): 153–68. https://doi.org/10.1111/jse.12450.

Logan, William Bryant. 2005. *Oak: Frame of Civilization*. New York: W. W. Norton.

Lotsy, Johannes Paulus. 1917. "La Quintessence de La Théorie Du Croisement."
*Archives Néerlandaises des Sciences Exactes et Naturelles. Série IIIB (Sciences Na-
turelles)* 3:351–53.

———. 1925. "Species or Linneon." *Genetica* 7 (5): 487–506. https://doi.org/10.1007
/BF01676287.

Lowry, David B. 2012. "Ecotypes and the Controversy over Stages in the Formation
of New Species." *Biological Journal of the Linnean Society* 106 (2): 241–57. https://
doi.org/10.1111/j.1095-8312.2012.01867.x.

Lucek, Kay, Hannah Augustijnen, and Marcial Escudero. 2022. "A Holocentric Twist
to Chromosomal Speciation?" *Trends in Ecology and Evolution* (April). https://doi
.org/10.1016/j.tree.2022.04.002.

Lumibao, Candice Y., Sean M. Hoban, and Jason McLachlan. 2017. "Ice Ages Leave
Genetic Diversity 'Hotspots' in Europe but Not in Eastern North America." *Ecology Letters* 20 (11): 1459–68. https://doi.org/10.1111/ele.12853.

Luo, Ao, Xiaoting Xu, Yunpeng Liu, Yaoqi Li, Xiangyan Su, Yichao Li, Tong Lyu,
et al. 2023. "Spatio-Temporal Patterns in the Woodiness of Flowering Plants."
Global Ecology and Biogeography 32 (3): 384–96. https://doi.org/10.1111/geb.13627.

Luterbacher, Jürg, Daniel Dietrich, Elena Xoplaki, Martin Grosjean, and Heinz Wan-
ner. 2004. "European Seasonal and Annual Temperature Variability, Trends, and
Extremes Since 1500." *Science* 303 (5663): 1499–1503. https://doi.org/10.1126
/science.1093877.

Lutzoni, François, Michael D. Nowak, Michael E. Alfaro, Valérie Reeb, Jolanta Mi-
adlikowska, Michael Krug, A. Elizabeth Arnold, et al. 2018. "Contemporaneous
Radiations of Fungi and Plants Linked to Symbiosis." *Nature Communications* 9
(1): 5451. https://doi.org/10.1038/s41467-018-07849-9.

Lynch, Michael, and John S. Conery. 2003. "The Origins of Genome Complexity."
Science 302 (5649): 1401–4. https://doi.org/10.1126/science.1089370.

Macciardi, Fabio, and Fabio Martini. 2022. "Chapter 6—The Neanderthal Brain: Biological and Cognitive Evolution." In *Updating Neanderthals*, edited by Francesca
Romagnoli, Florent Rivals, and Stefano Benazzi, 89–108. Cambridge, MA: Academic Press. https://doi.org/10.1016/B978-0-12-821428-2.00008-1.

MacDougal, D. T. 1907. "Hybridization of Wild Plants." *Botanical Gazette* 43 (1): 45–
58. https://doi.org/10.1086/329077.

MacDougall, Andrew S., Alisha Duwyn, and Natalie T. Jones. 2010. "Consumer-
Based Limitations Drive Oak Recruitment Failure." *Ecology* 91 (7): 2092–99.
https://doi.org/10.1890/09-0204.1.

Magallón, Susana, Sandra Gómez-Acevedo, Luna L. Sánchez-Reyes, and Tania

Hernández-Hernández. 2015. "A Metacalibrated Time-Tree Documents the Early Rise of Flowering Plant Phylogenetic Diversity." *New Phytologist* 207 (2): 437–53. https://doi.org/10.1111/nph.13264.

Magni, C. R., A. Ducousso, H. Caron, R. J. Petit, and A. Kremer. 2005. "Chloroplast DNA Variation of *Quercus rubra* L. in North America and Comparison with Other Fagaceae." *Molecular Ecology* 14 (2): 513–24.

Mahner, Martin, and Michael Kary. 1997. "What Exactly Are Genomes, Genotypes and Phenotypes? And What About Phenomes?" *Journal of Theoretical Biology* 186 (1): 55–63. https://doi.org/10.1006/jtbi.1996.0335.

Mallet, James. 1995. "A Species Definition for the Modern Synthesis." *Trends in Ecology and Evolution* 10 (7): 294–99. https://doi.org/10.1016/0169-5347(95)90031-4.

———. 2003. "Poulton, Wallace and Jordan: How Discoveries in *Papilio* Butterflies Led to a New Species Concept 100 Years Ago." *Systematics and Biodiversity* 1 (December): 441–52. https://doi.org/10.1017/S1477200003001300.

———. 2020. "Alternative Views of Biological Species: Reproductively Isolated Units or Genotypic Clusters?" *National Science Review* 7 (8): 1401–7. https://doi.org/10.1093/nsr/nwaa116.

Manchester, Steven R. 1994. "Fruits and Seeds of the Middle Eocene Nut Beds Flora, Clarno Formation, Oregon." *Palaeontographica Americana* 58:1–205.

———. 2011. "Fruits of Ticodendraceae (Fagales) from the Eocene of Europe and North America." *International Journal of Plant Sciences* 172 (9): 1179–87. https://doi.org/10.1086/662135.

———. 2014. "Revisions to Roland Brown's North American Paleocene Flora." *Acta Musei Nationalis Pragae, Series B, Historia Naturalis / Sborník Národního Muzea Řada B, Přírodní Vědy* (December): 153–210. https://doi.org/10.14446/AMNP.2014.153.

Manos, Paul S. 1997. "Systematics of *Nothofagus* (Nothofagaceae) Based on rDNA Spacer Sequences (ITS): Taxonomic Congruence with Morphology and Plastid Sequences." *American Journal of Botany* 84 (8): 1137–55. https://doi.org/10.2307/2446156.

Manos, Paul S., Charles H. Cannon, and Sang-Hun Oh. 2008. "Phylogenetic Relationships and Taxonomic Status of the Paleoendemic Fagaceae of Western North America: Recognition of a New Genus, *Notholithocarpus*." *Madroño* 55 (3): 181–90. https://doi.org/10.3120/0024-9637-55.3.181.

Manos, Paul S., Jeff J. Doyle, and Kevin C. Nixon. 1999. "Phylogeny, Biogeography, and Processes of Molecular Differentiation in *Quercus* Subgenus *Quercus* (Fagaceae)." *Molecular Phylogenetics and Evolution* 12 (3): 333–49. https://doi.org/10.1006/mpev.1999.0614.

Manos, Paul S., and David E. Fairbrothers. 1987. "Allozyme Variation in Populations of Six Northeastern American Red Oaks (Fagaceae: *Quercus* Subg. *Erythrobalanus*)." *Systematic Botany* 12 (3): 265–373.

Manos, Paul S., and Andrew L. Hipp. 2021. "An Updated Infrageneric Classification of the North American Oaks (*Quercus* Subgenus *Quercus*): Review of the Contribution of Phylogenomic Data to Biogeography and Species Diversity." *Forests* 12 (6): 786. https://doi.org/10.3390/f12060786.

Manos, Paul S., Kevin C. Nixon, and Jeff J. Doyle. 1993. "Cladistic Analysis of Restriction Site Variation within the Chloroplast DNA Inverted Repeat Region of Selected Hamamelididae." *Systematic Botany* 18 (4): 551–62. https://doi.org/10.2307/2419533.

Manos, Paul S., and Kelly P. Steele. 1997. "Phylogenetic Analyses of 'Higher' Hamamelididae Based on Plastid Sequence Data." *American Journal of Botany* 84 (10): 1407–19. https://doi.org/10.2307/2446139.

Marçais, Benoit, and Marie-Laure Desprez-Loustau. 2014. "European Oak Powdery Mildew: Impact on Trees, Effects of Environmental Factors, and Potential Effects of Climate Change." *Annals of Forest Science* 71 (6): 633–42. https://doi.org/10.1007/s13595-012-0252-x.

Martin, Paul S., and Byron E. Harrell. 1957. "The Pleistocene History of Temperate Biotas in Mexico and Eastern United States." *Ecology* 38 (3): 468–80. https://doi.org/10.2307/1929892.

Martinetto, Edoardo, Nareerat Boonchai, Friðger Grímsson, Paul Joseph Grote, Gregory Jordan, Marianna Kováčová, Lutz Kunzmann, et al. 2020. "Triumph and Fall of the Wet, Warm, and Never-More-Diverse Temperate Forests (Oligocene-Pliocene)." In *Nature through Time: Virtual Field Trips through the Nature of the Past*, edited by Edoardo Martinetto, Emanuel Tschopp, and Robert A. Gastaldo, 55–81. Cham, Germany: Springer International. https://doi.org/10.1007/978-3-030-35058-1_2.

Martins, Karina, Paul F. Gugger, Jesus Llanderal-Mendoza, Antonio González-Rodríguez, Sorel T. Fitz-Gibbon, Jian-Li Zhao, Hernando Rodríguez-Correa, et al. 2018. "Landscape Genomics Provides Evidence of Climate-Associated Genetic Variation in Mexican Populations of *Quercus rugosa*." *Evolutionary Applications* 11 (10): 1842–58. https://doi.org/10.1111/eva.12684.

Mason, Sarah L. R. 1992. "Acorns in Human Subsistence." PhD thesis, University College London.

Matsumura, Emi, Kenta Morinaga, and Kenji Fukuda. 2022. "Host Specificity and Seasonal Variation in the Colonization of *Tubakia* Sensu Lato Associated with Evergreen Oak Species in Eastern Japan." *Microbial Ecology* (July): 1–13. https://doi.org/10.1007/s00248-022-02067-9.

Mauri, Achille, Marco Girardello, Giovanni Strona, Pieter S. A. Beck, Giovanni Forzieri, Giovanni Caudullo, Federica Manca, et al. 2022. "EU-Trees4F, a Dataset on the Future Distribution of European Tree Species." *Scientific Data* 9 (1): 37. https://doi.org/10.1038/s41597-022-01128-5.

May, Michael R., Mitchell C. Provance, Andrew C. Sanders, Norman C. Ellstrand, and Jeffrey Ross-Ibarra. 2009. "A Pleistocene Clone of Palmer's Oak Persisting in

Southern California." *PLOS ONE* 4 (12): e8346. https://doi.org/10.1371/journal
.pone.0008346.

Mayr, Ernst. 1942. *Systematics and the Origin of Species, from the Viewpoint of a Zool-
ogist*. Garden City, NY: Dover Publications.

Maze, Jack. 1968. "Past Hybridization between *Quercus macrocarpa* and *Quercus gambelii*." *Brittonia* 20 (4): 321. https://doi.org/10.2307/2805689.

McCarthy, Diane M., and Roberta J. Mason-Gamer. 2016. "Chloroplast DNA-Based
Phylogeography of *Tilia americana* (Malvaceae)." *Systematic Botany* 41 (4): 865–
80. https://doi.org/10.1600/036364416X693964.

McCauley, Ross A., Aurea C. Cortés-Palomec, and Ken Oyama. 2019. "Species Diver-
sification in a Lineage of Mexican Red Oak (*Quercus* Section *Lobatae* Subsection
Racemiflorae)—the Interplay between Distance, Habitat, and Hybridization."
Tree Genetics and Genomes 15 (2): 27. https://doi.org/10.1007/s11295-019-1333-x.

McCormack, M. Luke, Katie P. Gaines, Melissa Pastore, and David M. Eissenstat.
2015. "Early Season Root Production in Relation to Leaf Production among Six
Diverse Temperate Tree Species." *Plant and Soil* 389 (1–2): 121–29. https://doi.org
/10.1007/s11104-014-2347-7.

McInerney, Francesca A., and Scott L. Wing. 2011. "The Paleocene-Eocene Thermal
Maximum: A Perturbation of Carbon Cycle, Climate, and Biosphere with Implications for the Future." *Annual Review of Earth and Planetary Sciences* 39 (1):
489–516. https://doi.org/10.1146/annurev-earth-040610-133431.

McIntyre, D.J. 1991. "Pollen and Spore Flora of an Eocene Forest, Eastern Axel Heiberg Island." *N.W.T. Geological Survey of Canada Bulletin* 403:83–97.

McIver, E. E., and J. F. Basinger. 1999. "Early Tertiary Floral Evolution in the
Canadian High Arctic." *Annals of the Missouri Botanical Garden* 86 (2): 523–45.
https://doi.org/10.2307/2666184.

McKenna, Malcolm C. 1975. "Fossil Mammals and Early Eocene North Atlantic
Land Continuity." *Annals of the Missouri Botanical Garden* 62 (2): 335–53. https://
doi.org/10.2307/2395200.

McLachlan, Jason S., James S. Clark, and Paul S. Manos. 2005. "Molecular Indicators
of Tree Migration Capacity under Rapid Climate Change." *Ecology* 86 (8): 2088–
98. https://doi.org/10.1890/04-1036.

McVay, John D., Duncan Hauser, Andrew L. Hipp, and Paul S. Manos. 2017. "Phylogenomics Reveals a Complex Evolutionary History of Lobed-Leaf White Oaks
in Western North America." *Genome* 60 (9): 733–42. https://doi.org/10.1139/gen
-2016-0206.

McVay, John D., Andrew L. Hipp, and Paul S. Manos. 2017. "A Genetic Legacy of Introgression Confounds Phylogeny and Biogeography in Oaks." *Proceedings of the
Royal Society B: Biological Sciences* 284 (1854): 20170300. https://doi.org/10.1098
/rspb.2017.0300.

Meding, Stephen Mercer. 2007. "The Function of Common Mycorrhizal Networks on the Transfer of Nutrients between Oak Woodland Plants of the Sierra Foothills, California." PhD thesis, University of California, Davis.

Menitsky, Yu L. 2005. *Oaks of Asia*. Boca Raton, FL: CRC Press.

Mensing, Scott. 2005. "The History of Oak Woodlands in California, Part I: The Paleoecologic Record." https://scholarworks.calstate.edu/downloads/05741w226.

———. 2015. "The Paleohistory of California Oaks." In *Proceedings of the Seventh California Oak Symposium: Managing Oak Woodlands in a Dynamic World; Gen. Tech. Rep. PSW-GTR-251*, edited by Richard B. Standiford and Kathryn L. Purcell, 35–47. Berkeley: U.S. Department of Agriculture, Forest Service, Pacific Southwest Research Station.

Merceron, Nastasia R., Thibault Leroy, Emilie Chancerel, Jeanne Romero-Severson, Daniel S. Borkowski, Alexis Ducousso, Arnaud Monty, et al. 2017. "Back to America: Tracking the Origin of European Introduced Populations of *Quercus rubra* L." *Genome* 60 (9): 778–90. https://doi.org/10.1139/gen-2016-0187.

Milne, Richard I., and Richard J. Abbott. 2002. "The Origin and Evolution of Tertiary Relict Floras." *Advances in Botanical Research*, 38:281–314. Cambridge, MA: Academic Press. https://doi.org/10.1016/S0065-2296(02)38033-9.

Mishra, Bagdevi, Bartosz Ulaszewski, Joanna Meger, Jean-Marc Aury, Catherine Bodénès, Isabelle Lesur-Kupin, Markus Pfenninger, et al. 2022. "A Chromosome-Level Genome Assembly of the European Beech (*Fagus sylvatica*) Reveals Anomalies for Organelle DNA Integration, Repeat Content and Distribution of SNPs." *Frontiers in Genetics* 12. https://doi.org/10.3389/fgene.2021.691058.

Mitrus, Sławomir. 2021. "Acorn Ants May Create and Use Two Entrances to the Nest Cavity." *Insects* 12 (10): 912. https://doi.org/10.3390/insects12100912.

Mogensen, H. Lloyd. 1970. "Syncotyly in *Quercus arizonica*." *American Journal of Botany* 57 (10): 1207–10. https://doi.org/10.2307/2441359.

———. 1975. "Ovule Abortion in *Quercus* (Fagaceae)." *American Journal of Botany* 62 (2): 160–65. https://doi.org/10.2307/2441590.

Mohler, C. L. 1990. "Co-Occurrence of Oak Subgenera: Implications for Niche Differentiation." *Bulletin of the Torrey Botanical Club* 117 (3): 247–55. https://doi.org/10.2307/2996693.

Moore, Jeffrey E., Amy B. McEuen, Robert K. Swihart, Thomas A. Contreras, and Michael A. Steele. 2007. "Determinants of Seed Removal Distance by Scatter-Hoarding Rodents in Deciduous Forests." *Ecology* 88 (10): 2529–40. https://doi.org/10.1890/07-0247.1.

Moore, Peter D. 1999. "Woodpecker Population Drills." *Nature* 399 (6736): 528–29. https://doi.org/10.1038/21078.

Moorjani, Priya, Sriram Sankararaman, Qiaomei Fu, Molly Przeworski, Nick Patterson, and David Reich. 2016. "A Genetic Method for Dating Ancient Genomes Provides a Direct Estimate of Human Generation Interval in the Last 45,000 Years." *Proceedings of the National Academy of Sciences* 113 (20): 5652–57. https://doi.org/10.1073/pnas.1514696113.

Moracho, E., G. Moreno, P. Jordano, and A. Hampe. 2016. "Unusually Limited Pollen Dispersal and Connectivity of Pedunculate Oak (*Quercus robur*) Refugial Populations at the Species' Southern Range Margin." *Molecular Ecology* 25 (14): 3319–31. https://doi.org/10.1111/mec.13692.

Morales-Saldaña, Saddan, Susana Valencia-Avalos, Ken Oyama, Efraín Tovar Sánchez, Andrew L. Hipp, and Antonio González-Rodríguez. 2022. "Even More Oak Species in Mexico? Genetic Structure and Morphological Differentiation Support the Presence of at Least Two Specific Entities within *Quercus laeta*." *Journal of Systematics and Evolution* 60 (5): 1124–39. https://doi.org/10.1111/jse.12818.

Moran, Emily V., John Willis, and James S. Clark. 2012. "Genetic Evidence for Hybridization in Red Oaks (*Quercus* Sect. *Lobatae*, Fagaceae)." *American Journal of Botany* 99 (1): 92–100. https://doi.org/10.3732/ajb.1100023.

Moricca, Salvatore, Beatrice Ginetti, and Alessandro Ragazzi. 2012. "Species- and Organ-Specificity in Endophytes Colonizing Healthy and Declining Mediterranean Oaks." *Phytopathologia Mediterranea* 51 (3): 587.

Morillas, Lourdes, María José Leiva, Ignacio M. Pérez-Ramos, Jesús Cambrollé, and Luis Matías. 2023. "Latitudinal Variation in the Functional Response of *Quercus suber* Seedlings to Extreme Drought." *Science of the Total Environment* 887 (August): 164122. https://doi.org/10.1016/j.scitotenv.2023.164122.

Morjan, Carrie L., and Loren H. Rieseberg. 2004. "How Species Evolve Collectively: Implications of Gene Flow and Selection for the Spread of Advantageous Alleles." *Molecular Ecology* 13 (6): 1341–56.

Muir, Graham, Colin C. Fleming, and Christian Schlötterer. 2000. "Species Status of Hybridizing Oaks." *Nature (London)* 405:1016.

Mukherjee, S. 2016. *The Gene: An Intimate History.* New York: Scribner.

Muller, Cornelius H. 1951. "The Significance of Vegetative Reproduction in *Quercus*." *Madroño* 11 (3): 129–37.

———. 1952. "Ecological Control of Hybridization in *Quercus*: A Factor in the Mechanism of Evolution." *Evolution* 6 (2): 147–61.

Müller-Wille, Staffan. 2007. "The Love of Plants." *Nature* 446 (7133): 268. https://doi.org/10.1038/446268a.

Munoz, Samuel E., David J. Mladenoff, Sissel Schroeder, and John W. Williams. 2014. "Defining the Spatial Patterns of Historical Land Use Associated with the Indigenous Societies of Eastern North America." *Journal of Biogeography* 41 (12): 2195–2210. https://doi.org/10.1111/jbi.12386.

Nagamitsu, Teruyoshi, Hajime Shimizu, Mineaki Aizawa, and Atsushi Nakanishi. 2019. "An Admixture of *Quercus dentata* in the Coastal Ecotype of *Q. mongolica* var. *crispula* in Northern Hokkaido and Genetic and Environmental Effects on Their Traits." *Journal of Plant Research* (January). https://doi.org/10.1007/s10265-018-01079-2.

Nagamitsu, Teruyoshi, and Kato Shuri. 2021. "Seed Transfer across Geographic Regions in Different Climates Leads to Reduced Tree Growth and Genetic Admixture in *Quercus mongolica* var. *crispula*." *Forest Ecology and Management* 482 (February): 118787. https://doi.org/10.1016/j.foreco.2020.118787.

Narango, Desirée L., Douglas W. Tallamy, and Kimberley J. Shropshire. 2020. "Few Keystone Plant Genera Support the Majority of Lepidoptera Species." *Nature Communications* 11 (1): 5751. https://doi.org/10.1038/s41467-020-19565-4.

Navarro-Cerrillo, Rafael, Antonio Cabrera-Ariza, Antonio Avaria, Guillermo Palacios-Rodríguez, and Rómulo Santelices-Moya. 2020. "Stand Structure, Regeneration and Seed Dispersal Patterns of *Nothofagus glauca* (Hualo) in Central Chile." *Southern Forests: A Journal of Forest Science* 82 (1): 75–85. https://doi.org/10.2989/20702620.2020.1733759.

Nicholls, James A., George Melika, and Graham N. Stone. 2017. "Sweet Tetra-Trophic Interactions: Multiple Evolution of Nectar Secretion, a Defensive Extended Phenotype in Cynipid Gall Wasps." *American Naturalist* 189 (1): 67–77. https://doi.org/10.1086/689399.

Nickerson, Megan N., Lillian P. Moore, and Jana M. U'Ren. 2023. "The Impact of Polyphenolic Compounds on the in Vitro Growth of Oak-Associated Foliar Endophytic and Saprotrophic Fungi." *Fungal Ecology* 62 (April): 101226. https://doi.org/10.1016/j.funeco.2023.101226.

Nixon, Kevin C. 1997. "*Quercus*." In *Flora of North America North of Mexico*, edited by Flora of North America Editorial Committee, 3:445–47. New York: Oxford University Press.

———. 2009. "An Overview of *Quercus*: Classification and Phylogenetics with Comments on Differences in Wood Anatomy." In *The Proceedings of the 2nd National Oak Wilt Symposium*, edited by Ronald F. Billings and David N. Appel, 13–25. International Society of Arboriculture—Texas Chapter. https://texasoakwilt.org/assets/studies/NOWS/conference_assets/conferencepapers/Nixon.pdf.

Núñez-Farfán, J., and C. D. Schlichting. 2001. "Evolution in Changing Environments: The 'Synthetic' Work of Clausen, Keck, and Hiesey." *Quarterly Review of Biology* 76 (4): 433–57. https://doi.org/10.1086/420540.

Nurk, Sergey, Sergey Koren, Arang Rhie, Mikko Rautiainen, Andrey V. Bzikadze, Alla Mikheenko, Mitchell R. Vollger, et al. 2022. "The Complete Sequence of a Human Genome." *Science* 376 (6588): 44–53. https://doi.org/10.1126/science.abj6987.

"The Oak Name Checklist." 2022. International Oak Society. http://www.oaknames.org.

O'Connor, Philip A., and Mark V. Coggeshall. 2011. "White Oak Seed Source Performance across Multiple Sites in Indiana through Age 16." In *Proceedings, 17th Central Hardwood Forest Conference; 2010 April 5–7, Lexington, KY, Gen. Tech. Rep. NRS-P-78*, edited by Songlin Fei, John M. Lhotka, Jeffrey W. Stringer, Kurt W. Gottschalk, and Gary W. Miller, 358–63. Newtown Square, PA: U.S. Department of Agriculture, Forest Service, Northern Research Station.

O'Donnell, Scott. 2023. "Evidence of Hybridization and Introgression in Two Distantly Related, Sympatric Californian White Oaks (*Quercus* Sect. *Quercus*)." PhD thesis, University of California, Los Angeles.

O'Donnell, Scott T., Sorel T. Fitz-Gibbon, and Victoria L. Sork. 2021. "Ancient Introgression between Distantly Related White Oaks (*Quercus* Sect. *Quercus*) Shows Evidence of Climate-Associated Asymmetric Gene Exchange." *Journal of Heredity*, no. esab053 (September). https://doi.org/10.1093/jhered/esab053.

Oh, Sang-Hun, and Paul S. Manos. 2008. "Molecular Phylogenetics and Cupule Evolution in Fagaceae as Inferred from Nuclear CRABS CLAW Sequences." *Taxon* 57:434–51.

Oikawa, Hana Londoño, and Paulo C. Pulgarín-R. 2019. "Is the Distribution of the Acorn Woodpecker (*Melanerpes formicivorus flavigula*) Associated with Oaks and Granaries? A Local Study in an Urban Area in Northern South America." bioRxiv .org. https://doi.org/10.1101/765487.

Oney-Birol, Signem, Sorel Fitz-Gibbon, Jin-Ming Chen, Paul F. Gugger, and Victoria L. Sork. 2018. "Assessment of Shared Alleles in Drought-Associated Candidate Genes among Southern California White Oak Species (*Quercus* Sect. *Quercus*)." *BMC Genetics* 19 (1): 88. https://doi.org/10.1186/s12863-018-0677-9.

Opedal, Øystein H., W. Scott Armbruster, Thomas F. Hansen, Agnes Holstad, Christophe Pélabon, Stefan Andersson, Diane R. Campbell, et al. 2023. "Evolvability and Trait Function Predict Phenotypic Divergence of Plant Populations." *Proceedings of the National Academy of Sciences* 120 (1): e2203228120. https://doi.org/10.1073/pnas.2203228120.

Ordonez, Alejandro, and John W. Williams. 2013. "Climatic and Biotic Velocities for Woody Taxa Distributions over the Last 16 000 Years in Eastern North America." *Ecology Letters* 16 (6): 773–81. https://doi.org/10.1111/ele.12110.

Ortego, Joaquín, Josep Maria Espelta, Dolors Armenteras, María Claudia Díez, Alberto Muñoz, and Raúl Bonal. 2023. "Demographic and Spatially Explicit Landscape Genomic Analyses in a Tropical Oak Reveal the Impacts of Late Quaternary Climate Change on Andean Montane Forests." *Molecular Ecology* 32 (12): 3182–99. https://doi.org/10.1111/mec.16930.

Ortego, Joaquín, Paul F. Gugger, Erin C. Riordan, and Victoria L. Sork. 2014. "Influence of Climatic Niche Suitability and Geographical Overlap on Hybridization Patterns among Southern Californian Oaks." *Journal of Biogeography* 41 (10): 1895–1908. https://doi.org/10.1111/jbi.12334.

Ortego, Joaquín, Paul F. Gugger, and Victoria L. Sork. 2015. "Climatically Stable Landscapes Predict Patterns of Genetic Structure and Admixture in the Californian Canyon Live Oak." *Journal of Biogeography* 42 (2): 328–38. https://doi.org/10.1111/jbi.12419.

———. 2018. "Genomic Data Reveal Cryptic Lineage Diversification and Introgression in Californian Golden Cup Oaks (Section *Protobalanus*)." *New Phytologist* 218 (2): 804–18. https://doi.org/10.1111/nph.14951.

Osman, Matthew B., Jessica E. Tierney, Jiang Zhu, Robert Tardif, Gregory J. Hakim,

Jonathan King, and Christopher J. Poulsen. 2021. "Globally Resolved Surface Temperatures since the Last Glacial Maximum." *Nature* 599 (7884): 239–44. https://doi.org/10.1038/s41586-021-03984-4.

Ostfeld, Richard S., Charles D. Canham, Kelly Oggenfuss, Raymond J. Winchcombe, and Felicia Keesing. 2006. "Climate, Deer, Rodents, and Acorns as Determinants of Variation in Lyme-Disease Risk." *PLOS Biology* 4 (6): e145. https://doi.org/10.1371/journal.pbio.0040145.

Overpeck, Jonathan T., Patrick J. Bartlein, and Thompson Webb. 1991. "Potential Magnitude of Future Vegetation Change in Eastern North America: Comparisons with the Past." *Science* 254 (5032): 692–95.

Owusu, Sandra A., Alexis R. Sullivan, Jaime A. Weber, Andrew L. Hipp, and Oliver Gailing. 2015. "Taxonomic Relationships and Gene Flow in Four North American *Quercus* Species (*Quercus* Section *Lobatae*)." *Systematic Botany* 40 (July): 510–21. https://doi.org/10.1600/036364415x688754.

Palcu, Dan V., and Wout Krijgsman. 2021. "The Dire Straits of Paratethys: Gateways to the Anoxic Giant of Eurasia." *Geological Society, London, Special Publications* (December): SP523–2021. https://doi.org/10.1144/SP523-2021-73.

Palcu, Dan Valentin, Irina Stanislavovna Patina, Ionuț Șandric, Sergei Lazarev, Iuliana Vasiliev, Marius Stoica, and Wout Krijgsman. 2021. "Late Miocene Megalake Regressions in Eurasia." *Scientific Reports* 11 (1): 11471. https://doi.org/10.1038/s41598-021-91001-z.

Palmer, Ernest J. 1948. "Hybrid Oaks of North America." *Journal of the Arnold Arboretum* 29 (1): 1–48.

Panchy, Nicholas, Melissa Lehti-Shiu, and Shin-Han Shiu. 2016. "Evolution of Gene Duplication in Plants." *Plant Physiology* 171 (4): 2294–2316. https://doi.org/10.1104/pp.16.00523.

Papper, Prahlada D. 2021. "Three Scales of Gene Flow in California White Oaks." PhD thesis, University of California, Berkeley.

Pascual, Gemma, Marisa Molinas, and Dolors Verdaguer. 2002. "Comparative Anatomical Analysis of the Cotyledonary Region in Three Mediterranean Basin *Quercus* (Fagaceae)." *American Journal of Botany* 89 (3): 383–92. https://doi.org/10.3732/ajb.89.3.383.

Paterson, Andrew H., Michael Freeling, Haibao Tang, and Xiyin Wang. 2010. "Insights from the Comparison of Plant Genome Sequences." *Annual Review of Plant Biology* 61 (1): 349–72. https://doi.org/10.1146/annurev-arplant-042809-112235.

Pearse, Ian S., and Andrew L. Hipp. 2009. "Phylogenetic and Trait Similarity to a Native Species Predict Herbivory on Non-Native Oaks." *Proceedings of the National Academy of Sciences* 106 (43): 18097–102. https://doi.org/10.1073/pnas.0904867106.

———. 2012. "Global Patterns of Leaf Defenses in Oak Species." *Evolution* 66 (7): 2272–86. https://doi.org/10.1111/j.1558-5646.2012.01591.x.

———. 2014. "Native Plant Diversity Increases Herbivory to Non-Natives." *Proceed-*

ings of the Royal Society B: Biological Sciences 281 (1794): 20141841. https://doi.org
/10.1098/rspb.2014.1841.

Pearse, Ian S., Walter D. Koenig, Kyle A. Funk, and Mario B. Pesendorfer. 2015. "Pollen Limitation and Flower Abortion in a Wind-Pollinated, Masting Tree." *Ecology* 96 (2): 587–93. https://doi.org/10.1890/14-0297.1.

Pearse, Ian S., Walter D. Koenig, and Dave Kelly. 2016. "Mechanisms of Mast Seeding: Resources, Weather, Cues, and Selection." *New Phytologist* 212 (3): 546–62. https://doi.org/10.1111/nph.14114.

Pearse, Ian S., Jalene M. LaMontagne, Michael Lordon, Andrew L. Hipp, and Walter D. Koenig. 2020. "Biogeography and Phylogeny of Masting: Do Global Patterns Fit Functional Hypotheses?" *New Phytologist* 227 (5): 1557–67. https://doi
.org/10.1111/nph.16617.

Pearse, Ian S., Andreas P. Wion, Angela D. Gonzalez, and Mario B. Pesendorfer. 2021. "Understanding Mast Seeding for Conservation and Land Management." *Philosophical Transactions of the Royal Society B: Biological Sciences* 376 (1839): 20200383. https://doi.org/10.1098/rstb.2020.0383.

Peattie, Donald Culross. 1991. *A Natural History of Trees of Eastern and Central North America*. Boston: Houghton Mifflin.

Penman, Donald E., and James C. Zachos. 2018. "New Constraints on Massive Carbon Release and Recovery Processes during the Paleocene-Eocene Thermal Maximum." *Environmental Research Letters* 13 (10): 105008. https://doi.org/10.1088
/1748-9326/aae285.

Pérez-Pedraza, Alberto, Hernando Rodríguez-Correa, Susana Valencia-Avalos, César Andrés Torres-Miranda, Maribel Arenas-Navarro, and Ken Oyama. 2021. "Effect of Hybridization on the Morphological Differentiation of the Red Oaks *Quercus acutifolia* and *Quercus grahamii* (Fagaceae)." *Plant Systematics and Evolution* 307 (3): 37. https://doi.org/10.1007/s00606-021-01757-0.

Pesendorfer, Mario B., Walter D. Koenig, Ian S. Pearse, Johannes M. H. Knops, and Kyle A. Funk. 2016. "Individual Resource Limitation Combined with Population-Wide Pollen Availability Drives Masting in the Valley Oak (*Quercus lobata*)." *Journal of Ecology* 104 (3): 637–45. https://doi.org/10.1111/1365-2745.12554.

Pesendorfer, Mario B., T. Scott Sillett, Walter D. Koenig, and Scott A. Morrison. 2016. "Scatter-Hoarding Corvids as Seed Dispersers for Oaks and Pines: A Review of a Widely Distributed Mutualism and Its Utility to Habitat Restoration." *Condor* 118 (2): 215–37. https://doi.org/10.1650/CONDOR-15-125.1.

Petit, Rémy J., C. Bodénès, A. Ducousso, G. Roussel, and A. Kremer. 2003. "Hybridization as a Mechanism of Invasion in Oaks." *New Phytologist* 161:151–64.

Petit, Rémy J., Simon Brewer, Sándor Bordács, Kornel Burg, Rachid Cheddadi, Els Coart, Joan Cottrell, et al. 2002. "Identification of Refugia and Post-Glacial Colonisation Routes of European White Oaks Based on Chloroplast DNA and Fossil Pollen Evidence." *Forest Ecology and Management* 156 (1): 49–74. https://doi.org
/10.1016/S0378-1127(01)00634-X.

Petit, Rémy J., U. Csaikl, S. Bordács, K. Burg, E. Coart, J. Cottrell, B. van Dam, et al. 2002. "Chloroplast DNA Variation in European White Oaks. Phylogeography and Patterns of Diversity Based on Data from over 2600 Populations." *Forest Ecology and Management* 156:5–26.

Petit, Rémy J., and Laurent Excoffier. 2009. "Gene Flow and Species Delimitation." *Trends in Ecology and Evolution* 24 (7): 386–93. https://doi.org/10.1016/j.tree.2009.02.011.

Petit, Rémy J., and Arndt Hampe. 2006. "Some Evolutionary Consequences of Being a Tree." *Annual Review of Ecology, Evolution, and Systematics* 37 (1): 187–214. https://doi.org/10.1146/annurev.ecolsys.37.091305.110215.

Petit, Rémy J., A. Kremer, and D. B. Wagner. 1993. "Geographic Structure of Chloroplast DNA Polymorphisms in European Oaks." *Theoretical and Applied Genetics* 87 (1): 122–28. https://doi.org/10.1007/BF00223755.

Petit, Rémy J., and Clément Larue. 2022. "Confirmation That Chestnuts Are Insect-Pollinated." *Botany Letters* 169 (3): 370–74. https://doi.org/10.1080/23818107.2022.2088612.

Peyrégne, Stéphane, Viviane Slon, and Janet Kelso. 2023. "More Than a Decade of Genetic Research on the Denisovans." *Nature Reviews Genetics* (September): 1–21. https://doi.org/10.1038/s41576-023-00643-4.

Pham, Kasey Khanh, Andrew L. Hipp, Paul S. Manos, and Richard C. Cronn. 2017. "A Time and a Place for Everything: Phylogenetic History and Geography as Joint Predictors of Oak Plastome Phylogeny." *Genome* 60 (April): 720–32. https://doi.org/10.1139/gen-2016-0191.

Phillips, Richard P., Edward Brzostek, and Meghan G. Midgley. 2013. "The Mycorrhizal-Associated Nutrient Economy: A New Framework for Predicting Carbon–Nutrient Couplings in Temperate Forests." *New Phytologist* 199 (1): 41–51. https://doi.org/10.1111/nph.12221.

Pielou, E. C. 2008. *After the Ice Age: The Return of Life to Glaciated North America.* Chicago: University of Chicago Press.

Pimm, Stuart L., and Lucas N. Joppa. 2015. "How Many Plant Species Are There, Where Are They, and at What Rate Are They Going Extinct?" *Annals of the Missouri Botanical Garden* 100 (3): 170–76. https://doi.org/10.3417/2012018.

Pina-Martins, Francisco, João Baptista, Georgios Pappas Jr., and Octávio S. Paulo. 2019. "New Insights into Adaptation and Population Structure of Cork Oak Using Genotyping by Sequencing." *Global Change Biology* 25 (1): 337–50. https://doi.org/10.1111/gcb.14497.

Piovesan, Gianluca, Michele Baliva, Lucio Calcagnile, Marisa D'Elia, Isabel Dorado-Liñán, Jordan Palli, Antonino Siclari, and Gianluca Quarta. 2020. "Radiocarbon Dating of Aspromonte Sessile Oaks Reveals the Oldest Dated Temperate Flowering Tree in the World." *Ecology* 101 (12): e03179. https://doi.org/10.1002/ecy.3179.

Piredda, Roberta, Guido W. Grimm, Ernst-Detlef Schulze, Thomas Denk, and Marco

Cosimo Simeone. 2021. "High-Throughput Sequencing of 5S-IGS in Oaks: Exploring Intragenomic Variation and Algorithms to Recognize Target Species in Pure and Mixed Samples." *Molecular Ecology Resources* 21 (2): 495–510. https://doi.org/10.1111/1755-0998.13264.

Plomion, Christophe, Jean-Marc Aury, Joëlle Amselem, Tina Alaeitabar, Valérie Barbe, Caroline Belser, Hélène Bergès, et al. 2016. "Decoding the Oak Genome: Public Release of Sequence Data, Assembly, Annotation and Publication Strategies." *Molecular Ecology Resources* 16 (1): 254–65. https://doi.org/10.1111/1755-0998.12425.

Plomion, Christophe, Jean-Marc Aury, Joëlle Amselem, Thibault Leroy, Florent Murat, Sébastien Duplessis, Sébastien Faye, et al. 2018. "Oak Genome Reveals Facets of Long Lifespan." *Nature Plants* 4 (7): 440–52. https://doi.org/10.1038/s41477-018-0172-3.

Polette, France, and David J. Batten. 2017. "Fundamental Reassessment of the Taxonomy of Five Normapolles Pollen Genera." *Review of Palaeobotany and Palynology* 243 (August): 47–91. https://doi.org/10.1016/j.revpalbo.2017.04.001.

Pons, Josep, and Juli G. Pausas. 2007. "Acorn Dispersal Estimated by Radio-Tracking." *Oecologia* 153 (4): 903–11. https://doi.org/10.1007/s00442-007-0788-x.

Pont, Caroline, Stefanie Wagner, Antoine Kremer, Ludovic Orlando, Christophe Plomion, and Jerome Salse. 2019. "Paleogenomics: Reconstruction of Plant Evolutionary Trajectories from Modern and Ancient DNA." *Genome Biology* 20 (1): 29. https://doi.org/10.1186/s13059-019-1627-1.

Poulton, Edward Bagnall. 1904. *What Is a Species?* London: Richard Clay & Sons.

Pound, Matthew J., and Ulrich Salzmann. 2017. "Heterogeneity in Global Vegetation and Terrestrial Climate Change during the Late Eocene to Early Oligocene Transition." *Scientific Reports* 7 (1): 43386. https://doi.org/10.1038/srep43386.

Preston, Stephanie D., and Lucia F. Jacobs. 2009. "Mechanisms of Cache Decision Making in Fox Squirrels (*Sciurus niger*)." *Journal of Mammalogy* 90 (4): 787–95. https://doi.org/10.1644/08-MAMM-A-254.1.

Prothero, Donald R. 1994. "The Late Eocene–Oligocene Extinctions." *Annual Review of Earth and Planetary Sciences* 22:145–65.

Prum, Richard O., Jacob S. Berv, Alex Dornburg, Daniel J. Field, Jeffrey P. Townsend, Emily Moriarty Lemmon, and Alan R. Lemmon. 2015. "A Comprehensive Phylogeny of Birds (Aves) Using Targeted Next-Generation DNA Sequencing." *Nature* 526 (7574): 569–73. https://doi.org/10.1038/nature15697.

"*Quercus alba*—White Oak." n.d. Eastern Oldlist (website). Accessed July 6, 2023. http://dendro.cnre.vt.edu/olds/detail.cfm?genus=Quercus&species=alba.

Rabarijaona, Arivoara, Stéphane Ponton, Didier Bert, Alexis Ducousso, Béatrice Richard, Joseph Levillain, and Oliver Brendel. 2022. "Provenance Differences in Water-Use Efficiency Among Sessile Oak Populations Grown in a Mesic Common Garden." *Frontiers in Forests and Global Change* 5:914199.

Racimo, Fernando, Sriram Sankararaman, Rasmus Nielsen, and Emilia Huerta-Sánchez. 2015. "Evidence for Archaic Adaptive Introgression in Humans." *Nature Reviews Genetics* 16 (6): 359–71. https://doi.org/10.1038/nrg3936.

Raff, Jennifer. 2021. "Genomes Reveal Humanity's Journey into the Americas." *Scientific American*, May 1. https://www.scientificamerican.com/article/genomes-reveal-humanitys-journey-into-the-americas/.

Ramírez-Valiente, José A., and Jeannine Cavender-Bares. 2017. "Evolutionary Trade-Offs between Drought Resistance Mechanisms across a Precipitation Gradient in a Seasonally Dry Tropical Oak (*Quercus oleoides*)." *Tree Physiology* (April): 1–13. https://doi.org/10.1093/treephys/tpx040.

Ramírez-Valiente, José A., Alyson Center, Jed Sparks, Kimberlee Sparks, Julie Etterson, Timothy Longwell, George Pilz, et al. 2017. "Population-Level Differentiation in Growth Rates and Leaf Traits in Seedlings of the Neotropical Live Oak *Quercus oleoides* Grown under Natural and Manipulated Precipitation Regimes." *Frontiers in Plant Science* 8. https://doi.org/10.3389/fpls.2017.00585.

Ramírez-Valiente, José A., Rosana López, Andrew L. Hipp, and Ismael Aranda. 2020. "Correlated Evolution of Morphology, Gas Exchange, Growth Rates and Hydraulics as a Response to Precipitation and Temperature Regimes in Oaks (*Quercus*)." *New Phytologist* 227 (3): 794–809. https://doi.org/10.1111/nph.16320.

Ramírez-Valiente, José A., David Sánchez-Gómez, Ismael Aranda, and Fernando Valladares. 2010. "Phenotypic Plasticity and Local Adaptation in Leaf Ecophysiological Traits of 13 Contrasting Cork Oak Populations under Different Water Availabilities." *Tree Physiology* 30 (5): 618–27. https://doi.org/10.1093/treephys/tpq013.

Ramos-Ortiz, S., K. Oyama, H. Rodríguez-Correa, and A. González-Rodríguez. 2016. "Geographic Structure of Genetic and Phenotypic Variation in the Hybrid Zone between *Quercus affinis* and *Q. laurina* in Mexico." *Plant Species Biology* 31 (3): 219–32. https://doi.org/10.1111/1442-1984.12109.

Rauschendorfer, James, Rebecca Rooney, and Carsten Külheim. 2022. "Strategies to Mitigate Shifts in Red Oak (*Quercus* Sect. *Lobatae*) Distribution under a Changing Climate." *Tree Physiology* 42 (12): 2383–2400. https://doi.org/10.1093/treephys/tpac090.

Reich, Peter B., R. O. Teskey, P. S. Johnson, and T. M. Hinckley. 1980. "Periodic Root and Shoot Growth in Oak." *Forest Science* 26 (4): 590–98.

Reid, Clement. 1895. "The Dispersal of Acorns by Rooks." *Nature* 53 (6): 6. https://doi.org/10.1038/053006a0.

———. 1899. *The Origin of the British Flora*. London: Dulau & Co.

Rellstab, Christian, Stefan Zoller, Lorenz Walthert, Isabelle Lesur, Andrea R. Pluess, René Graf, Catherine Bodénès, et al. 2016. "Signatures of Local Adaptation in Candidate Genes of Oaks (*Quercus* Spp.) with Respect to Present and Future Climatic Conditions." *Molecular Ecology* 25 (23): 5907–24. https://doi.org/10.1111/mec.13889.

Renault, Marion. 2021. "The Strange Quest to Save North America's Most Elusive Oak Tree." *New Republic*, December 20.

———. 2022. "Found: One Oak Tree, Famously Missing." *New Republic*, July 7.

Renner, S. S., Guido W. Grimm, Paschalia Kapli, and Thomas Denk. 2016. "Species Relationships and Divergence Times in Beeches: New Insights from the Inclusion of 53 Young and Old Fossils in a Birth–Death Clock Model." *Philosophical Transactions of the Royal Society B: Biological Sciences* 371 (1699): 20150135. https://doi.org/10.1098/rstb.2015.0135.

Retallack, Gregory J. 1997. "Neogene Expansion of the North American Prairie." *PALAIOS* 12 (4): 380. https://doi.org/10.2307/3515337.

———. 2001. "Cenozoic Expansion of Grasslands and Climatic Cooling." *Journal of Geology* 109 (4): 407–26. https://doi.org/10.1086/320791.

Reutimann, Oliver, Benjamin Dauphin, Andri Baltensweiler, Felix Gugerli, Antoine Kremer, and Christian Rellstab. 2023. "Abiotic Factors Predict Taxonomic Composition and Genetic Admixture in Populations of Hybridizing White Oak Species (*Quercus* Sect. *Quercus*) on Regional Scale." *Tree Genetics and Genomes* 19 (3): 22. https://doi.org/10.1007/s11295-023-01598-7.

Rey, María-Dolores, Mónica Labella-Ortega, Víctor M. Guerrero-Sánchez, Rômulo Carleial, María Ángeles Castillejo, Valentino Ruggieri, and Jesús V. Jorrín-Novo. 2023. "A First Draft Genome of Holm Oak (*Quercus ilex* subsp. *ballota*), the Most Representative Species of the Mediterranean Forest and the Spanish Agrosylvopastoral Ecosystem 'Dehesa.'" *Frontiers in Molecular Biosciences* 10 (October): 1242943. https://doi.org/10.3389/fmolb.2023.1242943.

Riahi, Keywan, Shilpa Rao, Volker Krey, Cheolhung Cho, Vadim Chirkov, Guenther Fischer, Georg Kindermann, et al. 2011. "RCP 8.5—A Scenario of Comparatively High Greenhouse Gas Emissions." *Climatic Change* 109 (1): 33. https://doi.org/10.1007/s10584-011-0149-y.

Rieseberg, Loren H. 2001. "Chromosomal Rearrangements and Speciation." *Trends in Ecology and Evolution* 16 (7): 351–58.

Rieseberg, Loren H., Chrystal Van Fossen, and Andree M. Desrochers. 1995. "Hybrid Speciation Accompanied by Genomic Reorganization in Wild Sunflowers." *Nature (London)* 375 (6529): 313–16.

Rivers, Descanso House, Megan Barstow, and Alejandra C. D. Fuentes. 2019. "*Ticodendron Incognitum*." *IUCN Red List of Threatened Species* e.T37468A128258819. https://doi.org/10.2305/IUCN.UK.2019-3.RLTS.T37468A128258819.en.

Robin, Vincent, Marie-Josée Nadeau, Pieter M. Grootes, Hans-Rudolf Bork, and Oliver Nelle. 2016. "Paleobotanical and Climate Data Support the Plausibility of Temperate Trees Spread into Central Europe during the Late Glacial." *New Phytologist* 212 (1): 19–21. https://doi.org/10.1111/nph.14148.

Ronquist, Fredrik. 1999. "Phylogeny, Classification and Evolution of the Cynipoidea." *Zoologica Scripta* 28 (1–2): 139–64. https://doi.org/10.1046/j.1463-6409.1999.00022.x.

Rosas, Antonio, Markus Bastir, and Antonio García-Tabernero. 2022. "Chapter 5—Neanderthals: Anatomy, Genes, and Evolution." In *Updating Neanderthals*, edited by Francesca Romagnoli, Florent Rivals, and Stefano Benazzi, 71–87. Cambridge, MA: Academic Press. https://doi.org/10.1016/B978-0-12-821428-2 .00007-X.

Rubio De Casas, Rafael, Pablo Vargas, Esther Pérez-Corona, Esteban Manrique, José Ramón Quintana, Carlos García-Verdugo, and Luis Balaguer. 2007. "Field Patterns of Leaf Plasticity in Adults of the Long-Lived Evergreen *Quercus coccifera*." *Annals of Botany* 100 (2): 325–34. https://doi.org/10.1093/aob/mcm112.

Rushton, B. S. 1993. "Natural Hybridization within the Genus *Quercus* L." Supplement. *Annales des Sciences Forestières* 50:73s–90s. https://doi.org/10.1051/forest: 19930707.

Sadowski, Eva-Maria, Jörg U. Hammel, and Thomas Denk. 2018. "Synchrotron X-Ray Imaging of a Dichasium Cupule of *Castanopsis* from Eocene Baltic Amber." *American Journal of Botany* 105 (12): 2025–36. https://doi.org/10.1002/ajb2.1202.

Sadowski, Eva-Maria, Alexander R. Schmidt, and Thomas Denk. 2020. "Staminate Inflorescences with in Situ Pollen from Eocene Baltic Amber Reveal High Diversity in Fagaceae (Oak Family)." *Willdenowia* 50 (3). https://doi.org/10.3372/wi.50 .50303.

Sáenz-Romero, Cuauhtémoc, Jean-Baptiste Lamy, Alexis Ducousso, Brigitte Musch, François Ehrenmann, Sylvain Delzon, Stephen Cavers, et al. 2017. "Adaptive and Plastic Responses of *Quercus petraea* Populations to Climate across Europe." *Global Change Biology* 23 (7): 2831–47. https://doi.org/10.1111/gcb.13576.

Saleh, Dounia, Jun Chen, Jean-Charles Leplé, Thibault Leroy, Laura Truffaut, Benjamin Dencausse, Céline Lalanne, et al. 2022. "Genome-Wide Evolutionary Response of European Oaks during the Anthropocene." *Evolution Letters* 6 (1): 4–20. https://doi.org/10.1002/evl3.269.

Sancho-Knapik, Domingo, Alfonso Escudero, Sonia Mediavilla, Christine Scoffoni, Joseph Zailaa, Jeannine Cavender-Bares, Tomás Gómez Álvarez-Arenas, et al. 2021. "Deciduous and Evergreen Oaks Show Contrasting Adaptive Responses in Leaf Mass per Area across Environments." *New Phytologist* 230 (2): 521–34. https://doi.org/10.1111/nph.17151.

Sanmartín, Isabel, and Fredrik Ronquist. 2004. "Southern Hemisphere Biogeography Inferred by Event-Based Models: Plant versus Animal Patterns." *Systematic Biology* 53 (2): 278–98. https://doi.org/10.1080/10635150490423430.

Satake, Akiko, and Dave Kelly. 2021. "Delayed Fertilization Facilitates Flowering Time Diversity in Fagaceae." *Philosophical Transactions of the Royal Society B: Biological Sciences* 376 (1839): 20210115. https://doi.org/10.1098/rstb.2021.0115.

Sato, Akie, Colm O'hUigin, Felipe Figueroa, Peter R. Grant, B. Rosemary Grant, Herbert Tichy, and Jan Klein. 1999. "Phylogeny of Darwin's Finches as Revealed by mtDNA Sequences." *Proceedings of the National Academy of Sciences* 96 (9): 5101–6. https://doi.org/10.1073/pnas.96.9.5101.

Sauquet, Hervé, Simon Y. W. Ho, Maria A. Gandolfo, Gregory J. Jordan, Peter Wilf, David J. Cantrill, Michael J. Bayly, et al. 2012. "Testing the Impact of Calibration on Molecular Divergence Times Using a Fossil-Rich Group: The Case of *Nothofagus* (Fagales)." *Systematic Biology* 61 (2): 289–313. https://doi.org/10.1093/sysbio /syr116.

Schatz, Michael C., Jan Witkowski, and W. Richard McCombie. 2012. "Current Challenges in de Novo Plant Genome Sequencing and Assembly." *Genome Biology* 13 (4): 243. https://doi.org/10.1186/gb-2012-13-4-243.

Scheffer, Theodore C., George H. Englerth, and Catherine G. Duncan. 1949. "Decay Resistance of Seven Native Oaks." *Journal of Agricultural Research* 78 (5–6): 129–52.

Schmid-Siegert, Emanuel, Namrata Sarkar, Christian Iseli, Sandra Calderon, Caroline Gouhier-Darimont, Jacqueline Chrast, Pietro Cattaneo, et al. 2017. "Low Number of Fixed Somatic Mutations in a Long-Lived Oak Tree." *Nature Plants* 3 (12): 926–29. https://doi.org/10.1038/s41477-017-0066-9.

Schoene, Blair, Kyle M. Samperton, Michael P. Eddy, Gerta Keller, Thierry Adatte, Samuel A. Bowring, Syed F. R. Khadri, et al. 2015. "U-Pb Geochronology of the Deccan Traps and Relation to the End-Cretaceous Mass Extinction." *Science* 347 (6218): 182–84. https://doi.org/10.1126/science.aaa0118.

Schulze, Ernst-Detlef, and Guido W. Grimm. 2022. "Alles Bastarde: Die Buche, ein eurasisches Art-Mosaik." *Biologie in unserer Zeit* 52 (4): 340–51. https://doi.org /10.11576/biuz-5864.

Schwilk, Dylan W., Maria S. Gaetani, and Helen M. Poulos. 2013. "Oak Bark Allometry and Fire Survival Strategies in the Chihuahuan Desert Sky Islands, Texas, USA." *PLOS ONE* 8 (11): e79285. https://doi.org/10.1371/journal.pone.0079285.

Scotese, Christopher R. 2021. "An Atlas of Phanerozoic Paleogeographic Maps: The Seas Come In and the Seas Go Out." *Annual Review of Earth and Planetary Sciences* 49 (1): 679–728. https://doi.org/10.1146/annurev-earth-081320-064052.

Scotese, Christopher R., Arthur J. Boucot, and Chen Xu. 2014. "Atlas of Phanerozoic Climatic Zones (Mollweide Projection)," Volumes 1–6, PALEOMAP Project PaleoAtlas for ArcGIS, PALEOMAP Project, Evanston, IL. https://doi.org/10 .13140/2.1.2757.8567. Unpublished.

Scotese, Christopher R., Haijun Song, Benjamin J. W. Mills, and Douwe G. Van Der Meer. 2021. "Phanerozoic Paleotemperatures: The Earth's Changing Climate during the Last 540 Million Years." *Earth-Science Reviews* 215 (April): 103503. https://doi.org/10.1016/j.earscirev.2021.103503.

Scotti-Saintagne, Caroline, Catherine Bodénès, Teresa Barreneche, Evangelista Bertocchi, Christophe Plomion, and Antoine Kremer. 2004. "Detection of Quantitative Trait Loci Controlling Bud Burst and Height Growth in *Quercus robur* L." *TAG. Theoretical and Applied Genetics. Theoretische Und Angewandte Genetik* 109 (8): 1648–59. https://doi.org/10.1007/s00122-004-1789-3.

Scotti-Saintagne, Caroline, Stephanie Mariette, Ilga Porth, Pablo G. Goicoechea, Teresa Barreneche, Catherine Bodénès, Kornel Burg, and Antoine Kremer. 2004. "Genome Scanning for Interspecific Differentiation between Two Closely Related Oak Species [*Quercus robur* L. and *Q. petraea* (Matt.) Liebl.]." *Genetics* 168 (3): 1615–26.

Segovia, Ricardo A., R. Toby Pennington, Tim R. Baker, Fernanda Coelho de Souza, Danilo M. Neves, Charles C. Davis, Juan J. Armesto, et al. 2020. "Freezing and Water Availability Structure the Evolutionary Diversity of Trees across the Americas." *Science Advances* 6 (19): eaaz5373. https://doi.org/10.1126/sciadv.aaz5373.

Serpell, Mick. 2013. "Guest Editorial." *British Journal of Pain* 7 (4): 161–161. https://doi.org/10.1177/2049463713507019.

Sharp, Ward M., and Henry H. Chisman. 1961. "Flowering and Fruiting in the White Oaks. I. Staminate Flowering through Pollen Dispersal." *Ecology* 42 (2): 365–72. https://doi.org/10.2307/1932087.

Sharp, Ward M., and Vance G. Sprague. 1967. "Flowering and Fruiting in the White Oaks. Pistillate Flowering, Acorn Development, Weather, and Yields." *Ecology* 48 (2): 243–51. https://doi.org/10.2307/1933106.

Shastry, Barkur S. 2009. "SNPs: Impact on Gene Function and Phenotype." In *Single Nucleotide Polymorphisms: Methods and Protocols*, edited by Anton A. Komar, 3–22. Methods in Molecular Biology™ 578. Totowa, NJ: Humana Press. https://doi.org/10.1007/978-1-60327-411-1_1.

Shepard, David A. 1993. "The Legitimacy of *Quercus ellipsoidalis* Based on a Populational Study of *Quercus coccinea* in Illinois." MS thesis, Western Illinois University.

———. 2009. "A Review of the Taxonomic Status of *Quercus ellipsoidalis* and *Quercus coccinea* in the Eastern United States." *International Oak Journal* 20:65–84.

Shimada, Takuya, Takashi Saitoh, Eiki Sasaki, Yosuke Nishitani, and Ro Osawa. 2006. "Role of Tannin-Binding Salivary Proteins and Tannase-Producing Bacteria in the Acclimation of the Japanese Wood Mouse to Acorn Tannins." *Journal of Chemical Ecology* 32 (6): 1165–80. https://doi.org/10.1007/s10886-006-9078-z.

Shirasawa, Kenta, Sogo Nishio, Shingo Terakami, Roberto Botta, Daniela Torello Marinoni, and Sachiko Isobe. 2021. "Chromosome-Level Genome Assembly of Japanese Chestnut (*Castanea crenata* Sieb. et Zucc.) Reveals Conserved Chromosomal Segments in Woody Rosids." *DNA Research* 28 (5): dsab016. https://doi.org/10.1093/dnares/dsab016.

Shirone, Bartolomeo, Federico Vessella, and Maria Carolina Varela. 2019. "EUFORGEN Technical Guidelines for Genetic Conservation and Use: Holm Oak, *Quercus ilex*." European Forest Genetic Resources Programme (EUFORGEN): European Forest Institute.

Shi Yong, Zhou Biao-Feng, Liang Yi-Ye, and Wang Baosheng. 2023. "Linked Selection and Recombination Rate Generate Both Shared and Lineage-Specific Genomic Islands of Divergence in Two Independent *Quercus* Species Pairs." *Journal of Systematics and Evolution*, 0. https://doi.org/10.1111/jse.13008.

Shubin, Neil, Cliff Tabin, and Sean Carroll. 2009. "Deep Homology and the Origins of Evolutionary Novelty." *Nature* 457 (7231): 818–23. https://doi.org/10.1038/nature07891.

Silvertown, Jonathan. 2005. *Demons in Eden: The Paradox of Plant Diversity*. Chicago: University of Chicago Press.

Simeone, Marco Cosimo, Simone Cardoni, Roberta Piredda, Francesca Imperatori, Michael Avishai, Guido W. Grimm, and Thomas Denk. 2018. "Comparative Systematics and Phylogeography of *Quercus* Section *Cerris* in Western Eurasia: Inferences from Plastid and Nuclear DNA Variation." *PeerJ* 6 (October): e5793. https://doi.org/10.7717/peerj.5793.

Simeone, Marco Cosimo, Guido W. Grimm, Alessio Papini, Federico Vessella, Simone Cardoni, Enrico Tordoni, Roberta Piredda, et al. 2016. "Plastome Data Reveal Multiple Geographic Origins of *Quercus* Group Ilex." *PeerJ* 4 (April): e1897. https://doi.org/10.7717/peerj.1897.

Simonti, Corinne N., Benjamin Vernot, Lisa Bastarache, Erwin Bottinger, David S. Carrell, Rex L. Chisholm, David R. Crosslin, et al. 2016. "The Phenotypic Legacy of Admixture between Modern Humans and Neandertals." *Science* 351 (6274): 737–41. https://doi.org/10.1126/science.aad2149.

Sims, Hallie J., Patrick S. Herendeen, and Peter R. Crane. 1998. "New Genus of Fossil Fagaceae from the Santonian (Late Cretaceous) of Central Georgia, U.S.A." *International Journal of Plant Sciences* 159 (2): 391–404. https://doi.org/10.1086/297559.

Sims, Hallie J., Patrick S. Herendeen, Richard Lupia, Raymond A. Christopher, and Peter R. Crane. 1999. "Fossil Flowers with Normapolles Pollen from the Upper Cretaceous of Southeastern North America." *Review of Palaeobotany and Palynology* 106 (3): 131–51. https://doi.org/10.1016/S0034-6667(99)00008-1.

Siniscalchi, Carolina M., Julian Correa-Narvaez, Heather R. Kates, Douglas E. Soltis, Pamela S. Soltis, Robert P. Guralnick, Steven R. Manchester, et al. 2023. "Fagalean Phylogeny in a Nutshell: Chronicling the Diversification History of Fagales." bioRxiv. https://doi.org/10.1101/2023.03.06.531381.

Sittaro, Fabian, Alain Paquette, Christian Messier, and Charles A. Nock. 2017. "Tree Range Expansion in Eastern North America Fails to Keep Pace with Climate Warming at Northern Range Limits." *Global Change Biology* 23 (8): 3292–3301. https://doi.org/10.1111/gcb.13622.

Slattery, Joshua, William Cobban, Kevin Mckinney, Peter Harries, and Ashley Sandness. 2013. "Early Cretaceous to Paleocene Paleogeography of the Western Interior Seaway: The Interaction of Eustasy and Tectonism." In *Wyoming Geological Association Handbook*, edited by Marron Bingle-Davis, 68:22–60. Casper, WY. https://doi.org/10.13140/RG.2.1.4439.8801.

Smith, Stephen A., and Michael J. Donoghue. 2008. "Rates of Molecular Evolution Are Linked to Life History in Flowering Plants." *Science* 322 (5898): 86–89. https://doi.org/10.1126/science.1163197.

Snell, Rebecca S., and Sharon A. Cowling. 2015. "Consideration of Dispersal Processes and Northern Refugia Can Improve Our Understanding of Past Plant Migration Rates in North America." *Journal of Biogeography* 42 (9): 1677–88. https://doi.org/10.1111/jbi.12544.

Sogo, Akiko, and Hiroshi Tobe. 2006. "Delayed Fertilization and Pollen-Tube Growth in Pistils of *Fagus japonica* (Fagaceae)." *American Journal of Botany* 93 (12): 1748–56. https://doi.org/10.3732/ajb.93.12.1748.

Sork, Victoria L., Judy Bramble, and Owen Sexton. 1993. "Ecology of Mast-Fruiting in Three Species of North American Deciduous Oaks." *Ecology* 74 (2): 528–41.

Sork, Victoria L., Shawn J. Cokus, Sorel T. Fitz-Gibbon, Aleksey V. Zimin, Daniela Puiu, Jesse A. Garcia, Paul F. Gugger, et al. 2022. "High-Quality Genome and Methylomes Illustrate Features Underlying Evolutionary Success of Oaks." *Nature Communications* 13 (1): 2047. https://doi.org/10.1038/s41467-022-29584-y.

Sork, Victoria L, Kirk A. Stowe, and Cris Hochwender. 1993. "Evidence for Local Adaptation in Closely Adjacent Subpopulations of Northern Red Oak (*Quercus rubra* L.) Expressed as Resistance to Leaf Herbivores." *American Naturalist* 142 (6): 928–36.

Speelman, E. N., M. M. L. Van Kempen, J. Barke, H. Brinkhuis, G. J. Reichart, A. J. P. Smolders, J. G. M. Roelofs, et al. 2009. "The Eocene Arctic *Azolla* Bloom: Environmental Conditions, Productivity and Carbon Drawdown." *Geobiology* 7 (2): 155–70. https://doi.org/10.1111/j.1472-4669.2009.00195.x.

Spellenberg, Richard, Jeffrey R. Bacon, and M. Socorro González-Elizondo. 1998. "Los Encinos (*Quercus*, Fagaceae) en un transecto sobre la Sierra Madre Occidental." *Boletín Del Instituto de Botánica de La Universidad de Guadalajara* 5:357–87.

Spinney, Laura. 2008. "Remnants of Evolution." *New Scientist* 198 (2656): 42–45. https://doi.org/10.1016/S0262-4079(08)61231-2.

Sridhar, Hari. 2021. "Revisiting Cavender-Bares et al. 2004." *Reflections on Papers Past* (blog), June 3.

Stairs, G. R. 1964. "Microsporogenesis and Embryogenesis in *Quercus*." *Botanical Gazette* 125 (2): 115–21.

Steane, Dorothy A., Karen L. Wilson, and Robert S. Hill. 2003. "Using matK Sequence Data to Unravel the Phylogeny of Casuarinaceae." *Molecular Phylogenetics and Evolution* 28 (1): 47–59. https://doi.org/10.1016/S1055-7903(03)00028-9.

Stebbins, G. Ledyard. 1950. *Variation and Evolution in Plants*. Columbia Biological Series. New York: Columbia University Press.

Stebbins, G. Ledyard, E. G. Matzke, and C. Epling. 1947. "Hybridization in a Population of *Quercus marilandica* and *Q. ilicifolia*." *Evolution* 1:79–88.

Steele, Michael A. 2021. *Oak Seed Dispersal: A Study in Plant-Animal Interactions*. Baltimore: Johns Hopkins University Press.

Steele, Michael A., Travis Knowles, Kenneth Bridle, and Ellen L. Simms. 1993. "Tannins and Partial Consumption of Acorns: Implications for Dispersal of Oaks by

Seed Predators." *The American Midland Naturalist* 130 (2): 229–38. https://doi.org/10.2307/2426123.

Steele, Michael A., Peter Smallwood, William B. Terzaghi, John E. Carlson, Thomas Contreras, and Amy McEuen. 2004. "Oak Dispersal Syndromes: Do Red and White Oaks Exhibit Different Dispersal Strategies?" In *Upland Oak Ecology Symposium: History, Current Conditions, and Sustainability. Gen. Tech. Rep. SRS–73.*, edited by Martin A. Spetich, 72–77. Asheville, NC: U.S. Department of Agriculture, Forest Service, Southern Research Station.

Stern, William Louis. 1973. "Development of the Amentiferous Concept." *Brittonia* 25 (4): 316–33. https://doi.org/10.2307/2805638.

Stevens, Peter F. 2001. "Angiosperm Phylogeny Website," Version 12, July 2012 [and updated more or less continuously since]. Accessed August 2012. http://www.Mobot.Org/MOBOT/Research/APweb/.

Stone, Graham N., Richard J. Challis, Rachel J. Atkinson, György Csóka, Alex Hayward, George Melika, Serap Mutun, et al. 2007. "The Phylogeographical Clade Trade: Tracing the Impact of Human-Mediated Dispersal on the Colonization of Northern Europe by the Oak Gallwasp *Andricus kollari.*" *Molecular Ecology* 16 (13): 2768–81. https://doi.org/10.1111/j.1365-294X.2007.03348.x.

Stone, Graham N., Antonio Hernandez-Lopez, James A. Nicholls, Erica di Pierro, Juli Pujade-Villar, George Melika, and James M. Cook. 2009. "Extreme Host Plant Conservatism during at Least 20 Million Years of Host Plant Pursuit by Oak Gallwasps." *Evolution* 63 (4): 854–69.

Stone, Graham N., and Karsten Schönrogge. 2003. "The Adaptive Significance of Insect Gall Morphology." *Trends in Ecology and Evolution* 18 (10): 512–22. https://doi.org/10.1016/S0169-5347(03)00247-7.

Stone, Graham N., Karsten Schönrogge, Rachel J. Atkinson, David Bellido, and Juli Pujade-Villar. 2002. "The Population Biology of Oak Gall Wasps (Hymenoptera: Cynipidae)." *Annual Review of Entomology* 47 (1): 633–68. https://doi.org/10.1146/annurev.ento.47.091201.145247.

Storey, Michael, Robert A. Duncan, and Carl C. Swisher. 2007. "Paleocene-Eocene Thermal Maximum and the Opening of the Northeast Atlantic." *Science* 316 (5824): 587–89. https://doi.org/10.1126/science.1135274.

Stringer, Chris, and Lucile Crété. 2022. "Mapping Interactions of *H. neanderthalensis* and *Homo sapiens* from the Fossil and Genetic Records." *PaleoAnthropology* 2022 (2). https://doi.org/10.48738/2022.iss2.130.

Stringham, Sydney A., Elisabeth E. Mulroy, Jinchuan Xing, David Record, Michael W. Guernsey, Jaclyn T. Aldenhoven, Edward J. Osborne, et al. 2012. "Divergence, Convergence, and the Ancestry of Feral Populations in the Domestic Rock Pigeon." *Current Biology* 22 (4): 302–8. https://doi.org/10.1016/j.cub.2011.12.045.

Stull, Gregory W. 2023. "Evolutionary Origins of the Eastern North American–

Mesoamerican Floristic Disjunction: Current Status and Future Prospects." *American Journal of Botany* 110 (3): e16142. https://doi.org/10.1002/ajb2.16142.

Sturtevant, Alfred Henry. 1913. "The Linear Arrangement of Six Sex-Linked Factors in *Drosophila*, as Shown by Their Mode of Association." *Journal of Experimental Zoology* 14 (1): 43–59. https://doi.org/10.1002/jez.1400140104.

Sun, Xi-Qing, Yi-Gang Song, Bin-Jie Ge, Xi-Ling Dai, and Gregor Kozlowski. 2021. "Intermediate Epicotyl Physiological Dormancy in the Recalcitrant Seed of *Quercus chungii* F.P.Metcalf with the Elongated Cotyledonary Petiole." *Forests* 12 (3): 263. https://doi.org/10.3390/f12030263.

Sun, Yanqing, Lianguang Shang, Qian-Hao Zhu, Longjiang Fan, and Longbiao Guo. 2022. "Twenty Years of Plant Genome Sequencing: Achievements and Challenges." *Trends in Plant Science* 27 (4): 391–401. https://doi.org/10.1016/j.tplants.2021.10.006.

Sun, Ye, Jianling Guo, Xiaorong Zeng, Risheng Chen, Yi Feng, Shuang Chen, and Kai Yang. 2021. "Chromosome-Scale Genome Assembly of *Castanopsis tibetana* Provides a Powerful Comparative Framework to Study the Evolution and Adaptation of Fagaceae Trees." *Molecular Ecology Resources* (October). https://doi.org/10.1111/1755-0998.13539.

Sundaram, Mekala, Ashley E. Higdon, Karl V. Wood, Connie C. Bonham, and Robert K. Swihart. 2020. "Mechanisms Underlying Detection of Seed Dormancy by a Scatter-Hoarding Rodent." *Integrative Zoology* 15 (2): 89–102. https://doi.org/10.1111/1749-4877.12417.

Suseela, Vidya, Nishanth Tharayil, Galya Orr, and Dehong Hu. 2020. "Chemical Plasticity in the Fine Root Construct of Quercus Spp. Varies with Root Order and Drought." *New Phytologist* 228 (6): 1835–51. https://doi.org/10.1111/nph.16841.

Sutikna, Thomas, Matthew W. Tocheri, Michael J. Morwood, E. Wahyu Saptomo, Jatmiko, Rokus Due Awe, Sri Wasisto, et al. 2016. "Revised Stratigraphy and Chronology for *Homo Floresiensis* at Liang Bua in Indonesia." *Nature* 532 (7599): 366–69. https://doi.org/10.1038/nature17179.

Svensen, Henrik, Sverre Planke, Anders Malthe-Sørenssen, Bjørn Jamtveit, Reidun Myklebust, Torfinn Rasmussen Eidem, and Sebastian S. Rey. 2004. "Release of Methane from a Volcanic Basin as a Mechanism for Initial Eocene Global Warming." *Nature* 429 (6991): 542–45. https://doi.org/10.1038/nature02566.

Swanston, Chris, Leslie A. Brandt, Maria K. Janowiak, Stephen D. Handler, Patricia Butler-Leopold, Louis Iverson, Frank R. Thompson III, et al. 2018. "Vulnerability of Forests of the Midwest and Northeast United States to Climate Change." *Climatic Change* 146 (1): 103–16. https://doi.org/10.1007/s10584-017-2065-2.

Swenson, Nathan G., Jeanne M. Fair, and Jeff Heikoop. 2008. "Water Stress and Hybridization between *Quercus gambelii* and *Quercus grisea*." *Western North American Naturalist* 68 (4): 498–507. https://doi.org/10.3398/1527-0904-68.4.498.

Szwed, John. 2004. *So What: The Life of Miles Davis*. New York: Simon and Schuster.

Takahashi, Masamichi, Else Marie Friis, Patrick S. Herendeen, and Peter R. Crane.

2008. "Fossil Flowers of Fagales from the Kamikitaba Locality (Early Coniacian; Late Cretaceous) of Northeastern Japan." *International Journal of Plant Sciences* 169 (7): 899–907. https://doi.org/10.1086/589933.

Takamatsu, Susumu, Uwe Braun, Saranya Limkaisang, Sawwanee Kom-un, Yukio Sato, and James H. Cunnington. 2007. "Phylogeny and Taxonomy of the Oak Powdery Mildew *Erysiphe alphitoides* Sensu Lato." *Mycological Research* 111 (7): 809–26. https://doi.org/10.1016/j.mycres.2007.05.013.

Tallamy, Douglas W. 2021. *The Nature of Oaks: The Rich Ecology of Our Most Essential Native Trees*. Portland, OR: Timber Press.

Tang, Ting, Naili Zhang, Franca J. Bongers, Michael Staab, Andreas Schuldt, Felix Fornoff, Hong Lin, et al. 2022. "Tree Species and Genetic Diversity Increase Productivity via Functional Diversity and Trophic Feedbacks." *eLife* 11 (November): e78703. https://doi.org/10.7554/eLife.78703.

Tang, Yang, Enzai Du, Hongbo Guo, Yang Wang, Josep Peñuelas, and Peter B. Reich. 2023. "Rapid Migration of Mongolian Oak into the Southern Asian Boreal Forest." *Global Change Biology*, e17002. https://doi.org/10.1111/gcb.17002.

Taylor, David Winship, Shusheng Hu, and Bruce H. Tiffney. 2012. "Fossil Floral and Fruit Evidence for the Evolution of Unusual Developmental Characters in Fagales." *Botanical Journal of the Linnean Society* 168 (4): 353–76. https://doi.org/10.1111/j.1095-8339.2012.01217.x.

Taylor, T. 2009. "Flowering Plants." In *Biology and Evolution of Fossil Plants*, 873–997. Elsevier. https://doi.org/10.1016/B978-0-12-373972-8.00022-X.

Tedersoo, Leho, Mohammad Bahram, and Martin Zobel. 2020. "How Mycorrhizal Associations Drive Plant Population and Community Biology." *Science* 367 (6480): eaba1223. https://doi.org/10.1126/science.aba1223.

Tekleva, Maria V., Svetlana Polevova, and Natalia N. Naryshkina. 2023. "Pollen Characteristics Used in Determination and Systematics of *Quercus* (Fagaceae): New Data and Verification of Previous Concepts." *Botanical Journal of the Linnean Society* (June): boad001. https://doi.org/10.1093/botlinnean/boad001.

Terborgh, John. 2020. "At 50, Janzen-Connell Has Come of Age." *BioScience* 70 (12): 1082–92. https://doi.org/10.1093/biosci/biaa110.

Thompson, Jamie B., and Santiago Ramírez-Barahona. 2023. "No Phylogenetic Evidence for Angiosperm Mass Extinction at the Cretaceous–Palaeogene (K-Pg) Boundary." *Biology Letters* 19 (9): 20230314. https://doi.org/10.1098/rsbl.2023.0314.

Thorne, Robert F. 1973. "The 'Amentiferae' or Hamamelidae as an Artificial Group: A Summary Statement." *Brittonia* 25 (4): 395–405. https://doi.org/10.2307/2805643.

Tiffney, Bruce H. 1985. "The Eocene North Atlantic Land Bridge: Its Importance in Tertiary and Modern Phytogeography of the Northern Hemisphere." *Journal of the Arnold Arboretum* 66 (2): 243–73.

Tingen, Paul. 2001. *Miles Beyond: The Electric Explorations of Miles Davis 1967–1991*. New York: Billboard Books.

———. 2017. "Miles Davis and the Making of Bitches Brew." *JazzTimes*, July 10.

Tipper, Edward T., Emily I. Stevenson, Victoria Alcock, Alasdair C. G. Knight, J. Jotautas Baronas, Robert G. Hilton, Mike J. Bickle, et al. 2021. "Global Silicate Weathering Flux Overestimated Because of Sediment–Water Cation Exchange." *Proceedings of the National Academy of Sciences* 118 (1): e2016430118. https://doi.org/10.1073/pnas.2016430118.

Toju, Hirokazu, Satoshi Yamamoto, Hirotoshi Sato, and Akifumi S. Tanabe. 2013. "Sharing of Diverse Mycorrhizal and Root-Endophytic Fungi among Plant Species in an Oak-Dominated Cool–Temperate Forest." *PLOS ONE* 8 (10): e78248. https://doi.org/10.1371/journal.pone.0078248.

Trelease, W. 1924. "The American Oaks." *Memoirs of the National Academy of Sciences* 20:1–255.

Trieff, Danny. 2002. "Composition of the Coleoptera and Associated Insects Collected by Canopy Fogging of Northern Red Oak (*Quercus rubra* L.) Trees in the Great Smoky Mountains National Park and The University of Tennessee Arboretum." MS thesis, University of Tennessee.

Truffaut, Laura, Emilie Chancerel, Alexis Ducousso, Jean Luc Dupouey, Vincent Badeau, François Ehrenmann, and Antoine Kremer. 2017. "Fine-Scale Species Distribution Changes in a Mixed Oak Stand over Two Successive Generations." *New Phytologist* 215 (1): 126–39. https://doi.org/10.1111/nph.14561.

Tucker, John M. 1961. "Studies in the *Quercus undulata* Complex. I. A Preliminary Statement." *American Journal of Botany* 48 (3): 202–8. https://doi.org/10.1002/j.1537-2197.1961.tb11626.x.

———. 1963. "Studies in the *Quercus undulata* Complex. III. the Contribution of *Q. arizonica*." *American Journal of Botany* 50 (7): 699–708. https://doi.org/10.1002/j.1537-2197.1963.tb12245.x.

———. 1970. "Studies in the *Quercus undulata* Complex. IV. The Contribution of *Quercus havardii*." *American Journal of Botany* 57 (1): 71–84. https://doi.org/10.1002/j.1537-2197.1970.tb09792.x.

———. 1971. "Studies in the *Quercus undulata* Complex. V. The Type of *Quercus undulata*." *American Journal of Botany* 58 (4): 329–41. https://doi.org/10.1002/j.1537-2197.1971.tb09981.x.

———. 1974. "Patterns of Parallel Evolution of Leaf Form in New World Oaks." *TAXON* 23 (1): 129–54. https://doi.org/10.2307/1218095.

Tucker, John M., Walter P. Cottam, and Rudy Drobnick. 1961. "Studies in the *Quercus undulata* Complex. II. The Contribution of *Quercus turbinella*." *American Journal of Botany* 48 (4): 329–39. https://doi.org/10.1002/j.1537-2197.1961.tb11647.x.

Tucker, John M., and J. R. Maze. 1966. "Bur Oak (*Quercus macrocarpa*) in New Mexico?" *Southwestern Naturalist* 11 (3): 402–5. https://doi.org/10.2307/3669480.

Ueno, S., G. Le Provost, V. Leger, C. Klopp, C. Noirot, J. M. Frigério, F. Salin, et al. 2010. "Bioinformatic Analysis of ESTs Collected by Sanger and Pyrosequencing Methods for a Keystone Forest Tree Species: Oak." *BMC Genomics* 11 (650): 650.

Ülker, Elif, Çağatay Tavşanoğlu, and Utku Perktaş. 2018. "Ecological Niche Modelling of Pedunculate Oak (*Quercus robur*) Supports the 'expansion-Contraction' Model of Pleistocene Biogeography." *Biological Journal of the Linnean Society* 123 (February): 338–247. https://doi.org/10.1093/biolinnean/blx154.

Upham, Nathan S., Jacob A. Esselstyn, and Walter Jetz. 2019. "Inferring the Mammal Tree: Species-Level Sets of Phylogenies for Questions in Ecology, Evolution, and Conservation." *PLOS Biology* 17 (12): e3000494. https://doi.org/10.1371/journal .pbio.3000494.

U'Ren, Jana M., and Naupaka B. Zimmerman. 2021. "Oaks Provide New Perspective on Seed Microbiome Assembly." *New Phytologist* 230 (4): 1293–95. https://doi .org/10.1111/nph.17305.

Valbuena-Carabana, M., S. C. González-Martinez, V. L. Sork, C. Collada, A. Soto, P. G. Goicoechea, and L. Gil. 2005. "Gene Flow and Hybridisation in a Mixed Oak Forest (*Quercus pyrenaica* Willd. and *Quercus petraea* (Matts.) Liebl.) in Central Spain." *Heredity* 95 (6): 457–65.

Valencia-Avalos, Susana. 2004. "Diversidad del género *Quercus* (Fagaceae) en México." *Boletín de la Sociedad Botánica de México* 75: 33–53.

———. 2020. "Species Delimitation in the Genus *Quercus* (Fagaceae)." *Botanical Sciences* 99 (1): 1–12. https://doi.org/10.17129/botsci.2658.

Valk, Tom van der, Patrícia Pečnerová, David Díez-del-Molino, Anders Bergström, Jonas Oppenheimer, Stefanie Hartmann, Georgios Xenikoudakis, et al. 2021. "Million-Year-Old DNA Sheds Light on the Genomic History of Mammoths." *Nature*, February, 1–5. https://doi.org/10.1038/s41586-021-03224-9.

Van Dersal, William R. 1940. "Utilization of Oaks by Birds and Mammals." *The Journal of Wildlife Management* 4 (4): 404–28. https://doi.org/10.2307/3796011.

Van Valen, Leigh. 1976. "Ecological Species, Multispecies, and Oaks." *Taxon* 25: 233–39.

Vargas-Rodriguez, Yalma L., William J. Platt, Lowell E. Urbatsch, and David W. Foltz. 2015. "Large Scale Patterns of Genetic Variation and Differentiation in Sugar Maple from Tropical Central America to Temperate North America." *BMC Evolutionary Biology* 15 (1): 257. https://doi.org/10.1186/s12862-015-0518-7.

Vargas-Rodriguez, Yalma L., Lowell E. Urbatsch, and Vesna Karaman-Castro. 2020. "Taxonomy and Phylogenetic Insights for Mexican and Central American Species of *Acer* (Sapindaceae)." *The Journal of the Torrey Botanical Society* 147 (1): 49. https://doi.org/10.3159/TORREY-D-19-00011.1.

Veblen, Thomas T., Robert S. Hill, and Jennifer Read. 1996. *The Ecology and Biogeography of Nothofagus Forests*. Yale University Press.

Vieira, Manuel, Reinhard Zetter, Friðgeir Grímsson, and Thomas Denk. 2023. "Niche Evolution versus Niche Conservatism and Habitat Loss Determine Persistence and Extirpation in Late Neogene European Fagaceae." *Quaternary Science Reviews* 300 (January): 107896. https://doi.org/10.1016/j.quascirev.2022.107896.

Visscher, Peter M, and Michael E. Goddard. 2019. "From R. A. Fisher's 1918 Paper to GWAS a Century Later." *Genetics* 211 (4): 1125–30. https://doi.org/10.1534/genetics.118.301594.

Vopson, Melvin M. 2021. "Estimation of the Information Contained in the Visible Matter of the Universe." *AIP Advances* 11 (10): 105317. https://doi.org/10.1063/5.0064475.

Wagner, Stefanie, Frédéric Lagane, Andaine Seguin-Orlando, Mikkel Schubert, Thibault Leroy, Erwan Guichoux, Emilie Chancerel, et al. 2018. "High-Throughput DNA Sequencing of Ancient Wood." *Molecular Ecology* 27 (5): 1138–54. https://doi.org/10.1111/mec.14514.

Walker, James W., and James A. Doyle. 1975. "The Bases of Angiosperm Phylogeny: Palynology." *Annals of the Missouri Botanical Garden* 62 (3): 664–723. https://doi.org/10.2307/2395271.

Wang, Chengshan, Xixi Zhao, Zhifei Liu, Peter C. Lippert, Stephan A. Graham, Robert S. Coe, Haisheng Yi, et al. 2008. "Constraints on the Early Uplift History of the Tibetan Plateau." *Proceedings of the National Academy of Sciences* 105 (13): 4987–92. https://doi.org/10.1073/pnas.0703595105.

Wang, H. George, Jennifer Wouk, Randi Anderson, and Robert J. Marquis. 2023. "Strong Influence of Leaf Tie Formation and Corresponding Weak Effect of Leaf Quality on Herbivory in Eight Species of *Quercus*." *Ecological Entomology* 48 (1): 69–80. https://doi.org/10.1111/een.13202.

Wang, Wen J., Frank R. Thompson III, Hong S. He, Jacob S. Fraser, William D. Dijak, and Todd Jones-Farrand. 2019. "Climate Change and Tree Harvest Interact to Affect Future Tree Species Distribution Changes." *Journal of Ecology* 107 (4): 1901–17. https://doi.org/10.1111/1365-2745.13144.

Wang, Yufei, Qingmin Han, Kaoru Kitajima, Hiroko Kurokawa, Takuya Shimada, Tamaho Yamaryo, Daisuke Kabeya, et al. 2023. "Resource Allocation Strategies in the Reproductive Organs of Fagaceae Species." *Ecological Research* 38 (2): 306–16. https://doi.org/10.1111/1440-1703.12350.

Wappler, Torsten, Ellen D. Currano, Peter Wilf, Jes Rust, and Conrad C. Labandeira. 2009. "No Post-Cretaceous Ecosystem Depression in European Forests? Rich Insect-Feeding Damage on Diverse Middle Palaeocene Plants, Menat, France." *Proceedings of the Royal Society B: Biological Sciences* 276 (1677): 4271–77. https://doi.org/10.1098/rspb.2009.1255.

Ward, Anna K. G., Robin K. Bagley, Scott P. Egan, Glen Ray Hood, James R. Ott, Kirsten M. Prior, Sofia I. Sheikh, et al. 2022. "Speciation in Nearctic Oak Gall Wasps Is Frequently Correlated with Changes in Host Plant, Host Organ, or Both." *Evolution* 76 (8): 1849-1867. https://doi.org/10.1111/evo.14562.

Ward, J., D. J. McCafferty, D. C. Houston, and G. D. Ruxton. 2008. "Why Do Vultures Have Bald Heads? The Role of Postural Adjustment and Bare Skin Areas in Thermoregulation." *Journal of Thermal Biology* 33 (3): 168–73. https://doi.org/10.1016/j.jtherbio.2008.01.002.

Warren, Robert J., Antoine Guiguet, Chloe Mokadam, John Tooker, and Andrew

Deans. 2022. "Oak Galls Exhibit Ant-Dispersal Convergent with Myrmecochorous Seeds." *American Naturalist* 200 (2): 292–301. https://doi.org/10.1086/720283.

Watson, G. W., A. M. Hewitt, M. Custic, and M. Lo. 2014. "The Management of Tree Root Systems in Urban and Suburban Settings II: A Review of Strategies to Mitigate Human Impacts." *Arboriculture and Urban Forestry* 40 (5): 249–71.

Weatherhead, Emily, Emily Lorine Davis, and Roger T. Koide. 2022. "Many Foliar Endophytic Fungi of *Quercus Gambelii* Are Capable of Psychrotolerant Saprotrophic Growth." *PLOS ONE* 17 (10): e0275845. https://doi.org/10.1371/journal.pone.0275845.

Webb, Sara L. 1986. "Potential Role of Passenger Pigeons and Other Vertebrates in the Rapid Holocene Migrations of Nut Trees." *Quaternary Research* 26 (3): 367–75. https://doi.org/10.1016/0033-5894(86)90096-7.

Weber, Marjorie G., and Kathleen H. Keeler. 2013. "The Phylogenetic Distribution of Extrafloral Nectaries in Plants." *Annals of Botany* 111 (6): 1251–61. https://doi.org/10.1093/aob/mcs225.

Weiss, Madeline C., Martina Preiner, Joana C. Xavier, Verena Zimorski, and William F. Martin. 2018. "The Last Universal Common Ancestor between Ancient Earth Chemistry and the Onset of Genetics." *PLOS Genetics* 14 (8): e1007518. https://doi.org/10.1371/journal.pgen.1007518.

Weissert, Helmut, and Elisabetta Erba. 2004. "Volcanism, CO_2 and Palaeoclimate: A Late Jurassic-Early Cretaceous Carbon and Oxygen Isotope Record." *Journal of the Geological Society* 161 (July): 695–702. https://doi.org/10.1144/0016-764903-087.

Wen, Jun, Ze-Long Nie, and Stefanie M. Ickert-Bond. 2016. "Intercontinental Disjunctions between Eastern Asia and Western North America in Vascular Plants Highlight the Biogeographic Importance of the Bering Land Bridge from Late Cretaceous to Neogene." *Journal of Systematics and Evolution* 54 (5): 469–90. https://doi.org/10.1111/jse.12222.

West, Ashley Marie. 2015. "Ecological Specialization of *Tubakia iowensis*, and Searching for Variation in Resistance to Bur Oak Blight." MS thesis, Iowa State University.

Westerhold, Thomas, Norbert Marwan, Anna Joy Drury, Diederik Liebrand, Claudia Agnini, Eleni Anagnostou, James S. K. Barnet, et al. 2020. "An Astronomically Dated Record of Earth's Climate and Its Predictability over the Last 66 Million Years." *Science* 369 (6509): 1383–87. https://doi.org/10.1126/science.aba6853.

White, JoAnn, and Charles E. Strehl. 1978. "Xylem Feeding by Periodical Cicada Nymphs on Tree Roots." *Ecological Entomology* 3 (4): 323–27. https://doi.org/10.1111/j.1365-2311.1978.tb00933.x.

Whittaker, Robert H. 1960. "Vegetation of the Siskiyou Mountains, Oregon and California." *Ecological Monographs* 30 (3): 279–338. https://doi.org/10.2307/1943563.

Whittaker, Robert H. 1969. "Evolution of Diversity in Plant Communities." In "Diversity and Stability in Ecological Systems." *Brookhaven Symposium in Biology* 22: 178–95.

Whittemore, Alan T., Ryan S. Fuller, Bethany H. Brown, Marlene Hahn, Linus Gog, Jaime A. Weber, and Andrew L. Hipp. 2021. "Phylogeny, Biogeography, and Classification of the Elms (*Ulmus*)." *Systematic Botany* 46 (3): 711–27. https://doi.org /10.1600/036364421X16312068417039.

Whittemore, Alan T., and Barbara A. Schaal. 1991. "Interspecific Gene Flow in Sympatric Oaks." *Proceedings of the National Academy of Sciences USA* 88: 2540–44.

Wiegand, Karl M. 1935. "A Taxonomist's Experience with Hybrids in the Wild." *Science* 81:161–66.

Wilf, Peter, Kevin C. Nixon, María A. Gandolfo, and N. Rubén Cúneo. 2019a. "Eocene Fagaceae from Patagonia and Gondwanan Legacy in Asian Rainforests." *Science* 364 (6444): eaaw5139. https://doi.org/10.1126/science.aaw5139.

———. 2019b. "Response to Comment on 'Eocene Fagaceae from Patagonia and Gondwanan Legacy in Asian Rainforests.'" *Science* 366 (6467): eaaz2297. https:// doi.org/10.1126/science.aaz2297.

Willeit, M., A. Ganopolski, R. Calov, and V. Brovkin. 2019. "Mid-Pleistocene Transition in Glacial Cycles Explained by Declining CO_2 and Regolith Removal." *Science Advances* 5 (4): eaav7337. https://doi.org/10.1126/sciadv.aav7337.

Williams, John W., Bryan N. Shuman, Thompson Webb, Patrick J. Bartlein, and Phillip L. Leduc. 2004. "Late-Quaternary Vegetation Dynamics in North America: Scaling from Taxa to Biomes." *Ecological Monographs* 74 (2): 309–34. https://doi .org/10.1890/02-4045.

Williams, Joseph H., William J. Boecklen, and Daniel J. Howard. 2001. "Reproductive Processes in Two Oak (*Quercus*) Contact Zones with Different Levels of Hybridization." *Heredity* 87 (6): 680–90. https://doi.org/10.1046/j.1365-2540.2001 .00968.x.

Wing, Scott L. 1987. "Eocene and Oligocene Floras and Vegetation of the Rocky Mountains." *Annals of the Missouri Botanical Garden* 74 (4): 748–84. https://doi .org/10.2307/2399449.

Wing, Scott L., and Ellen D. Currano. 2013. "Plant Response to a Global Greenhouse Event 56 Million Years Ago." *American Journal of Botany* 100 (7): 1234–54. https:// doi.org/10.3732/ajb.1200554.

Wing, Scott L., Fabiany Herrera, Carlos A. Jaramillo, Carolina Gómez-Navarro, Peter Wilf, and Conrad C. Labandeira. 2009. "Late Paleocene Fossils from the Cerrejón Formation, Colombia, Are the Earliest Record of Neotropical Rainforest." *Proceedings of the National Academy of Sciences* 106 (44): 18627–32. https://doi.org /10.1073/pnas.0905130106.

Wolfe, Jack A. 1975. "Some Aspects of Plant Geography of the Northern Hemisphere During the Late Cretaceous and Tertiary." *Annals of the Missouri Botanical Garden* 62 (2): 264–79. https://doi.org/10.2307/2395198.

———. 1977. "Paleogene Floras from the Gulf of Alaska Region." Geological Survey Professional Paper 997. Washington, DC: U.S. Government Printing Office. https://doi.org/10.3133/pp997.

———. 1987. "Late Cretaceous-Cenozoic History of Deciduousness and the Terminal Cretaceous Event." *Paleobiology* 13 (2): 215–26. https://doi.org/10.1017/S0094837300008769.

Wolff, Jerry O. 1996. "Population Fluctuations of Mast-Eating Rodents Are Correlated with Production of Acorns." *Journal of Mammalogy* 77 (3): 850–56. https://doi.org/10.2307/1382690.

Wright, Sewall. 1940. "Breeding Structure of Populations in Relation to Speciation." *The American Naturalist* 74 (752): 232–48. https://doi.org/10.1086/280891.

Wright, Jessica W., Christopher T. Ivey, Courtney Canning, and Victoria L. Sork. 2021. "Timing of Bud Burst Is Associated with Climate of Maternal Origin in *Quercus lobata* Progeny in a Common Garden." *Madroño* 68 (4): 443–49. https://doi.org/10.3120/0024-9637-68.4.443.

Wu, Yingtong, Alicia Brown, and Robert E. Ricklefs. 2022. "Host-specific Soil Microbes Contribute to Habitat Restriction of Closely Related Oaks (*Quercus* Spp.)." *Ecology and Evolution* 12 (12): e9614. https://doi.org/10.1002/ece3.9614.

Xia, Ke, William L. Harrower, Roy Turkington, Hong-Yu Tan, and Zhe-Kun Zhou. 2016. "Pre-Dispersal Strategies by *Quercus schottkyana* to Mitigate the Effects of Weevil Infestation of Acorns." *Scientific Reports* 6 (1): 37520. https://doi.org/10.1038/srep37520.

Xia, Ke, Hong-Yu Tan, Roy Turkington, Jin-Jin Hu, and Zhe-Kun Zhou. 2016. "Desiccation and Post-Dispersal Infestation of Acorns of *Quercus schottkyana*, a Dominant Evergreen Oak in SW China." *Plant Ecology* 217 (11): 1369–78. https://doi.org/10.1007/s11258-016-0654-1.

Xiang, Xiao-Guo, Wei Wang, Rui-Qi Li, Li Lin, Yang Liu, Zhe-Kun Zhou, Zhen-Yu Li, et al. 2014. "Large-Scale Phylogenetic Analyses Reveal Fagalean Diversification Promoted by the Interplay of Diaspores and Environments in the Paleogene." *Perspectives in Plant Ecology, Evolution and Systematics* 16 (3): 101–10. https://doi.org/10.1016/j.ppees.2014.03.001.

Xing, Yaowu, Renske E. Onstein, Richard J. Carter, Tanja Stadler, and H. Peter Linder. 2014. "Fossils and a Large Molecular Phylogeny Show That the Evolution of Species Richness, Generic Diversity, and Turnover Rates Are Disconnected." *Evolution* 68 (10): 2821–32. https://doi.org/10.1111/evo.12489.

Xu, He, Tao Su, Shi-Tao Zhang, Min Deng, and Zhe-Kun Zhou. 2016. "The First Fossil Record of Ring-Cupped Oak (*Quercus* L. Subgenus *Cyclobalanopsis* (Oersted) Schneider) in Tibet and Its Paleoenvironmental Implications." *Palaeogeography, Palaeoclimatology, Palaeoecology* 442 (January): 61–71. https://doi.org/10.1016/j.palaeo.2015.11.014.

Yacine, A., and F. Bouras. 1997. "Self- and Cross-Pollination Effects on Pollen Tube Growth and Seed Set in Holm Oak *Quercus ilex* L (Fagaceae)." *Annales Des Sciences Forestières* 54 (5): 447–62. https://doi.org/10.1051/forest:19970503.

Yang, Ying-Ying, Xiao-Jian Qu, Rong Zhang, Gregory W. Stull, and Ting-Shuang Yi. 2021. "Plastid Phylogenomic Analyses of Fagales Reveal Signatures of Conflict and Ancient Chloroplast Capture." *Molecular Phylogenetics and Evolution* 163 (October): 107232. https://doi.org/10.1016/j.ympev.2021.107232.

Yang, Ying-Ying, Gregory W. Stull, Xiao-Jian Qu, Lei Zhao, Yi Hu, Zhi-Heng Wang, Hong Ma, et al. 2023. "Genome Duplications, Genomic Conflict, and Rapid Phenotypic Evolution Characterize the Cretaceous Radiation of Fagales." bioRxiv. https://doi.org/10.1101/2023.06.11.544004.

Yi, Xianfeng, Andrew W. Bartlow, Rachel Curtis, Salvatore J. Agosta, and Michael A. Steele. 2019. "Responses of Seedling Growth and Survival to Post-Germination Cotyledon Removal: An Investigation among Seven Oak Species." *Journal of Ecology* 107 (4): 1817–27. https://doi.org/10.1111/1365-2745.13153.

Yi, Xianfeng, and Yueqin Yang. 2010. "Apical Thickening of Epicarp Is Responsible for Embryo Protection in Acorns of *Quercus variabilis*." *Israel Journal of Ecology and Evolution* 56 (2): 153–64. https://doi.org/10.1560/IJEE.56.2.153.

Yuan, Shuai, Yong Shi, Biao-Feng Zhou, Yi-Ye Liang, Xue-Yan Chen, Qing-Qing An, Yan-Ru Fan, et al. 2023. "Genomic Vulnerability to Climate Change in *Quercus acutissima*, a Dominant Tree Species in East Asian Deciduous Forests." *Molecular Ecology* 32 (7): 1639–1655. https://doi.org/10.1111/mec.16843.

Yücedağ, Cengiz, Markus Müller, and Oliver Gailing. 2021. "Morphological and Genetic Variation in Natural Populations of *Quercus vulcanica* and *Q. frainetto*." *Plant Systematics and Evolution* 307 (1): 8. https://doi.org/10.1007/s00606-020 -01737-w.

Zanazzi, Alessandro, Matthew J. Kohn, Bruce J. MacFadden, and Dennis O. Terry. 2007. "Large Temperature Drop across the Eocene–Oligocene Transition in Central North America." *Nature* 445 (7128): 639–42. https://doi.org/10.1038/nature 05551.

Zanetto, Anne, A. Kremer, G. Müller-Starck, and H. H. Hattemer. 1996. "Inheritance of Isozymes in Pedunculate Oak (*Quercus robur* L.)." *Journal of Heredity* 87 (5): 364–70. https://doi.org/10.1093/oxfordjournals.jhered.a023015.

Zanetto, Anne, G. Roussel, and A. Kremer. 1994. "Geographic Variation of Inter-Specific Differentiation between *Quercus robur* L. and *Quercus petraea* (Matt.) Liebl." *Forest Genetics* 1: 111–23.

Zanne, Amy E., David C. Tank, William K. Cornwell, Jonathan M. Eastman, Stephen A. Smith, Richard G. FitzJohn, Daniel J. McGlinn, et al. 2014. "Three Keys to the Radiation of Angiosperms into Freezing Environments." *Nature* 506 (7486): 89–92. https://doi.org/10.1038/nature12872.

Zeebe, Richard E., Andy Ridgwell, and James C. Zachos. 2016. "Anthropogenic Carbon Release Rate Unprecedented during the Past 66 Million Years." *Nature Geoscience* 9 (4): 325–29. https://doi.org/10.1038/ngeo2681.

Zhang, Qiuyue, Richard H. Ree, Nicolas Salamin, Yaowu Xing, and Daniele Silvestro. 2022. "Fossil-Informed Models Reveal a Boreotropical Origin and Divergent Evolutionary Trajectories in the Walnut Family (Juglandaceae)." *Systematic Biology* 71 (1): 242–58. https://doi.org/10.1093/sysbio/syab030.

Zhou, Biao-Feng, Shuai Yuan, Andrew Crowl, Yi-Ye Liang, Yong Shi, Xue-Yan Chen, Qing-Qing An, et al. 2022. "Phylogenomic Analyses Highlight Innovation and Introgression in the Continental Radiations of Fagaceae across the Northern Hemisphere." *Nature Communications* 13 (1): 1320. https://doi.org/10.1038/s41467-022-28917-1.

Zhou, Wenbin, and Qiu-Yun (Jenny) Xiang. 2022. "Phylogenomics and Biogeography of *Castanea* (Chestnut) and *Hamamelis* (Witch-Hazel)—Choosing between RAD-Seq and Hyb-Seq Approaches." *Molecular Phylogenetics and Evolution* 176 (November): 107592. https://doi.org/10.1016/j.ympev.2022.107592.

Zhou, Xia, Na Liu, Xiaolong Jiang, Zhikuang Qin, Taimoor Hassan Farooq, Fuliang Cao, and He Li. 2022. "A Chromosome-Scale Genome Assembly of *Quercus gilva*: Insights into the Evolution of *Quercus* Section *Cyclobalanopsis* (Fagaceae)." *Frontiers in Plant Science* 13.

Zimmer, Carl. 2019. *She Has Her Mother's Laugh: The Powers, Perversions, and Potential of Heredity.* London: Pan Macmillan.

Zorrilla-Azcué, Sofía, Antonio González-Rodríguez, Ken Oyama, Mailyn A. González, and Hernando Rodríguez-Correa. 2021. "The DNA History of a Lonely Oak: *Quercus humboldtii* Phylogeography in the Colombian Andes." *Ecology and Evolution* 11 (May): 6814–28. https://doi.org/10.1002/ece3.7529.

Index

Page numbers in italics refer to illustrations.

American Oak Group. See *Quercus* subgenus *Quercus*

amplified fragment length polymorphism (AFLP), 167

Anderson, Edgar, 72–74, 162

Angiosperm Phylogeny Group (APG), 102

angiosperms. *See* flowering plants (angiosperms)

annual acorns, 11–13, 188–89

anthers, 8–10, *9*

Anthropocene, 128

Antiquacupula, 106–7

ants, 189–90, 193

arbor vitae (*Thuja occidentalis*), 90–91

Archaefagacea, 106, 108, 228

Aristotle, 69

backcrossing, 73, 80, 82, 84, 86; evolution of Fagales and, 108; to *Homo sapiens*, 162

baobabs (*Adansonia* spp.), 91

Bartholomé, Jérôme, 55

Bartram, John, 72

Bartram oak, 71–72, 73, 75

bear oak (*Q. ilicifolia*), 73–74, 189

beech taxonomy. See Fagaceae (Beech Family); Fagales (Beech Order); *Fagus* (beeches)

Bering Land Bridge, 133, 146

Betulaceae (Birch Family), 98, 99, *100–101*, 102, 105, 112

biennial acorn maturation, 11–12, 188–89

biodiversity, 180, 181. See also diversity in oak communities

biological species concept, 69, 83–84, 223. *See also* reproductive isolation

Birch Family (Betulaceae), 98, 99, *100–101*, 102, 105, 112

birds: acorn woodpeckers, 21; in dinosaur clade, 94; eating cotyledons, 115; eating gall wasps, 193; grackles, 22; jays (*see* jays); passenger pigeons, 31, 115, 197; rooks, 19, 29–30

blackjack oak (*Q. marilandica*), 73–74, 142, 189

black oak (*Q. velutina*), 71, 83, 175, 189

blue jays, 19. *See also* jays

Bodénès, Catherine, 156–57, 159

Bourdeau, Philippe, 181

Brusatte, Steve, 103

buds: female flowers concealed in, 10; opening of, 7, 50

Burger, William, 77, 78

bur oak (*Q. macrocarpa*), 2; acorns' appearance, 65–66; acorns' winter dormancy, 18, 175; bark and branches of, 66; boundaries between other species and, 77–78, 83; chloroplast DNA of, 79–80; climate change now and, 198; gall wasp emerged from, *190*; genome of Gambel's oak and, 169; geographic range of, 41–42, 61, 198; interbreeding with other White Oak species, 66, 67, *76*, 77; introgressed with Gambel's oak, 74–75; leaves and catkins of, 8; long-lived, 40, 87; masting by, 24; nuclear genome and, 81; plastic and genetic variation combined in, 42; pollen shed by, 35; solitary on Great Plains, 184

bur oak blight, 187–88

California black oak (*Q. kelloggii*), 139, 184

California blue oak (*Q. douglasii*), 52

California scrub oak (*Q. berberidifolia*), 84, 165

California valley oak (*Q. lobata*): bud opening in warmer climates and, 50; climate change and, 61; declining population of, 199; evolution of,

communities of oaks, 195; convergently evolved traits and, 183; of different clades growing together, 183–84; evolutionary history of, 175–76; interactions shaping, 189; network of organisms in, 193–94; nutrient cycling in, 186–87, 241; solitary oaks compared to, 184. *See also* diversity in oak communities

communities of trees, latitudinal turnover in, 176–78

competition: Darwin on closely related species and, 181–82, 183; intraspecific, 57–58

Connell, Joseph, 187, 188

conservation of oak genes, 200

convergence, 180–84, 186

Core Fagales, 100–101, 105–6; diversifying, 112, 114. *See also* Normapolles

cork oak (*Q. suber*), 12, 50–51, 184

Cork Oak Group (*Q. sect. Cerris*): chloroplast genomes in, 135, 233; in communities, 184; convergence and, 180–81; Holly Oak Group and, 135; *Q. ithaburensis* acorn, *148*; reuniting with other sections in Eurasia, 146

cotyledonary petioles, *36*

cotyledons, 14; eaten by insects, 189; eaten by rodents and birds, 21–22, 115; in evolution of Fagaceae, 114–15; fungi in, 190; hidden in soil as seedlings grow, 14; nutrient content of, 14, 16, 21; photosynthetic, *15*, 115; pouring nutrients into developing root, 33

Coyne, Jerry, 83

Cretaceous: flowering plants in, 89–90; origins of Fagales in, 104–5; world of, 89

Cronquist, Arthur, 99, 101

crossing over, 153–54, 155

cross-pollination, and delayed fertilization, 11

cupules, 16; evolution of Fagales and, 106–7; in four-genus clade including *Quercus*, 117–19; of Gambel's oak, *13*; of *Lithocarpus*, *92*, *93*, 107, *111*, 119; of Nothofagaceae, *90*, 107, 109; spiny, *90*, 107, *116*, 118, *118*, 119

Cynipids. *See* gall wasps

Darwin, Charles, 42–46, 219; on classifications, 131; on competition between closely related species, 181–82, 183; on hybridization, 69, 72; on oaks, 45, 46–47, 85; on species concept, 69; on sub-species and varieties, 226; Tree of Life and, 43, 91–92, 94, 181. See also *Origin of Species* (Darwin)

Darwin, Leonard, 61

Davis, Miles, 151–52

deciduous species: in communities with evergreen species, 189; forests of, 131–33

Deer Oak Group (*Q. sect. Ponticae*), 129, 140, 146, 167–69, *168*, 184; *Q. pontica*, 147, *168*, 168–69. See also *Ponticae*

delayed fertilization, 11; facilitating hybridization, 170; in Golden-Cup Oak Group, 11, 16, 129; masting and, 24

Denisovans, 67–68, 73, 161, 197, 202, 237

Denk, Thomas, 106

Desprez-Loustau, Marie-Laure, 55

DeWald, Laura, 22

dinosaurs, nonavian: extinction of, 114, 120, 127, 197

diploid organisms, 53

disease resistance: genes helping trees with, 159, 160; Neanderthal alleles and, 161; in White Oaks, 179–80 diseases of trees: spread by humans, 199. *See also* fungi

diversity in oak communities: with both evergreen and deciduous species, 189; convergence and, 183, 184, 186; divergence among close relatives and, 184; ecological trade-offs and, 186, 241; as experimental grounds for alleles, 194–95; facilitation and, 186–87, 241; fungi and, 190, 191–92; hybridization possibilities and, 67; shared insect herbivores and, 188; and susceptibility to diseases and pests, 187–88; timing of germination and, 188–89; unique roles of clades in, 194

diversity of organisms: evolution and, 46; phylogeny and, 94

diversity of tree species, 194; in Fagales, 115

DNA, 53–54; allele frequencies and, 59; of chloroplasts, 79–80, 99, 101; clouds of variation of, 62; of Mexican species, 144; regulatory regions of, 153, 160, 165. *See also* genomes

DNA sequencing, 79–82; evolution and, 103–4; improved technology of, 166–67; of reference genomes, 158–61

Dobzhansky, Theodosius, 42

domesticated animals and plants, 43–44

dormancy: of acorns, 18, 32, 188; of pistillate flowers, 11

drought tolerance: adaptive introgression and, 162; convergent evolution and, 183; leaves and, 50–51, 181; trade-offs involving, 50–51, 186, 241. *See also* water availability

Ducousso, Alexis, 57–58

dwarf chestnut (*Castanea pumila*), 115

dwarf live oak (*Q. minima*), 23, 75

Early Eocene Climatic Optimum, 128, 131

East Asian oaks, 135, 146, 235; communities of, 184. *See also* Ring-Cupped Oak Group (*Q.* sect. *Cyclobalanopsis*)

eastern white oak (*Q. alba*): acorn germination of, 18, 66; acorns half-eaten by squirrels, 22; acorns producing taproots, 175; barely replacing itself, 199; characteristics of, 65–66; chemical protection against fungi, 218; chloroplast DNA of, 79–80; East Asian spread of ancestor of, 145–46; Eurasian White Oaks as sister to, 167, 169; factors maintaining distinct species of, 83; gall wasp emerged from, 192; heritable variation in, 47; hybridization of, 75–76, 76; as king in eastern North American forests, 66, 147; long-lived, 40, 87; nuclear genome and, 81; potentially hybridized with bur oak, 67; reproduction crisis in, 200; roots of, 7; stamens of, 9; symbiotic fungi and, 40

ecological niche: adaptive alleles and, 172; adaptive zone of, 77–78; boundaries between species and, 83, 84; introgressed genes and, 195. *See also* phylogenetic niche conservatism

ecosystem: of each oak tree, 189; role of every clade in, 94

ectomycorrhizal fungi. *See* mycorrhizal fungi

egg cell: genes shuffled in creation of, 54, 153; sperm cell merging with, 1, 13, 53

embryo, 13–14, 16; acorn's defenses of, 21–22; of *Lithocarpus*, 93

endangered and threatened tree species, 199

endosperm, 13, 14

Engelmann, George, 71, 223

Engelmann's oak (*Q. engelmannii*), 84, 145, 164–65, 170

Engler, Adolph, 98, 99, 102

environment: in common gardens, 47; elevation differences in Mexico, 144; genes fitting organisms to, 157; natural selection and, 46, 50–51, 60–61; plasticity and, 40–41, 42; plasticity and local adaptation combined in, 52; protein-coding genes triggered by, 154; reciprocal transplant experiment and, 49; shared regions of genome and, 164. *See also* drought tolerance

Eocene cooling, 131, 232

Erysiphe alphitoides, 55. *See also* powdery mildews

Eurasian chinkapin. *See* chinkapin, Eurasian (*Castanopsis*)

Eurasian Oak Group. See *Quercus* subgenus *Cerris*

evergreen species: benefits and risks of, 131; in communities with deciduous species, 189; early evolution of Fagales and, 109; growth rate vs. drought tolerance in, 241; *Lithocarpus* in forests of, 119; Ring-Cupped Oaks in forests of, 134, 178

evolution: convergent, 180–84, 186; ecological interactions and, 181; evolutionary novelty in balance with phylogenetic niche conservatism, 179; genome duplications and, 159; hybridization and, 72–73, 75; Neanderthals as force in, 162; of new species, 52–53; of plasticity, 51–52; within populations, 52–53; statistical models of, 93–94; variation and, 42, 43. *See also* natural selection

extinctions, 92, 94; background rate of, 198–99; climate at beginning of Oligocene and, 140; end-Cretaceous meteor and, 114; expectations for oaks' future, 197–99; of nonavian dinosaurs, 114, 120, 127, 197; of oaks not always leaving fossils, 148; of seed plant species since 1700s, 200

facilitation, 186–87, 241

Fagaceae (Beech Family), 5, 11, 14, 89, 95, 96; aments and, 98, 99, 102, 107; ancient gene flow in, 148; "big-seeded," 100–101, 115–17; biotic seed dispersal in, 115; closest relatives of, 102; end-Cretaceous explosion of species in, 114; evolution of acorn in, 117–20; evolution of cotyledons and, 114–15; gall wasps on, 192–93; insect-pollinated genera of, 119; phylogenetic tree of, 99, 100–101; after splitting off of Nothofagaceae, 110, 112, 114; twelve pairs of chromosomes in, 172

Fagales (Beech Order), 90; biotic seed dispersal in, 115; conjectured ancestry of, 107–9; early evolution of, 109–10, 112, 114; in Engler's classification, 98, 99, 102; with families in place by end of Paleocene, 120; formal recognition of, 102–3; fossil record and, 103, 106–8; insect-pollinated genera of, 107; at Paleocene-Eocene boundary, 123; phylogenetic tree of, 99, 100–101. *See also* Core Fagales

Fagus (beeches): cupules of, 107; evolution of, 110, 112; *F. crenata* seedling, 15; *F. grandifolia*, 112; not evolving acorns, 117; photosynthetic cotyledons of, 15, 115

migration of organisms at PETM, 124–25

mitochondrial genome of oaks, 238

modern humans, 68, 70, 161. See also *Homo sapiens*

Mongolian oak (*Q. mongolica*), 25–26, 237

Morgan, Thomas, 156

mountain chestnut oak (*Q. montana*), 146

Muller, Cornelius, 74, 75, 83, 170

mutations in genome, 53–54; in different branch tips of same tree, 160

mycorrhizal fungi, 190–91, 242; root tips colonized by, 177

natural selection, 43, 44–46; acting on groups of genes, 56; adaptation to climate and, 50; adaptive introgression and, 165; attributes of oaks beneficial for, 57; common gardens and, 49; diversity of oak species and, 186; environment and, 46, 50–51, 60–61; Eurasian White Oaks and, 169; evolution of Fagales and, 108; in French national forests, 60–61; genes involved in, 56–57; imperfect results of, 176; integrity of species and, 77–78, 84; intraspecific competition and, 57–58; introgressed genes and, 163, 195; Neanderthal alleles and, 161, 162; in oaks, 57–61; origin of oaks and, 128; requirements for, 45–46; trade-offs involving, 50–51. See also evolution

Neanderthals, 67–68, 70, 73, 161–62, 197, 202, 237

netleaf oak (*Q. rugosa*), 57

niche. See ecological niche; phylogenetic niche conservatism

Normapolles, 105–6, 116, 134, 228; fossilized pollen of, *104*, 105

Normapolles Province, 105

North Atlantic Land Bridge, 134, 146, 232, 235

Northern Hemisphere: oaks as pioneers in, 128; oaks spread across, 66–67; range of White Oak Group and, 147

northern red oak (*Q. rubra*), 7, 65; adapted to nearby microsites, 49; crossed with willow oak, 72; current climate change and, 198; expanding range of, 25; fine root plasticity and, 218; genome of Hill's oak and, 57, 162; germinating acorn of, 26, 30; hybridization of, 71; linkage map of, 155; range in eastern North America, 66; seedlings of, *36*, 175; squirrels eating acorns of, 21, 22; thriving across the Atlantic, 32

Nothofagaceae, 102, 105, 107, 108; gall wasps on, 192; split off from Fagales, 109–10

Nothofagus, 109; *N. alpina*, 90

Notholithocarpus, 117–20, 139, 148, 184; *N. densiflorus* acorn, *120*

nuclear genome, 79, 80–82, 225; ribosomal, 81, 166. See also genomes, of oaks

Nudopollis terminalis pollen grain, *104*

nutlets, 105, 107, 117, *118*

nutrient cycling, in mixed-species communities, 186–87, 241

nuts, in order Fagales, 102

oak names, 4, 207–11

oaks: billions worldwide, 217; characteristics of, 1, 2, 121, 176; as a clade, 94; classification of, 128–31; Darwin on, 45, 46–47; disciplines that are studying, 3; distant ancestors of, 89–90; first known fossils of, 125–27; future of, 197–202; gradual

oaks (*continued*)
  birth of, 128; as individuals, 1, 154,
    172–73; limited fossil record of, 103;
    monetary value to US, 194; rarity
    of chromosomal rearrangements
    in, 172. See also *Quercus*; sections of
    oaks; species of oaks; subsections
    of oaks
oak tree of life, 2, 4, 94; dead-end
    branches of, 202; disparate themes
    of, 169–70; genome and, 166–70;
    hybridization and, 86–87; stem of,
    128; traits and habitat preferences
    on, 182–83
oak wilt fungus, 179, 187, 241
Oglethorpe oak (*Q. oglethorpensis*), 75
order, taxonomic, 96
Origin of Species (Darwin), 42–46, 69,
    131, 218; Tree of Life illustrated in,
    91–92, 181–82
Orr, H. Allen, 83
ovary, 10, 13
ovules, 1, 8, 13, 13, 231
Oxford Oak, 58

Paleocene-Eocene Thermal Maximum
    (PETM), 123–29, 198, 230–31
Palmer, Ernest J., 74
Palmer's oak (*Q. palmeri*), 40; acorn
    of, 141
Paratethys Sea, 142, 234
passenger pigeons, 31, 115, 197
Pearse, Ian, 188
pedunculate oak (*Q. robur*), 32, 34; as-
    sembled reference genome of, 159,
    237; Darwin on, 45; extended range
    since LGM, 61; gene flow with ses-
    sile oak, 80, 156–57, 163; genetically
    distinct from sessile oak, 81–82,
    85, 170; geographic range of, 66; in
    group of four species exchanging

genes, 163–64; mycorrhizal fungi
    and, 191; nearly to Arctic Circle,
    184; North American White Oaks
    and, 167; powdery mildew on, 55;
    single-nucleotide mutations in, 53;
    trait heritability in, 219
periodical cicadas, 25
Petit, Rémy, 173
"Pharaoh's Dance," 151–52, 154, 156,
    159, 161, 172
phenotype, 53–56
photosynthetic cotyledons, 15, 115
photosynthetic rates, 50–52
phylogenetic niche conservatism, 178–
    79, 182. *See also* ecological niche
phylogenetics, 94, 96
pin oak (*Q. palustris*), 78, 198
pistillate (female) flowers, 7–8, 10–11,
    13
plasticity, 39–41, 42; Darwin on oaks
    and, 45; evolution of, 51–52; of
    flowering time, 52; genetic varia-
    tion and, 47, 61–62; of leaves from a
    single tree, 39–40, 41; in reciprocal
    transplant experiments, 49, 51;
    seemingly counter to adaptation, 52
Pleistocene, 26–27, 75, 143, 160, 197
Plomion, Christophe, 159
pollen: climate change and, 61; com-
    petition in fertilization by, 157;
    distance traveled by, 10; fertilization
    of different species, 157; genes shuf-
    fled in creation of, 54; germination
    of, 10, 13; of hybrids, 83; meiosis
    in creation of, 153–54; released by
    anthers, 9, 9–10; in sediment cores
    from lakes, 27–28. *See also* fossils,
    of pollen; wind pollination
pollen tubes, 10–11; boundaries be-
    tween species and, 83, 157; releasing
    two sperm cells, 13

polyploidy, 170

Ponticae, 129, 140, 146, 168, 184. See also Deer Oak Group (Q. sect. Ponticae)

populations, 34, 37; adjusting to environment, 46; variation within, 49, 50, 55–56, 57

post oak (Q. stellata), 75, 76, 79, 81, 142

powdery mildews, 55–56, 60

protein-coding genes, 152–53, 154, 159

proteins: functions of, 53; genotype with instructions for, 53; isozymes, 78–79, 81–82, 225; mutations with no effect on, 54

Proteus and Menelaus, 200

Protobalanus. See Golden-Cup Oak Group (Q. sect. Protobalanus); Intermediate Oak Group (Q. sect. Protobalanus)

Protofagacea, 106–7, 117, 228

provenance trials, 47

pubescent oak (Q. pubescens), 45, 184

Pyrenean oak (Q. pyrenaica), 40

Quercus, 2, 96; four-genus clade including, 117–20. See also oaks; sections of oaks; subsections of oaks; White Oak Group (Q. sect. Quercus); and individual species

Quercus sect. Cerris. See Cork Oak Group (Q. sect. Cerris)

Quercus sect. Cyclobalanopsis. See Ring-Cupped Oak Group (Q. sect. Cyclobalanopsis)

Quercus sect. Ilex. See Holly Oak Group (Q. sect. Ilex)

Quercus sect. Lobatae. See Red Oak Group (Q. sect. Lobatae)

Quercus sect. Ponticae. See Deer Oak Group (Q. sect. Ponticae)

Quercus sect. Protobalanus. See Golden-Cup Oak Group (Q. sect. Protobala-

nus); Intermediate Oak Group (Q. sect. Protobalanus)

Quercus sect. Quercus. See White Oak Group (Q. sect. Quercus)

Quercus sect. Virentes. See Southern Live Oak Group (Q. sect. Virentes)

Quercus subgenus Cerris, 129, 130, 134–36, 147, 166; gall wasps of, 193; phylogenetic tree of, 124–25

Quercus subgenus Quercus, 129, 147, 166; phylogenetic tree of, 124–25

Q. acutissima, 34, 61, 165

Q. agrifolia, 70, 189

Q. alba. See eastern white oak (Q. alba)

Q. aliena, 34

Q. alnifolia, leaves, 132

Q. berberidifolia, 84, 165

Q. bicolor, 10, 65–67, 76, 77, 80, 83

Q. cerris, 32, 48

Q. chrysolepis, 41, 184

Q. coccifera, 51–52, 184

Q. coccinea, 34, 84–85

Q. conzattii, 85

Q. cupreata, 144

Q. douglasii, 52

Q. dumosa, 180

Q. ellipsoidalis. See Hill's oak (Q. ellipsoidalis)

Q. engelmannii, 84, 145, 164–65, 170

Q. gambelii. See Gambel's oak (Q. gambelii)

Q. garryana, 199, 200

Q. geminata, 84

Q. gravesii, 143–44, 186

Q. grisea, 74, 83, 162, 186

Q. havardii, 40

Q. hinckleyi, 40

Q. humboldtii, 21, 66, 145, 184

Q. hypoxantha, 144

Q. ilex. See Holm oak (Q. ilex)

Q. ilex, acorns, 133

170, 225. *See also* biological species concept

resprouting: after fire, 40, 182, 183, 186; from stumps, 39–40, 116

restoration projects, facilitated by jays, 20

restriction site–associated DNA sequencing (RAD-seq), 166, 167

rhizomes, 40, 182, 183

Rhoiptelea, 105–6, 120

Ribicoff, Gabe, 42

Ring-Cupped Oak Group (*Q.* sect. *Cyclobalanopsis*), 130, 134–35, 139; in communities, 184; complicated history of, 147–48; delayed fertilization in, 11; Eurasian lineage of, 146; niche conservatism of, 178; *Q. glauca* acorns, 136; *Q. glauca* leaves, 137; timing of crops in, 22–23

Roble Colombiano. See *Q. humboldtii*

Rocky Mountains, 139–40, 142

rodents, 17, 115. *See also* mice, eating acorns; squirrels

rooks, 19, 29–30

roots: acorn germination with development of, 18, *30*, 32–33; drought stress and, 50; mycorrhizal fungi colonizing, *177*, 190–91; pathogens of, 60; phenology of, in *Q. alba* and *Q. rubra*, 7; plasticity of, 40, 218; as systems surviving for millennia, 40; taproots, 18, *30*, 33, 175

root tips: inside cotyledon, 16; mycorrhizal fungi colonizing, *177*; nipped off by a squirrel, 18; tannin concentration in, 21

Rydberg, Per Axel, 180

Saleh, Dounia, 59–60

sawtooth oak (*Q. acutissima*), 34, 61, 165

scarlet oak (*Q. coccinea*), 34, 84–85

scatter-hoarding, 17–20, 23

Schaal, Barbara, 79–81

sections of oaks, 4–5; classification of, 129–30; distribution map of, *126–27*; phylogenetic tree of, *124–25*; uniting in Eurasia, 146

sediment cores from lakes, 27–28

seedlings, 32–33; with cotyledons remaining in soil, 14; of *Fagus crenata*, *15*; few becoming trees, 35, 37; fungi in soil and, 191–92; growing with only part of cotyledon, 22; of northern red oak (*Q. rubra*), *36*, 175; overwintering, 175

seeds of oaks: encased in acorns, 1, 13; in two-seeded acorns, 13–14, 107, 214

segregation distortion, 157

self-pollination, 10–11, 57

sessile oak (*Q. petraea*): adapted to climates, 50; climate change and, 61; Darwin on, 45; in French national forests, 58, 59; gene flow with pedunculate oak, 80, 156–57, 163; genetically distinct from pedunculate oak, 81–82, 85, 170; geographic range of, 25, 66; in group of four species exchanging genes, 163–64; long-lived, 40, 58; in massive common garden experiment, 49–50; plastic leaf traits in, 52; powdery mildew on, 55; trait heritability in, 219

shrubby oak species, 1, 176; as adaptation to drought, 181; as clones, 40–41; co-occurrent with tree species, 189; hybridization of shrub white oaks, 170

Soepadmoa, Engkik, 106

Soepadmoa cupulata, 106–7, 117

soil, microorganisms in, 187, 191

solar radiation, and glaciation, 27

solitary oaks, 184

Sork, Victoria, 49

South America, only oak species in, 145, 178

Southern Beeches. *See* Nothofagaceae

southern live oak (*Q. virginiana*), 79, 81, 84

Southern Live Oak Group (*Q.* sect. *Virentes*), 129, 142, 145, 166; in communities, 182–83; convergence and, 180; Deer Oaks and, 168; oak wilt and, 187

speciation events: evolution within populations and, 34–35, 52–53; on Fagales tree of life, 100–101; of the oaks, 128–29; rapid in Mexican mountains, 144; on Tree of Life, 91–92

species concepts, 68–70, 83–84, 224; Darwin on, 45; Tree of Life and, 96

species of oaks, 1–2, 37; 91 in North America, 85; 425 recognized today, 2, 67, 96, 119; as clouds of variation, 62, 62–63; genome regions differentiating, 157–58; maintaining their distinctness, 67, 75, 82–84, 157, 170, 172; recognized for human uses, 67. *See also* "good species"; hybridization

sperm cells: fertilization and, 1, 53; genes shuffled in creation of, 54, 153–54; two in a pollen grain, 10, 13, 154

spiderworts (*Tradescantia*), introgression in, 73

sprouting. *See* resprouting; seedlings

squirrels, 17–19, 22, 29, 201

stamens, 8–9, 9. *See also* male (staminate) flowers

Stebbins, G. Ledyard, 72, 73–74, 75, 83

Steele, Kelly, 102

stem of oak clade, 128

stigmas, 10–11, 13, 14, 16, 83

stomata, and shared genes, 164

stone oaks. See *Lithocarpus* (stone oaks)

St. Pankraz pollen, 125–26, 128

stumps, sprouts from, 39–40, 116

Sturtevant, Alfred, 156

styles, 10–11, 13, 16, 83, 157

subgenera of oaks, 4–5, 129–30. See also *Quercus* subgenus *Cerris*; *Quercus* subgenus *Quercus*

subsections of oaks, 130, 233

swamp chestnut oak (*Q. michauxii*), 79, 81, 146, 167

swamp white oak (*Q. bicolor*), 10, 65–67, 76, 77, 80, 83

syngameon, 76–77, 224–25, 239

tannins, 20–22; jays favoring acorns with, 215; oak galls enriched in, 192

taproots, 18, 30, 33, 175

taxonomy: clades named in, 96, 102; consensus in, 85; of oaks, 129–31; Tree of Life and, 98. *See also* classifications

telomeres, 153

Theophrastus, 95

thickets of clones, 40

Ticodendraceae, 112, 114, 116

Ticodendron incognitum, 112, 114

Tradescantia (spiderworts), introgression in, 73

transposable elements, 153, 158

Tree of Life, 2, 3, 4–5, 37; Darwin on, 43, 91–92, 94, 181; in diverse cultures, 90–91; environment and, 176; hybridization in, 86–87; inferring, 93–94; stem of oak clade on, 128; taxonomy and, 98

trees: latitudinal turnover in, 176–78; not defining a clade, 94

Trelease, William, 180

Trigonobalanus, 112, 115, 117; *T. excelsa* inflorescence, *113*

trilliums, 98
Truffaut, Laura, 57–58
truffles, and Holm oaks, 135
Tubakia, 187–88
tubers, 33
Tucker, John, 74–75, 83, 169
Turgai Sea, 134, 135
turkey oak (*Q. cerris*), 32, 48
tyloses, 179

valves, 117–18, *118*
Van Valen, Leigh, 77–78, 80
variation, 43–47; complex patterns
    of gene flow and, 74; distance
    between populations and, 220; sex
    as wellspring of, 54. *See also* genetic
    variation; plasticity
varieties, 69, 85
vessels in trees: freezing and, 177–
    78; narrower in White Oaks, 179;
    tyloses growing into, 179. *See also*
    drought tolerance
volcanic activity, 114, 123
vultures, 98–99

Walnut Family (Juglandaceae), 105,
    108, 112
water availability: adaptation to, 49–
    50, 157. *See also* drought tolerance
weather: masting and, 24–25; pollina-
    tion and, 9, 24
weevil-infested acorns: development of
    weevils in, 17, 189; on early smaller
    acorns, 22–23; feeding at acorn
    base, 21–22; germinating despite
    some loss of cotyledons, 22; left
    behind by jays, 19–20; after seeking
    tree that has acorns, 214; squirrels'
    assessment of, 17–18
Western inland sea (North America),
    89, 105
white cedar (*Thuja occidentalis*), 90–91

White Oak Group (*Q.* sect. *Quercus*),
    129; acorns eaten by squirrels, 18;
    annual acorns in, 12–13, 188; Cen-
    tral American populations of, 144–
    45; chloroplast studies of, 79–80;
    clouds of DNA variation in, *62*; in
    communities, 182–84; complicated
    history of, 147–48; convergence
    and, 180–81, 183; disproportionate
    successes of, 179–80; divided into
    eastern and western clades, 140;
    early evolution of, 136–37, 139–
    40; East Asian species in, 146, 184;
    Eurasian, 146–47, 167–70, 179, 184,
    193; fossil leaves and, 136–37, *138*;
    four European species exchanging
    genes, 163–64; fungal diseases
    and, 187–88; gall wasps of, 32, 193;
    growing with Red Oak species, 181,
    183–84; hybridization and, 70,
    74, 75–77, *76*, 83–84, 223; insect
    herbivory and, 188; isozymes of, 78;
    leaf litter beneath, 186–87; Mexi-
    can populations of, 143–45
Whittaker, Robert, 181
Whittemore, Alan, 79–81
willow oak (*Q. phellos*), 72
wind pollination, 1, 8, 10; evolution
    of, 107; inefficiency of, 35; map of
    reproduction and, 33–34; natural
    selection and, 57; in order Fagales,
    102; species diversity of oaks and,
    119; weather and, 24
woodpeckers, 21
Wright, Sewall, 195

xylem rays, 176. *See also* vessels in trees

Yggdrasil, Norse Tree of Life, 90

Zawinul, Joe, 151, 152
zygote, 13